AF812564

Compendium of Plant Genomes

Series Editor

Chittaranjan Kole, Raja Ramanna Fellow, Department of Atomic Energy, Government of India, ICAR-National Institute for Plant Biotechnology, New Delhi, India

Whole-genome sequencing is at the cutting edge of life sciences in the new millennium. Since the first genome sequencing of the model plant *Arabidopsis thaliana* in 2000, whole genomes of about 100 plant species have been sequenced and genome sequences of several other plants are in the pipeline. Research publications on these genome initiatives are scattered on dedicated web sites and in journals with all too brief descriptions. The individual volumes elucidate the background history of the national and international genome initiatives; public and private partners involved; strategies and genomic resources and tools utilized; enumeration on the sequences and their assembly; repetitive sequences; gene annotation and genome duplication. In addition, synteny with other sequences, comparison of gene families and most importantly potential of the genome sequence information for gene pool characterization and genetic improvement of crop plants are described.

More information about this series at https://link.springer.com/bookseries/11805

Chittaranjan Kole
Editor

The *Catharanthus* Genome

Springer

Editor
Chittaranjan Kole
Raja Ramanna Fellow
Department of Atomic Energy
Government of India
ICAR-National Institute for Plant
Biotechnology
New Delhi, India

ISSN 2199-4781 ISSN 2199-479X (electronic)
Compendium of Plant Genomes
ISBN 978-3-030-89268-5 ISBN 978-3-030-89269-2 (eBook)
https://doi.org/10.1007/978-3-030-89269-2

© The Editor(s) (if applicable) and The Author(s), under exclusive license to Springer Nature Switzerland AG 2022
This work is subject to copyright. All rights are solely and exclusively licensed by the Publisher, whether the whole or part of the material is concerned, specifically the rights of translation, reprinting, reuse of illustrations, recitation, broadcasting, reproduction on microfilms or in any other physical way, and transmission or information storage and retrieval, electronic adaptation, computer software, or by similar or dissimilar methodology now known or hereafter developed.
The use of general descriptive names, registered names, trademarks, service marks, etc. in this publication does not imply, even in the absence of a specific statement, that such names are exempt from the relevant protective laws and regulations and therefore free for general use.
The publisher, the authors and the editors are safe to assume that the advice and information in this book are believed to be true and accurate at the date of publication. Neither the publisher nor the authors or the editors give a warranty, expressed or implied, with respect to the material contained herein or for any errors or omissions that may have been made. The publisher remains neutral with regard to jurisdictional claims in published maps and institutional affiliations.

This Springer imprint is published by the registered company Springer Nature Switzerland AG
The registered company address is: Gewerbestrasse 11, 6330 Cham, Switzerland

This book series is dedicated to my wife Phullara and our children Sourav and Devleena

Chittaranjan Kole

Preface to the Series

Genome sequencing has emerged as the leading discipline in the plant sciences coinciding with the start of the new century. For much of the twentieth century, plant geneticists were only successful in delineating putative chromosomal location, function and changes in genes indirectly through the use of a number of "markers" physically linked to them. These included visible or morphological, cytological, protein and molecular or DNA markers. Among them, the first DNA marker, the RFLPs, introduced a revolutionary change in plant genetics and breeding in the mid-1980s, mainly because of their infinite number and thus potential to cover maximum chromosomal regions, phenotypic neutrality, absence of epistasis and codominant nature. An array of other hybridization-based markers, PCR-based markers and markers based on both facilitated construction of genetic linkage maps, mapping of genes controlling simply inherited traits and even gene clusters (QTLs) controlling polygenic traits in a large number of model and crop plants. During this period, a number of new mapping populations beyond F2 were utilized and a number of computer programs were developed for map construction, mapping of genes, and for mapping of polygenic clusters or QTLs. Molecular markers were also used in the studies of evolution and phylogenetic relationship, genetic diversity, DNA fingerprinting and map-based cloning. Markers tightly linked to the genes were used in crop improvement employing the so-called marker-assisted selection. These strategies of molecular genetic mapping and molecular breeding made a spectacular impact during the last one and a half decades of the twentieth century. But still they remained "indirect" approaches for elucidation and utilization of plant genomes since much of the chromosomes remained unknown, and the complete chemical depiction of them was yet to be unraveled.

Physical mapping of genomes was the obvious consequence that facilitated the development of the "genomic resources" including BAC and YAC libraries to develop physical maps in some plant genomes. Subsequently, integrated genetic–physical maps were also developed in many plants. This led to the concept of structural genomics. Later on, emphasis was laid on EST and transcriptome analysis to decipher the function of the active gene sequences leading to another concept defined as functional genomics. The advent of techniques of bacteriophage gene and DNA sequencing in the 1970s was extended to facilitate sequencing of these genomic resources in the last decade of the twentieth century.

As expected, sequencing of chromosomal regions would have led to too much data to store, characterize and utilize with the-then available computer software could handle. But the development of information technology made the life of biologists easier by leading to a swift and sweet marriage of biology and informatics, and a new subject was born—bioinformatics.

Thus, the evolution of the concepts, strategies and tools of sequencing and bioinformatics reinforced the subject of genomics—structural and functional. Today, genome sequencing has traveled much beyond biology and involves biophysics, biochemistry and bioinformatics!

Thanks to the efforts of both public and private agencies, genome sequencing strategies are evolving very fast, leading to cheaper, quicker and automated techniques right from clone-by-clone and whole-genome shotgun approaches to a succession of second-generation sequencing methods. The development of software of different generations facilitated this genome sequencing. At the same time, newer concepts and strategies were emerging to handle sequencing of the complex genomes, particularly the polyploids.

It became a reality to chemically—and so directly—define plant genomes, popularly called whole-genome sequencing or simply genome sequencing.

The history of plant genome sequencing will always cite the sequencing of the genome of the model plant Arabidopsis thaliana in 2000 that was followed by sequencing the genome of the crop and model plant rice in 2002. Since then, the number of sequenced genomes of higher plants has been increasing exponentially, mainly due to the development of cheaper and quicker genomic techniques and, most importantly, the development of collaborative platforms such as national and international consortia involving partners from public and/or private agencies.

As I write this preface for the first volume of the new series "Compendium of Plant Genomes," a net search tells me that complete or nearly complete whole-genome sequencing of 45 crop plants, eight crop and model plants, eight model plants, 15 crop progenitors and relatives and three basal plants is accomplished, the majority of which are in the public domain. This means that we nowadays know many of our model and crop plants chemically, i.e., directly, and we may depict them and utilize them precisely better than ever. Genome sequencing has covered all groups of crop plants. Hence, information on the precise depiction of plant genomes and the scope of their utilization are growing rapidly every day. However, the information is scattered in research articles and review papers in journals and dedicated web pages of the consortia and databases. There is no compilation of plant genomes and the opportunity of using the information in sequence-assisted breeding or further genomic studies. This is the underlying rationale for starting this book series, with each volume dedicated to a particular plant.

Plant genome science has emerged as an important subject in academia, and the present compendium of plant genomes will be highly useful to both students and teaching faculties. Most importantly, research scientists involved in genomics research will have access to systematic deliberations on the plant genomes of their interest. Elucidation of plant genomes is of interest not only for the geneticists and breeders, but also for practitioners of an array of plant science disciplines, such as taxonomy, evolution, cytology, physiology, pathology, entomology, nematology, crop production, biochemistry and obviously bioinformatics. It must be mentioned that information regarding each plant genome is ever-growing. The contents of the volumes of this compendium are, therefore, focusing on the basic aspects of the genomes and their utility. They include information on the academic and/or economic importance of the plants, description of their genomes from a molecular genetic and cytogenetic point of view, and the genomic resources developed. Detailed deliberations focus on the background history of the national and international genome initiatives, public and private partners involved, strategies and genomic resources and tools utilized, enumeration on the sequences and their assembly, repetitive sequences, gene annotation and genome duplication. In addition, synteny with other sequences, comparison of gene families, and, most importantly, the potential of the genome sequence information for gene pool characterization through genotyping by sequencing (GBS) and genetic improvement of crop plants have been described. As expected, there is a lot of variation of these topics in the volumes based on the information available on the crop, model or reference plants.

I must confess that as the series editor, it has been a daunting task for me to work on such a huge and broad knowledge base that spans so many diverse plant species. However, pioneering scientists with lifetime experience and expertise on the particular crops did excellent jobs editing the respective volumes. I myself have been a small science worker on plant genomes since the mid-1980s, and that provided me the opportunity to personally know several stalwarts of plant genomics from all over the globe. Most, if not all, of the volume editors are my longtime friends and colleagues. It has been highly comfortable and enriching for me to work with them on this book series. To be honest, while working on this series I have been and will remain a student first, a science worker second and a series editor last. And I must express my gratitude to the volume editors and the chapter authors for providing me the opportunity to work with them on this compendium.

I also wish to mention here my thanks and gratitude to the Springer staff, particularly Dr. Christina Eckey and Dr. Jutta Lindenborn for the earlier set of volumes and presently Ing. Zuzana Bernhart for all their timely help and support.

I always had to set aside additional hours to edit books beside my professional and personal commitments—hours I could and should have given to my wife, Phullara, and our kids, Sourav and Devleena. I must mention that they not only allowed me the freedom to take away those hours from them but also offered their support in the editing job itself. I am really not sure whether my dedication of this compendium to them will suffice to do justice to their sacrifices for the interest of science and the science community.

Chittaranjan Kole

Preface

Catharanthus roseus (L.) G. Don, a tropical perennial medicinal subshrub of the family Apocynaceae, is a source of pharmaceutically important indole alkaloids. *C. roseus*, commonly known as Madagascar periwinkle, Cape periwinkle, rosy periwinkle, sadabahar (in Hindi) and Chang Chung Hua (in Chinese), is a common ornamental in tropical and subtropical regions found growing in humid dry, hot and cool locations. It is now becoming increasingly popular system for the molecular and physiological studies specially for its secondary metabolism. This plant also enjoys the distinction of a taxon possessing the largest number of alkaloids in the plant kingdom. However, it is widely known as a miraculous medicinal plant containing an array of potentially bioactive phytochemicals including flavonoids, saponins and alkaloids of which the alkaloids are the main therapeutic chemical constituents. A large number, claimed to be over 100 to 400 in different reports, of alkaloids are present in Catharanthus, which have pharmaceutical, food additive and agrochemical properties. Of these, vinblastine and vincristine are well-known source of commercial anticancer drugs. The alkaloid ajmalicine present in the root is used in the treatment of high blood pressure, and other leaf alkaloids have antineoplastic activities. This subshrub has been used for hundreds of years to treat a variety of ailments and was a favorite ingredient of medical component within the middle ages because of the multifarious activities of its components including antitumor, antibacterial, antihyperglycemic, antihypertensive, diabetic wound healing activity, asthma, constipation and menstrual problems. Large demand, low yields of alkaloids and rapid life cycle of the species are prompting significant research efforts on *C. roseus*.

Catharanthus has been a leading plant in researches in several disciplines including basic botany, genetics, genomics, biochemistry, physiology, medicine and horticulture. Some of the useful features of *Arabidopsis* are shared by *C. roseus,* and since latter is most studied for its chemistry and pharmacology, it is an apparent choice for studies on the functional genetics of alkaloids. However, an array of novel concepts and strategies of the post-genomics era has facilitated identification of the therapeutically important genes and their functions and more importantly their monitoring and manipulations aiming at developing genomically designed plant types containing quantitatively and qualitatively improved anticancer alkaloids. This book on The Catharanthus genome will hopefully present the basic and applied information on this "plant for health."

This book contains ten chapters covering various aspects. Chapter 1 of this book enumerates the taxonomy, distribution, nomenclature, habitat, morphology, anatomy, cytology, chemical constituents and their medicinal properties of Catharanthus. Chapter 2 discusses the plant-based cancer treatment with special reference to Catharanthus and different alkaloids present in the plant mainly vincristine and vinblastine and regulatory aspects of these anticancer herbal drugs. Chapter 3 deliberates on the systematic genetic improvement efforts in Catharanthus since its first identification in the Madagascar island forests, to its first cultivation at France and London, and its wide spread to tropical world. Driven by high commercial boost in development of ornamental cultivars for the landscape industry by the public sector plant breeders to its ongoing but limited continuous efforts in medicinal breeding for high concentration of monoterpene alkaloids supplemented with it targeted trait genetics and accomplishments. Somatic embryogenesis is a popular strategy of mass propagation particularly in horticultural plants. However, its success depends on the type of explants, growth promoters used and other regulating factors. These are explained in Chapter 4 along with details on somatic embryo development, structure and progress, i.e., maturation, germination of embryo. Chapter 5 presents an account of the effects of various media components and *in vitro* culture conditions governing alkaloid synthesis and a vivid depiction of alkaloid biosynthetic pathway in Catharanthus that could be made commercially viable.

The two major anticancer terpenoid indole alkaloids, vinblastine and vincristine, are naturally produced by Catharanthus. These are synthesized from vindoline via the catalysis of deacetylvindoline 4-*O*-acetyltransferase (DAT) in the last reaction of terpenoid indole alkaloid pathway. Chapter 6 presents the results of *Agrobacterium*-mediated genetic transformation for developing transgenic lines carrying the C. roseus DAT (CrDAT) gene to increase vincristine and alkaloid biosynthesis in the transgenic plants. Molecular breeding has emerged as a routine strategy to compliment traditional breeding. For this purpose, genetic linkage maps are constructed using different molecular markers and mapping populations, and genes and quantitative trait loci of interest are mapped to detect tightly linked molecular markers. Chapter 7 deliberates on these efforts made in Catharanthus along with prospects of marker-assisted breeding for improvement of the content of anticancer alkaloids. The chapter also deals with different omics approaches including transcriptomics, metabolomics and phenomics that could facilitate effective genomics-aided breeding.

Catharanthus is one of the few medicinal plants for which whole-genome sequencing has been accomplished. Information from genome sequencing not only provides the number and structure of genes but also depicts their phylogenetic relationships and evolutionary pathways. Besides, useful information on the function of the genes pave the way for formulation of strategies for their quantitative and qualitative improvement. The Chapter 8 enunciates the achievements of sequencing of nuclear and plastid genomes of Catharanthus and its various fundamental and applied implications. Chapter 9 deliberates on the biological pathways and regulatory genes and transcription factors involved in biosynthesis of various terpenoid indole alkaloids in Catharanthus, functional genes and transcription factors involved in transport of terpenoid indole alkaloids and on the functional genomic tools for engineering the biosynthetic pathways in Catharanthus. Chapter 10 enumerates the achievements of genetic and genomics researches made so far in Catharanthus and highlights on the potential future research areas.

These chapters have been contributed by 40 eminent scientists from three countries including India, Mexico and Vietnam. I express my thanks and gratitude to them for their contributions.

Catharanthus has a special place in my heart. Firstly, I grew up watching this plant just beside a plant of holy basil, *Ocimum sanctum*, my mother used to worship! Secondly, it is the plant on which I had started my formal "genetic" work in a center of the Central Institute of Medicinal and Aromatic Plants of the Council of Scientific & Industrial Research of India at Bengaluru, India as an amateurish "researcher"! This is the source of my confidence to dare to contemplate editing this book and ultimately complete it. Hence, I will feel blessed if the readers find this book useful.

Chittaranjan Kole

Contents

Contributors

Rohit Sam Ajee Independent Researcher, Sagar, India

Apoorva Department of Botany, Center of Advance Study in Botany, Institute of Science, Banaras Hindu University (BHU), Varanasi, India

Yashika Bansal Cellular Differentiation and Molecular Genetics Section, Department of Botany, New Delhi, India

Ha Thi Bui Department of Biology, Thai Nguyen University of Medicine and Pharmacy, Thai Nguyen, Viet Nam

Mau Hoang Chu School of Biology, Thai Nguyen University of Education, Thai Nguyen, Viet Nam

Shweta Dwivedi Division of Plant Breeding and Genetic Resource Conservation, CSIR- Central Institute of Medicinal Plants (CIMAP), Lucknow, India

Bushra Ejaz Cellular Differentiation and Molecular Genetics Section, Department of Botany, New Delhi, India

Tripta Jhang Division of Plant Breeding and Genetic Resource Conservation, CSIR- Central Institute of Medicinal Plants (CIMAP), Lucknow, India

Shuchi Kaushik State Forensic Science Laboratory, Sagar, Madhya Pradesh, India

Sushil Kumar SKA Institute for Research, Education and Development (SKAIRED), New Delhi, India

Moien Qadir Malik Cellular Differentiation and Molecular Genetics Section, Department of Botany, New Delhi, India

Maneesha Mall CSIR-Central Institute of Medicinal and Aromatic Plants, Lucknow, India

Jyoti Mamgain Cellular Differentiation and Molecular Genetics Section, Department of Botany, New Delhi, India

Daya N. Mani CSIR-Central Institute of Medicinal and Aromatic Plants, Lucknow, India

A. Mujib Cellular Differentiation and Molecular Genetics Section, Department of Botany, New Delhi, India

Lan Thi Ngoc Nguyen School of Biology, Thai Nguyen University of Education, Thai Nguyen, Viet Nam

Quan Huu Nguyen School of Biology, Thai Nguyen University of Education, Thai Nguyen, Viet Nam

Krishna Kumar Rai Department of Botany, Center of Advance Study in Botany, Institute of Science, Banaras Hindu University (BHU), Varanasi, India;
Dr. Rajendra Prasad Central Agricultural University, Samastipur, Bihar, India

Sanjay Kumar Rai Dr. Rajendra Prasad Central Agricultural University, Samastipur, Bihar, India

Shashi Pandey Rai Department of Botany, Center of Advance Study in Botany, Institute of Science, Banaras Hindu University (BHU), Varanasi, India

Neetu Rani Department of Botany, Nirmala College, Ranchi University, Ranchi, Jharkhand, India

Saumya Shah Biotechnology Division, CSIR-Central Institute of Medicinal and Aromatic Plants, Lucknow, Uttar Pradesh, India

Neha Sharma Amity Institute of Biotechnology, Amity University, Noida, India

Sarika Sharma Division of Plant Breeding and Genetic Resource Conservation, CSIR- Central Institute of Medicinal Plants (CIMAP), Lucknow, India

Ajit K. Shasany CSIR-Central Institute of Medicinal and Aromatic Plants, Lucknow, India

Ajit Kumar Shasany ICAR-National Institute for Plant Biotechnology (NIPB), New Delhi, India

Ashutosh K. Shukla CSIR-Central Institute of Medicinal and Aromatic Plants, Lucknow, India

Indranil Singh Independent Researcher, Sagar, India

Pooja Singh CSIR-Central Institute of Medicinal and Aromatic Plants, Lucknow, India

Velusamy Sundaresan CSIR-Central Institute of Medicinal and Aromatic Plants, Research Centre, Bengaluru, India

Rukaya Syeed Cellular Differentiation and Molecular Genetics Section, Department of Botany, New Delhi, India

Divya Vashisth Biotechnology Division, CSIR-Central Institute of Medicinal and Aromatic Plants, Lucknow, Uttar Pradesh, India

Felipe A. Vázquez-Flota Unidad de Bioquímica y Biología Molecular de Plantas, Centro de Investigación Científica de Yucatán, Mérida, Yucatán, México

Lien Thi Kim Vu Institute of Theoretical and Applied Research, Duy Tan University, Hanoi, Viet Nam;
Faculty of Natural Sciences, Duy Tan University, Da Nang, Viet Nam

Seema Yadav CSIR-Central Institute of Medicinal and Aromatic Plants, Lucknow, India

Abbreviations

10HGO	10-Hydroxygeraniol oxidoreductase
16OMT	16-Hydroxytabersonine-16-O-methyltransferase
2,4-D	2,4-Dichlorophenoxyacetic acid
2D-DIGE	Two-dimensional difference gel electrophoresis
7DLS	7-Deoxyloganetic acid synthase
A	Ajmalicine
AACT	Acetoacetyl-CoA thiolase
ABA	Abscisic acid
ABC	ATP-binding cassette
ADME	Absorption, dispersion, metabolism and excretion
AFLP	Amplified fragment length polymorphism
AJM	Ajmalicine
Akt	Ak strain transforming
AP	Activator protein
AP2/ERF	APETALA2/ethylene-responsive factor
APM	Alkaloid producing media
AS	Anthranilate synthase
ATP	Adenosine triphosphate
AVLB	α-3’,4’-Anhydrovinblastine
BAP	6-Benzylaminopurine
BCE	Before Common Era
bHLH	Basic helix-loop-helix protein
BIS	bHLH iridoid synthesis (transcription factor)
BSA	Bulked segregant analysis
BSLMM	Bayesian sparse linear mixed model
C	Catharanthine
cAMP	Cyclic adenosine monophosphate
CC BY	Creative Commons Attribution
cDNA-AFLP	Complementary DNA-AFLP
CHMIS-C	China’s Comprehensive Herbal Medicine Information System for Cancer
CMK	4-(cytidine 5’-diphospho)-2-C-methyl-D-erythritol kinase
COX-2	Cyclooxygenase 2 gene
CPR	Cytochrome P450 reductase
CR	Cathenamine reductase

CrDAT	*Catharanthus roseus* deacetylvindoline 4-O-acetyltransferase
CrGBF	*C. roseus* G box binding factor
CRISPR	Clustered regularly interspaced short palindromic repeats
croFGD	*C. roseus* Functional Genomics Database
CYP450	Cytochrome P 450
D17H	Desacetoxyvindoline-17-hydroxylase
D4H	Desacetoxyvindoline-4-hydroxylase
DAT	Deacetylvindoline-4-*O*-acetyltransferase
DL7H	7-Deoxyloganic acid-7-hydroxylase
DLGT	7-Deoxyloganetic acid glucosyltransferase
dsRNA	Double-stranded RNA
DVB	Double vindoline-bonded
DXPS	1-Deoxy-D-xylulose-5-phosphate synthase
DXR	1-Deoxy-D-xylulose-5-phosphate reductoisomerase
E. coli DH5α	*Escherichia coli* DH5 Alpha
EGFR	Epidermal growth factor receptor
EHRS	European Health Record System
EMMA	Efficient mixed-model association
EMMAX	Efficient mixed-model association eXpedite
EMS	Ethylmethane sulfonate
ERK	Extracellular-signal-regulated kinase
EST	Expressed sequence tag
F1	First filial (generation)
FaST-LMM	Factored spectrally transformed linear mixed model
FC	Flow cytometry
FDA	Food and Drug Administration (USA)
FM	Functional marker
G10H	Geraniol-10-hydroxylase
GA_3	Gibberellic acid
GACP	Guidelines of Good Agricultural and Collection Practices
GATA	GATA-binding protein
GC	Gas chromatography
GEMMA	Genome-wide efficient mixed model
GES	Geraniol-by-geraniol synthase
GES	Geraniol synthase
GI	Gradient index
GO	Gene ontology
GPPS	Geranyl pyrophosphate synthase
GTP	Guanosine triphosphate
GUS	β-Glucuronidase
GWAS	Genome-wide association studies
HDR	Hydroxymethylbutenyl-4-diphosphate reductase
HDS	Hydroxymethylbutenyl-4-diphosphate synthase
HIV	Human immunodeficiency virus
HMGR	Hydroxymethylglutaryl-CoA reductase
HMGS	Hydroxymethylglutaryl-CoA synthase

HPLC	High-pressure liquid chromatography
HRM	High-resolution melt analysis
IBA	Indole-3-butyric acid
IC_{50}	Half-maximal inhibitory concentration
IDI	Isopentenyl diphosphate isomerase
IDS	Iridoid synthase
IGR	Intergenic region
InDel	Insertion/deletion
IPAP	Internal phloem associated parenchyma
IPP	Isopentenyl diphosphate
IR	Inverted repeat
ISSR	Inter-simple sequence repeat
JAK-STAT	Janus kinase-STAT
Kn	Kinetin
LAMP	Loop-mediated isothermal amplification
LAMT	Loganic acid *O*-methyltransferase
LC	Liquid chromatography
LMM-Lasso	Linear mixed-model-Lasso
lncRNA	Long non-coding RNA
LSC	Large single copy
MAE	Microwave-assisted extraction
MAPK-ERK	Mitogen-activated protein kinase-ERK
MAS	Marker-assisted selection
MAT	Minovincine-O-acetyltransferase
MAT	Minovincinine-19-*O*-acetyltransferase
MATE	Multidrug and toxic compound extrusion
MCT	MEP cytidylyltransferase
MDR	Multidrug resistance
MECS	2-C-methyl-D-erythritol-2,4-cyclodiphosphate synthase
MeJA	Methyl jasmonate
MEP	Mevalonic acid pathway or methyl-D-erythritol-4 phosphate
MEP	2-C-Methyl-D-erythritol 4-phosphate
MFA	Metabolic flux analysis
MIA	Monoterpene indole alkaloid
miRNA	microRNA
MLMA	Model-based association analysis
MLMM	Multi-locus mixed model
MS	Murashige and Skoog (medium)
MS	Mass spectroscopy
MTIA	Monoterpene indole alkaloid
MTMM	Multi-trait mixed model
mTOR	Mammalian target of rapamycin
MVA	Mevalonate
NAA	Naphthalene acetic acid
NAC	No apical meristem

NCBI	National Center for Biotechnology Information
NCIM	National Collection of Industrial Microorganisms
NEU	N-Nitroso-N-ethyl urea
NF-κB	Nuclear factor kappa-light-chain-enhancer of activated B cells
nm	Nearly median
NMR	Nuclear magnetic resonance
NMT	N-methyltransferase
NMT	16-methoxy-2,3-dihydro-3-hydroxy-tabersonine-*N*-methyltransferase
NPF2.9	Nitrate/peptide family strictosidine transporter 2.9
nsm	Nearly submedian
nst	Nearly subtelocentric
nt	Nucleotide
OMT	O-Methyl transferase
ORCA	Octadecanoid-responsive Catharanthus AP2-domain protein
P3D	Population parameters previously determined
PCR	Polymerase chain reaction
PDR	Pleiotropic drug resistance
PEG	Polyethylene glycol
PEP	Phosphoenolpyruvate
PF	Pink flower variety
P-gp	P-glycoprotein
PGR	Plant growth regulator
PI3K	Phosphoinositide-3-kinase
PIF	Phytochrome interacting factor
PnWBP	Peanut witches' broom
PRX	Peroxidase
PTP-1B	Protein tyrosine phosphatase-1B
QTL	Quantitative trait locus
QTLs	Quantitative trait loci
RAPD	Random amplified of polymorphic DNA
rCrDAT	Recombinant CrDAT
RFLP	Restriction fragment length polymorphism
RILs	Recombinant inbred lines
RNAi	RNA interference
rRNA	Ribosomal ribonucleic acid
SAC	Spindle assembly checkpoint
SDG	Strictosidine degluocosidase
SE	Somatic embryogenesis
SGD	Strictosidine β-D-glucosidase
SI	Symmetry index
SLS	Secologanin synthase
SNP	Single-nucleotide polymorphism
SSC	Small single copy
SSR	Simple sequence repeat

STAT	Signal transducer and activator of transcription
STMS	Sequence-tagged microsatellite site
STR	Strictosidine synthase
T11H	Tabersonine-11-hydroxylase
T16H	Tabersonine-16-hydroxylase
T16H	Tabersonine-16-hydroxylase
T16H2	Tabersonine-16-hydroxylase 2
T3O	Tabersonine-3-oxygenase
T3R	Tabersonine-3-reductase
TCI	Total chromatin index
TCL	Total chromatin length
TDC	Tryptophan decarboxylase
TF	Total form
TF	Transcription factor
THAS	Tetrahydroalstonine synthase
TIA	Terpenoid indole alkaloid
TLC	Thin-layer chromatography
TM	Traditional medicine
TMIS	Traditional medicine information system
tRNA	Transfer RNA
TS	Transverse section
V	Vindoline
VBL	Vinblastine
VDS	Vindesine
VIGS	Virus-induced gene silencing
VIN	Vindoline
VRL	Vinorelbine
WF	White flower variety
WGBS	Whole-genome bisulfite sequencing
WHO	World Health Organization
WRKY	WRKY domain protein
WT	Non-transgenic plant
X^{ta}	Chiasmata
ZCT	Zinc finger transcription factor (from *Catharanthus*)
ZnO	Zinc oxide

Botanical Features of Periwinkle, *Catharanthus roseus*

1

Neetu Rani

Abstract

Periwinkle is the common name of *Catharanthus.* The scientific name of Madagascar Periwinkle and tiny periwinkle is *Catharanthus roseus* L. G. Don and *Catharanthus pusillus* (Murray) G. Don, respectively. *Catharanthus is* a genus of eight species out of which seven are indigenous to Madagascar and one (*C. pusillus*) is restricted to India and Sri Lanka. Periwinkle is an economically important perennial or annual medicinal plant having diploid chromosome number $2n = 2x = 16$ that belongs to the family Apocynaceae. To discuss the genomics of periwinkle there is necessity to have an insight related to the basic botanical features of periwinkle. This chapter provides a comprehensive account regarding its origin, distribution, nomenclature, taxonomic hierarchy, morphology, anatomy, cytology, chemical constituents, and medicinal and ayurvedic properties. This chapter also includes many original photographs and data to provide specific information related to periwinkle.

1.1 Introduction

The genus *Catharanthus* is a medicinally important plant that belongs to the family Apocynaceae. It occupies an enviably esteemed position in the realm of botany. The name *Catharanthus* is derived from the Greek word, meaning "Pure Flower". It is a handsome garden plant that flowers throughout the year and consequently it is called 'sadabahar' in Hindi. The plant is absolutely hardy and is popular for ornamental planting.

1.1.1 Species of *Catharanthus*

Catharanthus is a genus with eight species of herbaceous perennial or annual plants. These eight species are –

Catharanthus coriaceus Markgr.
Catharanthus Lanceus (Bojerex A. Dc.) Pichon.
Catharanthus longifolius (Pichon).
Catharanthusovalis ovalis (Markgr).
Catharanthus trichophyllus (Baker) Pichon.
Catharanthus scitulus (Pichon) Pichon.
Catharanthus roseus (L.) G. Don.
Catharanthus pusillus (Murray) G. Don.

N. Rani (✉)
Department of Botany, Nirmala College, Ranchi University, Ranchi, Jharkhand 834001, India

© The Author(s), under exclusive license to Springer Nature Switzerland AG 2022
C. Kole (ed.), *The Catharanthus Genome*, Compendium of Plant Genomes,
https://doi.org/10.1007/978-3-030-89269-2_1

1.1.2 Origin

Catharanthus is originally a native of Madagascar.

1.1.3 Distribution

Catharanthus is distributed globally. It is endemic to Madagascar but has now naturalized in all tropical and subtropical worlds. Among eight species seven are endemic to the island of Madagascar and the eighth, *Catharanthus pusillus* is from India and Sri Lanka. However, in India, this genus is represented by two species, i.e., *C. roseus* (L.) G. Don and *C. Pusillus* (Murray) G. Don. In India, *Catharanthusroseus* is cultivated as an ornamental plant, but it is invasive in Asia, Africa, and America.

Fig. 1.1 White variety of *C. roseus*

1.1.4 Nomenclature

The species *Catharanthus roseus* was formerly known as *Vinca rosea* or *Lochnera rosea*. *Catharanthus roseus* had been described under the name *Vinca rosea* by Swedish Naturalist Carl Von Linne in 1759. Then in 1828, the German botanist Heinrich Gotfilieb Ludwig Reichenbach proposed the genus name *Lochnera* mentioning the taxonomic disparities existing between the species *Vinca rosea* and the other species in the same genus (Vinca minor and Vinca Major). The species was renamed *Lochnera rosea* by an Austrian botanist Stephan Ldislaus Endlicher in 1838.

Catharanthus roseus (L.) G. Don was included in *Vinca rosea* L. by Bentham and Hooker (1876). Schumann (1895) Separated *Vinca rosea* from genus *Vinca* and placed it in *Lochnera* as *Lochnera rosea* Reichb on the basis of floral morphology. Bowden (1945) supported the separation of *Lochnera rosea* from *Vinca* on a cytological basis. Now the name *Lochnera rosea* has been declared illegitimate and *Catharanthus roseus* (L.) G. Don has been retained because it is an earlier name (Stearn, Loloydia 29: 169–200, 1966).

In India *Catharanthus roseus* (L.) G. Don (Figs. 1.1 and 1.2) has been recognized by a host of attractive names viz. Sadabahar in Hindi and Gujrati, Nayantara in Bengali, Kasikanigalu in Kannada, Ushamalari in Malayalam, Sadaphul in Marathi, Ainskati in Oriya Sudukadu Mallivai in Tamil, Billaganneru in Telgu and Ratanjot in Punjabi (Jain 1996).

Catharanthus pusillus (Fig. 1.3) is also recognized by many vernacular names in India such as Sangkhi, Ran–kel, or Tiloni in Marathi, Milakaych Chavkalatti in Tamil, Chupa–vela in Malayalam, Kanupoolaku in Telgu, and Bili Kaasi Kanigalu in Kannada.

Fig. 1.2 Pink variety of *C. roseus*

Fig. 1.3 C. Pusillus

Taxonomic position/hierarchy

Bentham and Hooker		APG Classification	
Kingdom:	Plantae	Kingdom:	Plantae
Division:	Magnoliophyta	Clade:	Angiospermae
Class:	Dicotyledonea	Clade:	Eudicots
Sub class:	Gamopetalae	Clade:	Asterides
Series:	Bicarpellatae	Order:	Gentianales
Order:	Gentianales	Family:	Apocynaceae
Family:	Apocynaceae	Genus:	*Catharanthus* G. Don
Genus:	*Catharanthus* G. Don	Species:	roseus (L) G. Don
Spices:	*roseus* (L.) G. Don		

1.1.5 Habitat

As an ornamental plant, it is appreciated for its hardiness in dry and nutritionally deficient conditions. It is tolerant to abiotic stress and can survive in various habitats like sand, dry land, roadsides and rocky soil, forest, riverbanks, and coastal areas. In India, *C. roseus* is not endangered and invasive but in Asia, Africa America, and Oceania it is listed as an invasive plant.

1.1.6 Morphology of *Catharanthus*

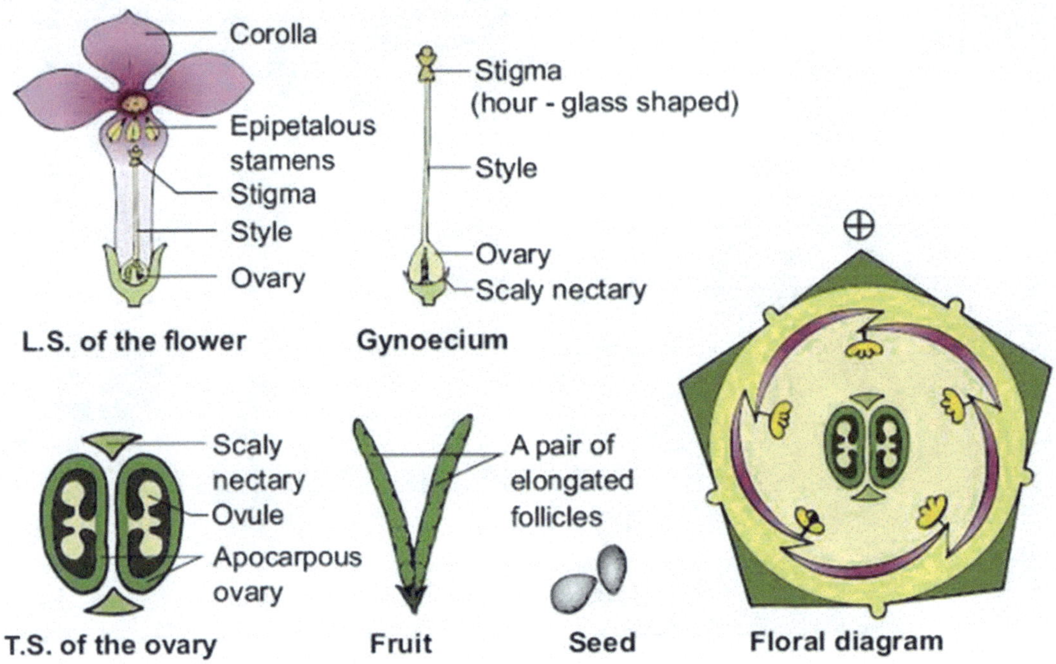

Floral formula Ebr.,Ebrl.,$\oplus$,⚥,$K_{(5)}$,$\overparen{C_{(5)},A_5}$,$\underline{G}_2$

1.1.6.1 Vegetative Characters

Habit: A perennial ornamental herb, flowering throughout the year, plant height is 1m. tall.

Root: Tap root, up to 70 cm long.

Stem: Erect, Cylindrical, Solid, nodes and internodes distinct, internodal region short, herbaceous, branched, whitish-green with a pinkish tinge.

Leaf: Leaf is simple, both cauline and ramal, exstipulate, petiolate, base of petiole glandular, opposite decussate, elliptic-oblong or obovate, dorsiventral, unicostate, reticulate, entire, glabrous, deep green, polished. Leaves 2.5–9 cm in length and 3.5cm in-breath petiole is 1–8 cm long.

1.1.6.2 Floral Characters

Inflorescence: Axillary, either solitary or a pair of two flowers.

Flower: Ebracteate, Complete, bisexual, regular, hypogynous, pedicelate, pentamerous, dichlamydeous rosy red, white, pink, and other flower varieties are also found.

Accessory Whorl

Calyx: Sepals 5, gamosepalous, bell-shaped inferior, persistent, accrescent, valvate, glandular, sepal linear, acute, glabrous.

Corolla: Petal 5, gamopetalous, hypocrateriform, inferior, twisted, corlla–tube–cylindrical, elongated greenish, swollen to urn shaped, its inner surface is hairy forming corona like structure. Corolla tube is 2.5–3 cm. long and corolla is 2–5 cm. in diameter.

Essential Whorl

Androecium: Stamen 5, epipetalous, inserted alternating with the corolla lobes. Anther is dithecous, sessile, sagittate, connivent round the stigmatic head, yellowish, adnate with the throat of corolla tube.

Gynoecium: Ovary 2, free below but united at the junction of ovaries and style, unilocilar with several ovules. Placentation marginal 2 large linear gland present which alternate the carpels. Style–simple, terminal, elongated but inserted within the corolla tube, white, cylindrical. Stigma–Dumb bell shaped.

Fruit: A pair of elongated slender follicles diverging from each other. Fruit is 2–4 cm. long and 3mm broad.

Seeds: Seeds are numerous, small in size (2–3 mm long), and black in color.

Pollination: Self-pollination is common.

1.1.6.3 Propagation

C. roseus is a fast growing perennial plant that produces numerous small seeds and is easily dispersed by ants, wind, and water. So they spread principally by seeds. Stem cuttings are also used for vegetative propagation.

1.1.6.4 Identifying Character

- Actinomorphic and Pentamerous flower
- Leaves opposite and decussate
- Stamens epipetalous
- Ovary superior and bicarpellary
- Fruit—a pair of follicles.

1.1.6.5 C. *Pusillus*

Habit: Tiny periwinkle is small (10–20 cm. tall) erect annual herb. It is native to India and Sri Lanka.

Root: tap root

Stem: Herbaceous, acutely quadrangular, branches spreading from the base.

Leaves: Opposite decussate, acute, lanceolate acuminate with minutely scabrid edges, otherwise glabrous, stipular glands subulate 3–8 cm. long, petiole 2–25" slender rough, margin.

Flower: Pentamerous , Bisexual, actinomorphic, hypogynous, white, small, singly, or in pairs in the upper leaf axils.

Calyx: Sepals 5, green 15" filiform.

Corolla: petals 5, white, flat, corolla tube 3-4" long, swollen base, valvate aestivation.

Androecium: stamens 5, epipetalous, dithecous.

Gynoecium: ovary 2, superior, bicarpellary style long.

Fruit: pair of follicles, pod is very slender 3–5 cm. long, straight pointed.

Seeds: black and small.

1.1.7 Anatomy

Leaves of *Catharanthus* are dorsiventral. T. S. of leaf exhibits that the upper epidermis is formed of a signal layer of small epidermal cells of larger size. Mesophyll is formed of a single layer of long palisade cells along with spongy parenchyma cells. Lower epidermis is formed at a single layer of small lignified cells which is covered with thin cuticles. Epidermal hair is absent in *Catharanthus.*

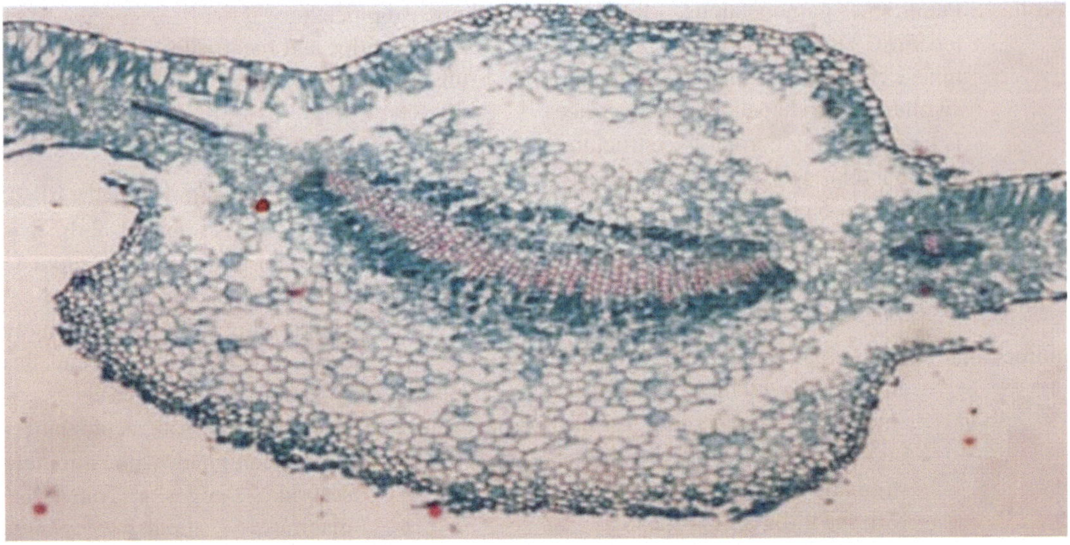

Fig. 1.4 T. S. of leaf of *Catharanthus roseus*

In midrib region, the upper epidermis is followed by 7–8 layers of collenchyma and 8 layers of parenchyma. A crescent-shaped bicollateral vascular bundle is present in the center. Outer and inner phloems are formed of sieve tube cells, companion cells, and phloem parenchyma. Xylem consists of radial vessels. Meta xylem is located at lower region while protoxylem in the upper region and are separated by xylem parenchyma. There are 4–5 layers of collenchyma that are followed by 10–11 layers of parenchyma, above the lower epidermis (Fig. 1.4).

1.1.7.1 Epidermal Anatomy

Epidermal anatomy includes stomatal index, shape of epidermal cell and the size of stomata provides character of taxonomic significance. This is due to the fact that genera and even families slow great constancy for their possession of a particular complex. The study of stomatal complexes becomes very useful to understand the evolutionary relationship of plants where classical cytology and genetics cannot be applied. Stomata of *Catharanthus* are dicytic, tricytic, tetracytic, pentacytic, and hexacytic with bean shaped guard cells (Figs. 1.5 and 1.6). However, the occurrence frequency of hexacytic stomata is less as compared to others. The epidermal cells of the stomata are irregular in shape on dorsal surface and regular on ventral surface (Figs. 1.7 and 1.8). There are slight differences in the stomatal index of both dorsal and ventral surfaces. It is very interesting to note that the dorsal, as well as ventral surface of white flower of Catharanthus, manifested maximum length as compared to pink flower and dorsal surface of pink variety showed maximum width to a white variety of *Catharanthus* whereas ventral surfaces of white variety exhibited maximum width as compared to pink variety. In white flower stomatal index is high in the ventral surface of the apex, middle, and base portion in sharp contrast to dorsal surface of pink flower variety (Table 1.1).

1.1.8 Cytology of Catharanthus

1.1.8.1 Mitotic Studies

The cytological studies on *Catharanthus* revealed that the Chromosomes of both the varieties of *Catharanthus roseus* and *Catharanthus pusillus* showed the same diploid chromosome number, i.e., $2n = 2x = 16$ (Figs. 1.9, 1.10 and 1.11).

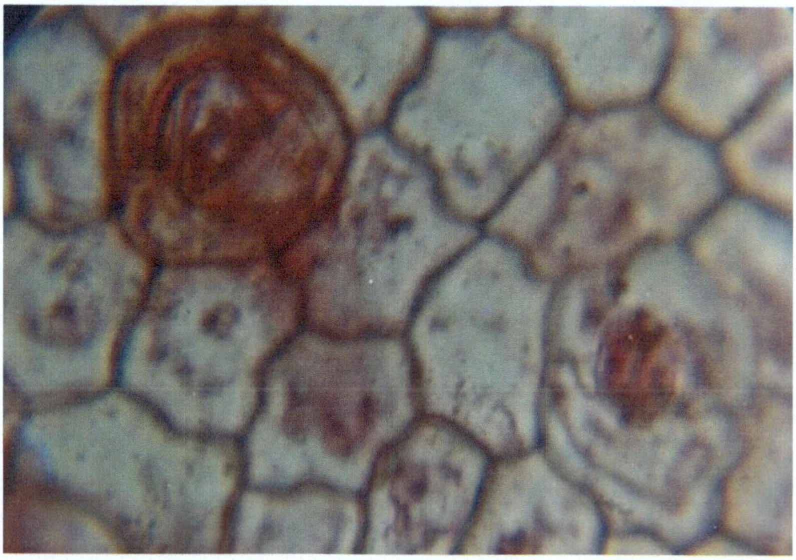

Fig. 1.5 Microphotographs showing dicytic and hexacytic stomata

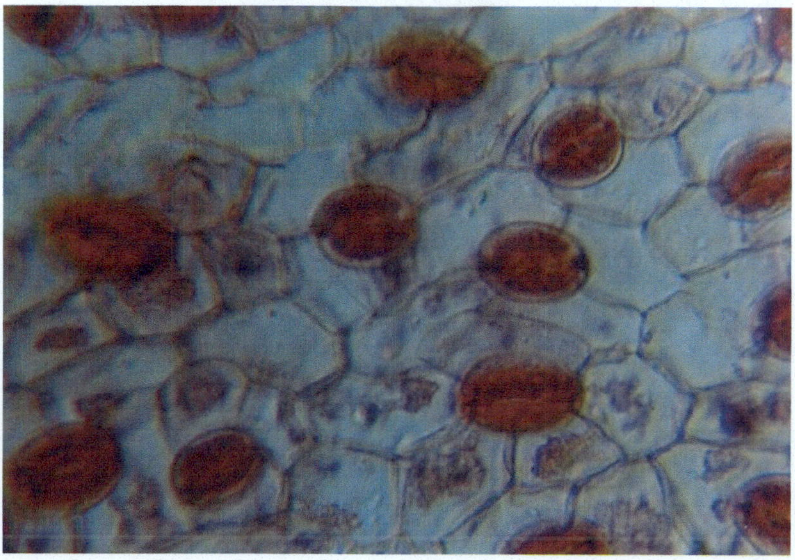

Fig. 1.6 Microphotographs showing tetracytic and pentacytic Stomata

The haploid complement in *C. roseus* has a median, nearly median, nearly sub-median, and nearly sub-terminal chromosomes. The chromosomes complement are small in size and could be classified into three distinct morphological types, i.e., long chromosome, medium sized chromosome, and small sized chromosome.

In *C. pusillus*, all the eight chromosomes are nearly sub-median type. Secondary constrictions are present and restricted to short arm only, while

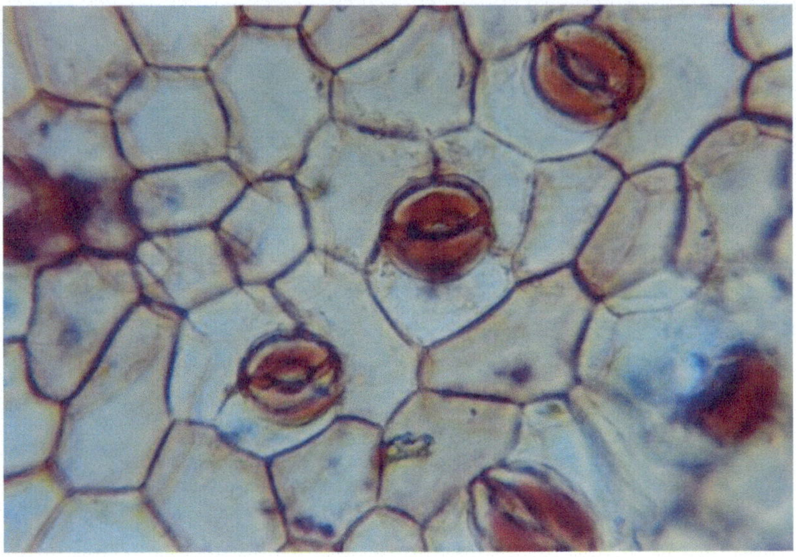

Fig. 1.7 Microphotographs showing regular epidermal cells on ventral surface

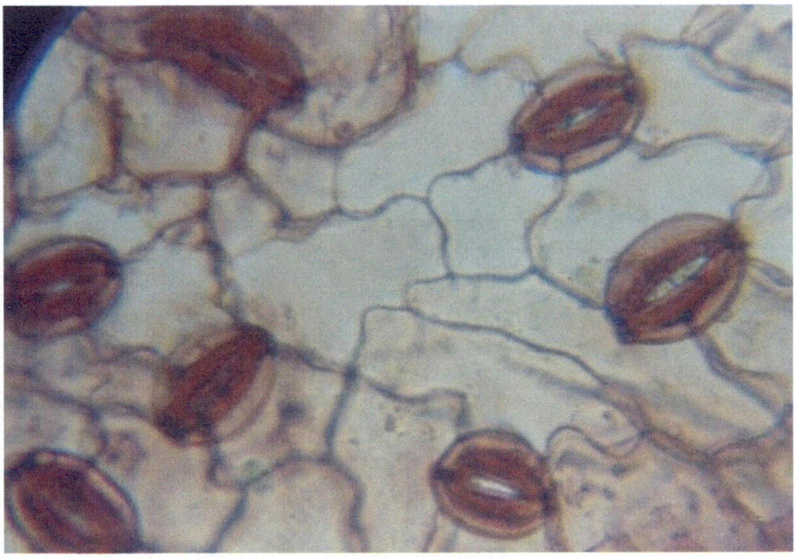

Fig. 1.8 Microphotographs showing Irregular epidermal cells on dorsal surface

it is completely absent in *Catharanthus roseus* (Table 1.2).

Karyotype formulae

A. *C. roseus*

1. White flower: 2n = 2x = 16 2(nst) + 10 (nsm) + 4(nm)
2. Pink flower: 2n = 2x = 16 14(nsm) + 2 (nm)

B. *C. pusillus*: 2n = 2x = 16 16(nm)

Regarding the total Chromatin length, the species having greater Chromatin length were supposed to be primitive whereas, species with lesser Chromatin length were treated as advanced (Stebbin 1971). This clearly indicates that *C. pusillus* is primitive having the highest Chromatin length while white variety of *C. roseus* is advanced. In the white variety of *C. roseus* three

Table 1.1 Range of leaf length and width of stomata (in μ) on the dorsal and ventral surface in both varieties of *Catharanthus*

Surface of leaf	Variety	Length (μ)	Width (μ)
Dorsal	PF	34.30–37.10 ± ± 1.71 1.40	28.00–29.40 ± ± 1.10 0.85
	WF	38.50–40.60 ± ± 1.92 1.78	26.60–28.00 ± ± 0.85 1.10
Ventral	PF	36.40–38.50 ± ± 1.78 1.92	31.50–32.20 ± ± 0.00 1.31
	WF	38.50–39.90 ± ± 1.56 1.40	28.70–32.90 ± ± 0.70 0.85

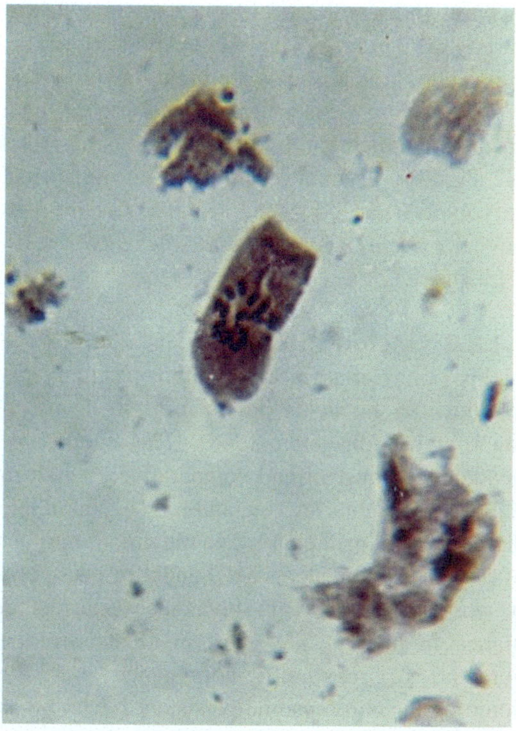

Fig. 1.9 Microphotographs showing mitotic metaphase in white variety of *C. roseus*

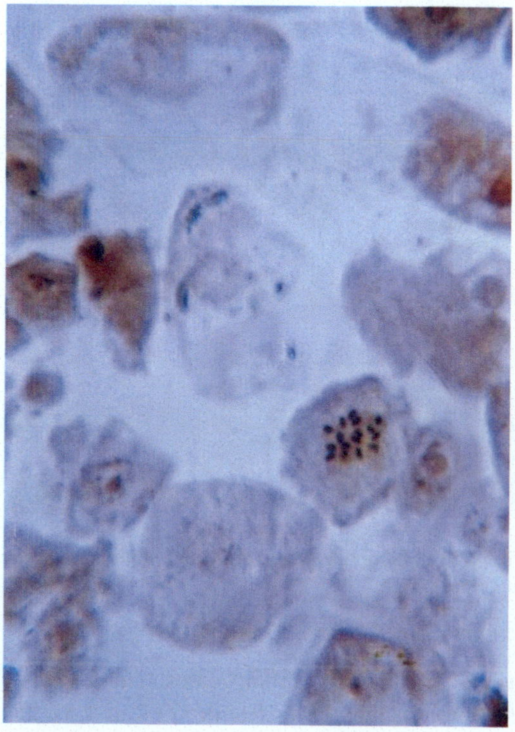

Fig. 1.10 Microphotographs showing mitotic metaphase in pink variety of *C. roseus*

are there pairs of long chromosomes, three pairs of medium sized chromosomes, and two pairs of small sized chromosomes. The length of the chromosome varies between 4.10 and 1.30 μ. Pink flower exhibits one pair of long chromosomes, five pairs of medium sized chromosomes, and 2 pairs of small sized chromosomes. The chromosome length ranges from 3.65 μ to 1.09 μ.

C. pusillus has four pairs of long chromosomes, 3 pairs of medium chromosomes, and one

Table 1.2 Data related to karyotype of *Catharanthus*

Species and varieties	TCL (μ)	T. F %	G. I %	S. I%
A. *C. roseus*				
i. White flower	21.25	27.29	31.70	37.54
ii. Pink flower	16.84	32.80	29.86	48.82
B. *C. pusillus*	28.07	44.49	28.03	80.73

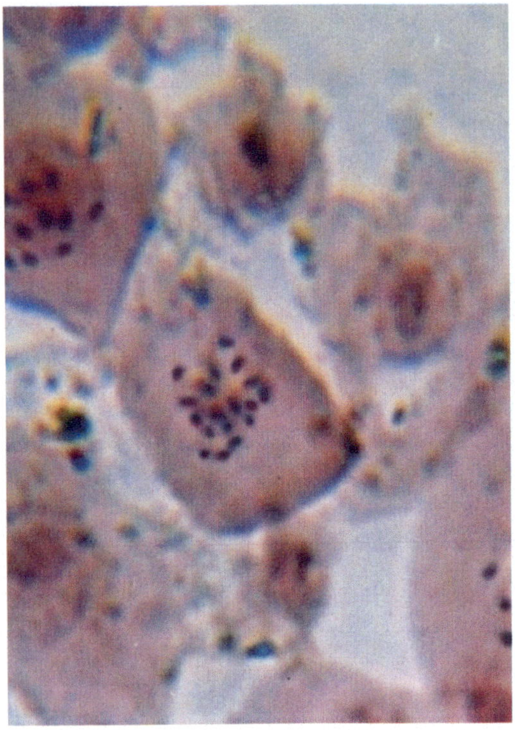

Fig. 1.11 Microphotographs showing mitotic metaphase in *C. pusillus*

pair of small sized chromosomes. The length of the chromosome varies from 5.60 μ to 1.57 μ.

All the chromosomes are nearly median in *C. pusillus*. The maximum (7 pairs) nearly sub-median chromosomes are found in pink flowers. The nearly sub-terminal chromosomes are observed only in white flowers (1 pair) and are found to be absent in other species and varieties ..

The relative length is maximum 100.00 μ in chromosome number I in both *C. roseus* and *C. pusillus* secondary constrictions are recorded in chromosomes I, II, and IV of *C. pusillus*. They are observed in the short arm only. The length of secondary constriction ranged from 0.17 μ to 0.35 μ.

Total chromatin index (T. C. I.) is found to be maximum in chromosome I and minimum in chromosome VIII in *Catharanthus roseus and Catharanthus pusillus.*

The higher value of total form percentage in *Catharanthus* indicates that the level of symmetry is higher except white flower variety of *C. roseus*. T. F.% is lower and considered asymmetrical. Regarding the gradient index value (G. I.), white flow variety is moderately asymmetrical (Fig. 1.12).

The genus *Catharanthus* shows high symmetrical index value (S.I.) except white flower variety of *C. roseus*. The high symmetrical index indicated that karyotypes are highly symmetrical in *Catharanthus* except for white variety which has low symmetrical index value and is regarded as slightly asymmetrical. The asymmetrical karyotype is an advanced feature according to Levitsky (1931) formulation (Figs. 1.9, 1.10, 1.11 and 1.12).

1.1.8.2 Meiotic Studies

Meiosis in *Catharanthus* exhibits regular formation of & bivalents (Furusato 1940; Danyansagar and Sudhakaran 1968). The number of bivalents is constant and comparatively small in size. The bivalents are ring and rod shaped in late diakinesis and metaphase. The maximum number of ring and rod bivalents are 7 and 4 respectively. It was found that ring bivalents occurred in higher frequency. Thy ring bivalents are the outcome of maximum homogeneity of the homologous chromosome persistent nucleolus is present only in pink variety while it is absent in white variety (Figs. 1.13 and 1.14).

Mean X^{ta} is 13.34 and 13.66 in pink variety and white variety respectively. Mean square between nuclei is 0.07 in both varieties of *C. roseus*. Mean square within nuclei is 11.58 and 12.17 in pink variety and white variety respectively. There is a negative correlation between bivalents in relation to inter chromosomal

SPECIS AND VARIETIES UNDER CONSIDERATION	MEDIAN	NEARLY MEDIAN	NEARLY SUB-MEDIAN	NEARLY SUB-TERMINAL
A. *Catharanthus roseus* 1. White flower		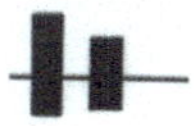		
2. Pink flower		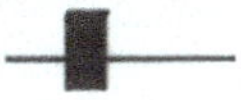		
B.Catharanthus pusillus				

Fig. 1.12 Karyotype analysis in Catharanthus

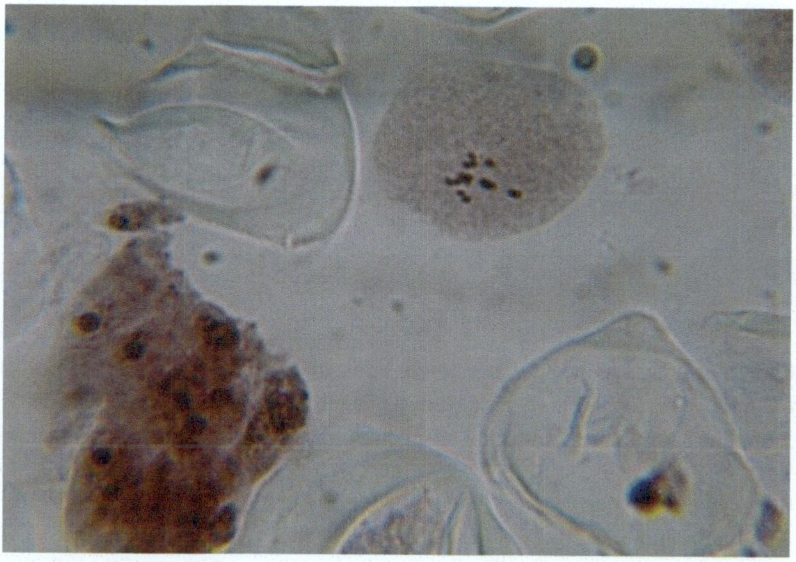

Fig. 1.13 Microphotographs showing meiotic metaphase in white variety of *C. roseus*

distribution of chiasmata. The mean chiasmata did not differ significantly. The intra plant variance is more than inter plant variance. There is variation in size, nature, and overall karyotype of bivalent in the same pollen mother cell (Table 1.3).

1.1.9 Chemical Constituents

1.1.9.1 C. *roseus*

More than a hundred alkaloids have been extracted from different parts of the plant *C. roseus* contains alkaloids, saponins,

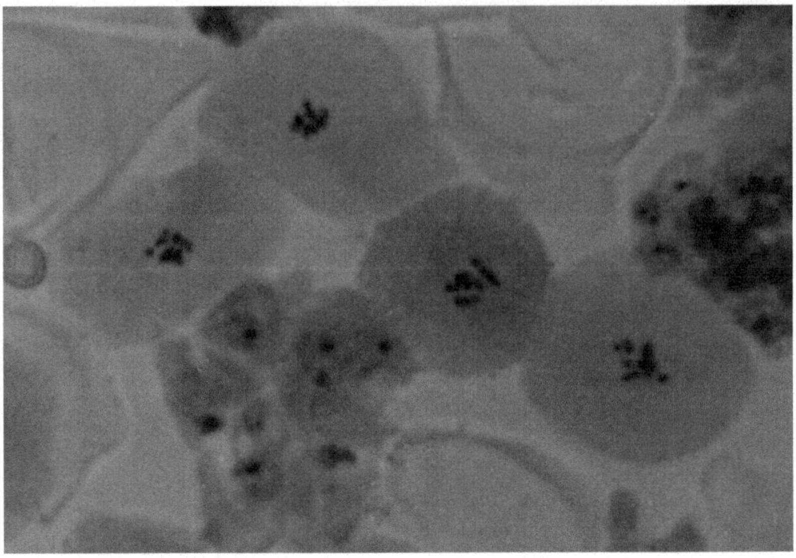

Fig. 1.14 Microphotographs showing mitotic metaphase in pink variety of *C. roseus*

Table 1.3 Meiotic analysis and chiasmata frequency in two varieties of *Catharanthus roseus* (L.) G. Don

Meiotic analysis	Varieties	No. of cell observed	No. of rings	No. of rods	Mean X^{ta}	½ X^{ta}	Mean square between nuclei	Mean square within nuclei	Variance ratio	Correlation coefficient
	Pink	100	1068	266	13.34	0.83	0.07	11.58	0.006	–0.01
	White	100	1132	234	13.66	0.85	0.07	12.17	0.005	–0.01
Chiasma Frequency	Varieties	No. of cell observed	No. of bivalent with Chiasmata in cell studied				Total chiasmata	Average chiasmata per nucleus	Chiasmata per bivalent	Terminalization coefficient
			No. of rings		No. of rods					
	PINK	100	1068		266		1334	13.34	1.67	80.06
	WHITE	100	1132		234		1366	13.66	1.70	82.86

carbohydrates, flavonoids, etc. It contains many essential phytochemical constituents such as leurosine, ajmalicine, raubasin, reserpine, vincoline, rosicine, leurosidine, vincolinine, vinacardine, leurocristine, vindoliscine, pleurosin, catharanthamine, deoxyvinblastine, vincardine, tabersonine, roseadin, etc. These chemical, constituents have significant pharmacological activities.

C. roseus assumed more importance particularly after the isolation vincristine and vinblastine from the leaves of the plant. The positive and significant anticancerous properties of these alkaloids propelled *C. roseus* to high pedestal in the management of solid tumors and lukemias (Hodes et al. 1961). The discovery of both of these active substances can be traced to a large scale effort of the United States National Cancer Institute in the 1950s to screen plants for anticancer properties. In the contemporary phase of science these alkaloids or now commercially produced and marketed by leading pharmaceutical industry houses.

The period of crowning glory came for *C. roseus* (L.) when it was aptly so by virtue of its unique medicinal properties to cure blood cancer. Extracts from Madagascar Periwinkle have been shown to be effective in the treatment of various

Fig. 1.15 Chemical constituents present in *C. roseus*

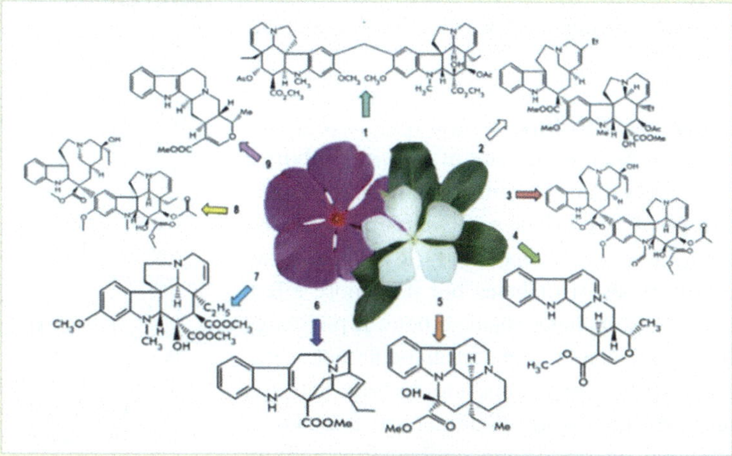

Alkaloids produced by Catharanthus roseus (1) vindolicine ($C_{51}H_{64}N_4O_{12}$, 925.08 g/mol); (2) anhydrovinblastine ($C_{46}H_{56}N_4O_8$, 792.97 g/mol); (3) vincristine ($C_{46}H_{56}N_4O_{10}$, 824.95 g/mol); (4) ajmalicine ($C_{21}H_{24}N_2O_3$, 352.43 g/mol); (5) tabersonine ($C_{21}H_{24}N_2O_2$, 336.44 g/mol); (6) catharanthine ($C_{21}H_{24}N_2O_2$, 336.42); (7) vindoline ($C_{25}H_{32}N_2O_6$, 456.53 g/mol); (8) vinblastine ($C_{46}H_{58}N_4O_9$, 810.97 g/mol)

kinds of leukemia, skin cancer, lymph cancer, breast cancer, and Hodgkin's disease. Vincristine Sulphate is being marketed under the trade name ONCOVIN, which is used against acute leukemia and vinblastine Sulphate as VELBE to cure Hodgkin's disease and other lymphomas and choriocinomas (Sharma 2004) roots of *C. roseus* have more ajmalicine and serpentine than even *Rauwolfia serpentina* (Jain 1996) (Fig. 1.15).

1.1.9.2 C. *Pusillus*

Catharanthus pusillus plant contains alkaloids—pusiline and pusilinine, leurosine, lochnerinine venoterpine, vindorosine, and vincapusine (isolated from leaves) and lochnericine (isolated from roots).

1.1.10 Medicinal Properties

1.1.10.1 C. *roseus*

Catharanthus has a fascinating global history in terms of its application for a host of medicinal purposes in diverse nature of ailments. It is used worldwide as panacea for different diseases (Duke 1985). It is used to treat wasp sting, diabetes, chest ailments, laryngitis, Menorrhagia rheumatism, asthma, flatulence, high blood pressure, and malaria (Chopra et al. 1956). Apart from anticancerous properties, it has medicinal importance for its hypotensive, sedative, tranquilizing and antispasmodic properties due to the presence of ajmalicine, serpentine, and reserpine alkaloid. Reserpine is used for circulatory disease and hypersensitivity. Due to its powerful herbal effect, the French called it "Violet of the Sorcerers".

1.1.10.2 C. *Pusillus*

The decoction of dried plant of *C. pusillus* boiled in oil is rubbed on the loins in case of lumbago. Whole plant is used in paralysis, epilepsy, and ulcers (Singh and Kumar 2007). *C. pusillus* can cure both types of diabetes but it is more effective in type–2 diabetes management. It possesses both hypoglycemic and antidiabetic activities. Apart from hypoglycaemic antidiabetic activity, it has anesthetic, anti-inflammatory, anticancer, antimalarial, antiviral, and antibacterial activity (Navitha et al. 2012).

1.1.11 Ayurvedic Properties

The science of Ayurveda mentions that the periwinkle possesses the properties of Tikta Rasa (in bitter taste), Rukshan Guna (i.e., dry qualities), and Ushna Virya (i.e., hot potency). Owing to the inherent properties, periwinkle (sadabahar) alleviates the Vata (i.e., air) and Kapha (i.e., earth and water) Doshas. The ayurvedic properties of the herb bestow the following effects on the body including Kashayaphaha (i.e., improves exhausted body tissues and treats chronic respiratory conditions like tuberculosis), Kshataksheenahara (i.e., helps in wound healing), Kapha—hara (expectorant), Madakari (hypnotic), Varnya and Agni-vriddhikar prabhava (i.e., effects on the body).

References

Bentham G, Hooker JD (1876) Genera plantarum, vol II. L. Reeve and Co. London, UK

Bowden WM (1945) (a) A list of chromosome number in higher plants I Acanthancea to Myrtaceae. Am J Bot 32:81–92

Chopra RN, Nayar SL, Chopra IC (1956) Glossary of Indian Medicinal plants. Council of Scientific and Industrial Research, New Delhi, India.

Danyansagar VR, Sudhakaran IV (1968) Meiotic studies in *Vince rosea* Linn. Cytologia 33:453–464

Duke JA (1985) Hand book of medicinal herbs. CRC Press, Boca Raton, FL, pp 105–107

Furusato K (1940) Polyploid plants induced by colchicine. Bot Zool 8:1303–1311

Hodes MR, Rohn RJ, Bond WH (1961) Vincaleucoblastin II. Further experiences in solid tumor and leukemias. In: Canadian cancer conference proceedings, vol 3. Academic, New York, pp 389–398

Jain SK (1996) Medicinal plants. National Book Trust India, New Delhi, India, pp 47–48

Levitsky GA (1931) The Karyotype in systematics. Bull Appl Bot Genet Plant Breed 27:19–174

Mustafa MA, Ahlam SEltahir and Hatil El- Kamali (2016) Comparative morphanatomical leaf character of *Nerium olender* and *Catharanthus roseus* family apocynaceae. Eur J Basic Appl Sci 3(3):68–73

Navitha A, Helen Sheeba DA, Sartaj RC, Banu M (2012) Hypoglycemic and antidiabetic activity of ethanolic extract of *Catharanthus pusillus* (murray) G. Don. ISOR J Pharm 2(4):17–21

Schuman K (1895) In: Engler A, Prantl KA Die Natürlichen Pflanzenfamilien 4.2:115,f.57

Singh G, Kumar J (2007) Few Medicinal plants of family apocynaceae in Ranchi. System Link, Ranchi, India, pp 22–25

Stearn T (1966) *Catharanthus roseus*, the correct name for the Madagascar periwinkle. Loloydia 29:196–200

Stebbins GL (1971) Chromosomal evolution in higher plants. Edward Arnold Ltd, London, UK

Catharanthus Roseus: A Source of Anticancer Phytomedicines

2

Neha Sharma, Indranil Singh, Rohit Sam Ajee, and Shuchi Kaushik

Abstract

Catharanthus rosesus, widely known as the Madagascar periwinkle and the Cape periwinkle, is a flowering plant believed to have potent medicinal and curative properties. Decoctions containing the extract of the plant have been used since older times as a potential cure to diabetes and the dreadful disease, cancer. However, to test the validity of the claims extraction of the valuable compounds in order to make an informed statement regarding the plant is required. As of late, with an inevitably detailed comprehension of the structure and function of components, a progression of new advancements and strategies have been applied to the improvement of therapeutic plants. In the event that we can set up a speedy and helpful pathway by which to first precisely anticipate an enormous number of substance mixtures and afterward, in light of these outcomes, act in vivo and in vitro pharmacological tests for check, this system will altogether improve the effectiveness of assessing the compound exercises of medicinal plants.

2.1 Basic Introduction to Cancer, Its Types and Plant Based Cancer Treatment

Cell division is the fundamental phenomenon in growth and development of an individual. However, when this basic process gets unregulated it results in a severe metabolic syndrome commonly known as cancer. Cancer is the uncontrolled symmetrical division of cells that leads to abnormal growth and causes an adverse effect on the general health and wellbeing of the body (Shahriyari and Komarova 2013). It is a complex disease that proves to be a global burden and one of the major medical challenges of the world. Despite advancement in medical diagnosis and treatment strategies, it is still one of the leading causes of mortality and morbidity worldwide and the number of cases is constantly increasing estimated to be 21 million by Ferlay et al. (2020) (Shahriyari and Komarova 2013; Fridlender et al. 2015; Bai et al. 2017; Rayan et al. 2017; Tay et al. 2019).

Cancer is due to accumulation of genetic alterations in cells such as mutations in DNA repair genes, tumor suppressor genes, proto-oncogenes, and genes involved in cell growth and metabolism. These instabilities transform

N. Sharma
Amity Institute of Biotechnology, Amity University, Madhya Pradesh, Noida, India

I. Singh · R. S. Ajee
Independent Researcher, Sagar, India

S. Kaushik (✉)
State Forensic Science Laboratory, Sagar, Madhya Pradesh, India

© The Author(s), under exclusive license to Springer Nature Switzerland AG 2022
C. Kole (ed.), *The Catharanthus Genome*, Compendium of Plant Genomes,
https://doi.org/10.1007/978-3-030-89269-2_2

normal cells into malignant cells and can be caused due to both external and internal factors (Krishnamurthi 2007). Cancer can be of different forms and in different parts of the body, among which lung cancer is reported the most in males followed by breast cancer in females (Sung et al. 2021).

The treatment strategies for cancer include a number of processes such as conventional treatments with surgery, radiation therapy, chemotherapy, hormone therapy, and/or in combination with highly targeted therapies such as immunotherapy and monoclonal antibody therapy. The drugs mainly work on inducing the intrinsic or extrinsic signaling pathways of apoptosis. The major drawback of using chemotherapeutics is the severe side effects on general health of the patient as these drugs are cytotoxic in nature and kill not only the cancer cells but the healthy ones also. The other adverse effects of these treatment procedures include limited bioavailability, toxicity, nonspecificity, and promotion of recurrence or metastasis (Gupta et al. 2013).

A lot of work has been done in order to minimize the side effects of cancer therapeutics, for example, reducing the adverse impacts on the close by cells and tissues, expanding drug aggregation and adequacy in the wound, creating novel drug delivery systems, and focusing on the targets (Iqbal et al. 2017). In this manner, widespread investigation has been coordinated to translate additional promising alternatives for better adequacy and less harmful malignancy treatments (Shim and Liu 2014).

Almost 80% of the total population rely upon routine medications and over 60% of clinically endorsed anticancer medications are obtained from these therapeutic plants (Cotto et al. 2018; Wishart et al. 2018). As indicated by extensive research work, there are numerous anticancer drugs clinically recognized and are suggested for malignancy therapy (Stark et al. 2006; Berman et al. 2000).

2.1.1 Transition from Traditional to Modern Medicine

The compilation of information, proficiency, and practice for the maintenance of health and to prevent, diagnose and treat various diseases traditional medicines prove to be a medical boon to society (Wachtel-Galor et al. 2011; Smith et al. 2014). Herbal drugs are the main component of traditional medical system (Wachtel-Galor et al. 2011) and are currently the oldest and the most frequently used medicine system in the world (Ogirima et al. 2015).

Contrasted with the non-customary medication, the Traditional Medicine (TM) follows an alternate methodology in the treatment measure (Solati 2016, 2017; Hasanpour-Dehkordi et al. 2017). The traditional medicine clinical information has a specific detailed account of patient's illness and the formulation of herbal medicine (Ansani et al. 2003; Kim et al. 2013; Carvalho et al. 2014). Accordingly, the coordinated information system satisfactorily records the TM's data. Subsequently, it prevents the repetition of various clinical tests and the information acquired through TMs data can be analyzed for providing better treatment (Wiesner and Knöss 2014; Walker 2015).

As a developing practice throughout the world, Traditional Medicine (TM) is encountering its quick progressions in time dependent health technology. The advanced and innovation based procedures are imperative for controlling and keeping up the nature of the native herbal medications (Sen and Chakraborty 2017). The formation of a data set for putting away an enormous bit of the information into helpful information has successfully added to the advancement of the Traditional Medicine Information System (TMIS) (Noraziah et al. 2011; Ogirima et al. 2015). China's Comprehensive Herbal Medicine Information System for Cancer (CHMIS-C) has filled in as a suitable data asset for the TM specialists (Fang et al. 2005).

Similarly, Web-based Decision Support System for Prescription in Herbal Medicine in Nigeria has assumed a huge part in controlling the nature of the homegrown medications remedy (Ogirima et al. 2015). Created for discussing with patients in the e-wellbeing framework, E-Health Record System (EHRS) in Australia has effectively helped the TM's experts in the treatment management (Bjering et al. 2011).

Despite the high idea given to the TM practice all over the world, it seems to go up against confusing troubles. The main drawback is the absence of a reference standard for deciding the valid measurement of the natural medication for the patients. This subsequently has achieved the development of misguided and deficient information about the characteristic prescriptions (Aghebati et al. 2014; Ogirima et al. 2015). There are no clear instructions regarding quality assurance, also no universal unified terminology, lack of evaluation of the safety as well as the lack of patient information system in the TM-centered clinics or hospitals are among the other challenges (Evans 2008; Zhou et al. 2010; Janamian et al. 2011; Niemeyer et al. 2013; Snow 2016). The local prescriptions are consistently acknowledged to have no adverse impacts on health so various buyers regard them better than the chemical drugs. Understand that the senseless use of these drugs would be connected with extended threats endangering the treatment process's success (Mukherjee et al. 2015). The viability of TM as a complicated framework has been convincingly tried and approved in various logical investigations.

2.1.2 *Current Cancer Therapy* Via *Phytochemicals: A Novel Approach*

It has been reported by World Health Organization (WHO) that nearly 70–95% of the world population use plant based drugs for meeting their healthcare requirements (WHO 2004; Jivad et al. 2016; Parsaei et al. 2016; Rouhi-Boroujeni et al. 2016; Sen and Chakraborty 2017). World Health Organization (WHO) report showed that more than 30% of all plant species have at some time been utilized for medicinal purposes (WHO 2013). The logical investigation of conventional therapeutic plants is vital for human wellbeing.

Distinguishing and anticipating the pharmacological premise of the action of customary therapeutic plants are significant for the objective of modernizing their utilization. Since the synthetic constituents of therapeutic plants are perplexing and fluctuate, explaining the particular compound segments in such plants and their major bioactive components is an intricate assignment.

The customary model of medicinal plant exploration can for the most part be isolated into the following steps: first, the extraction of compound monomers or divisions, trailed by their subjective and quantitative distinguishing proof, and afterward an assortment of pharmacological trials, for example, in vitro examinations and infusion or taking care of these array in individual at that point performing compelling estimation. As a rule, the entire examination measure is tedious and costly. In any case, a few pieces of this basic methodology can be changed to improve proficiency.

As of now, the science of the most broadly utilized medicinal plants, like ginseng and licorice, has been studied completely. Many synthetic mixtures have been extricated from these herbs. However, the vast majority of these mixtures have not to be examined whether they have likely no natural action. It would be a huge and tedious venture to methodically assess the exercises of these components by regular strategies.

2.2 Brief Review on *Catharanthus*

Catharanthus roseus (L.) is an important medicinal plant of the Apocynaceae family. Periwinkle and Nayantara are the common names for a pair of perennial flowering. The Latin name for this herb is periwinkle. There are about two common cultivars of *C. roseus* which are known on the basis of their flower color that is the pink flowered "Rosea" and also the white flowers "Alba". Periwinkle is found to be a species of *Catharanthus* native and also endemic to Madagascar. The synonyms of the plant name include Periwinkle, *Ammocallis rosea,* and *Lochnera rosea*, other English names occasionally used for the plant include periwinkle, Rosy Periwinkle, and "Old Maid" (Retna and Ethalsha 2013; Sain and Sharma 2013).

2.2.1 Classification

Kingdom: Plantae.
Order: Gentianales.
Family: Apocynaceae.
Genus: *Catharanthus.*
Species: *roseus.*
Binomial name: *Catharanthus roseus.*
Botanical Name: *Vinca rosea* (Erdogrul 2002).

2.2.2 The Wonderful Plant with Multifarious Activities

2.2.2.1 Morphology

Catharanthus roseus is an evergreen herb that is up to 1 m tall. The leaves are oval in shape with 2.5–9.0 cm. long and 1–3.5 cm. broad glossy green hairless pale midrib. It has a brief petiole about 1–1.8 cm. long and is arranged in the opposite pairs. The flowers are white to dark pink with a red center color. It has a basal tube about 2.5–3 cm. long and a corolla about 2–5 cm. diameter with five petal like lobes. The fruit maybe a pair of follicles about 2–4 cm. long and three mm broad (Sain and Sharma 2013).

2.2.2.2 Geographical Distribution

Catharanthus roseus is an inhabitant of the Indian Ocean Island of Madagascar. It is commercially cultivated in the United States, Spain, China, Africa, Australia, Southern Europe, and India for its medicinal uses. In the past, it was found to be under endangered plants and the main reason for their decline is the habitat devastation by the slash and burn agriculture. Now it is commonalty found in tropical and subtropical regions of the world including the Southern United states (Lata 2007; Sain and Sharma 2013).

2.2.2.3 Growth Conditions, Mode of Propagation, Culture

Catharanthus roseus is a native of Madagascar. It is cultivated in tropical areas as an ornamental plant. It is best grown as an annual plant in sandy loams. It needs regular moisture but avoids excessive watering. Cutting may be taken from plants in late summer for overwintering so as to provide a stock the following spring. Many cultivators have been selected for variation in flower color (white, mauve, peach, scarlet and reddish orange) and also for tolerance of cooler growing conditions in temperate regions (Muhammad et al. 2009; Sain and Sharma 2013).

2.2.2.4 Extraction of Vinca Alkaloids

For the optimal recovery of bioactive compounds from *C. roseus* several drying methods are employed prior to the actual extraction process. Drying the raw plant material was shown to greatly increase the yield of the compounds in the extraction phase. Freeze drying and thermal drying are some methods used for the preparation of the dry *C. roseus* sample. (Pham et al. 2017). After this, one of the various extraction techniques can be used to obtain the desired compounds. Conventional extraction methods such as soaking the sample in solvents and letting the mix rest overnight can be used to extract compounds like vincristine, vinblastine, catharanthine, vindoline, etc. Ultrasonic extraction techniques are advanced extraction processes that have greater yield as well as shorter extraction times compared to conventional techniques. Microwave assisted extraction (MAE), enzyme assisted extraction, supercritical extraction, etc. are all alternate methods that have been employed in the case of *C. roseus* samples.

However, regardless of the efficiency of the extraction process, the amount of Vinca Alkaloids obtained at the end of the process depends primarily on the content of the compound in the plant tissue. This is a matter of concern as Vinca

Alkaloids are present in quite low amounts within *C. roseus*. To remedy this, various studies are being carried out to increase the yields of the compounds in *C. roseus*. Two of the common methods used to increase the yield of Vinca Alkaloids are:

1. In vitro cultivation methods are used to optimize and maximize the production and extraction of Vinca Alkaloids.
2. Recombinant DNA technology is used to increase the production of the enzymes involved in the production of Vinca Alkaloids, which ultimately results in the increased production of the bioactive compounds (Collu et al. 2001).

2.2.3 Use in Traditional and Modern Medicine

Catharanthus roseus has many potentially active chemical constituents. It possesses carbohydrate, flavinoid, saponin, and alkaloids. Alkaloids are the main active chemical constituents of *C. roseus*. More than 400 alkaloids are present in the plant, which have pharmaceuticals, food additives, and agrochemicals properties. The alkaloids like actineo plastidemeric, Vinblastin, Vincrestine, Vindesine, Vindeline Tabersonine, etc. are mainly present in aerial parts whereas ajmalicine, vinceine, vineamine, raubasin, reserpine, catharanthine, etc. are present in roots and basal stem. Rosindin is an anthocyanin pigment found in the flower of *C. roseus* (Sain and Sharma 2013). The herb has been used for hundreds of years to treat a variety of ailments and was a favorite ingredient of medical components within the middle ages because of the multifarious activities of its components such as antitumor, antibacterial, antihyperglycemic, antihypertensive, diabetic wound healing activity, asthma, constipation and menstrual problem (Sain and Sharma Sain and Sharma 2013; Retna and Ethalsha 2013).

The genus *Catharanthus* consists of eight species of which seven are native to Madagascar and one, *C. pusillus* is from India. Periwinkle is one of the pharmacological plants that have an extended history of therapeutic journey from Mesopotamian folklore of 2600 BCE till today playing a significant role as an herbal and traditional medicine of various diseases (Das and Sharangi 2017), which is mentioned in Table 2.1.

2.3 Alkaloids Present in *Catharanthus*

C. roseus has been found to contain a variety of alkaloids having anticancer properties including vinblastine, vincristine, vindoline, vindolidine, vindolicine, vindolinine, and vindogentianine (Cragg and Newman 2003; Tiong et al. 2013, 2015). They hinder cell expansion by changing the microtubular elements, which prompts apoptosis (Almagro et al. 2015). Of these, vinblastine and vincristine were the principal plant-determined anticancer specialists conveyed for clinical use (Cragg and Newman 2003). Vinblastine sulfate has been used for therapy of Hodgkin's infection, lymphosarcoma, choriocarcinoma, neuroblastoma, carcinoma of bosom and lungs, and different organs in intense and persistent leukemia. Vincristine sulfate, an oxidized type of vinblastine (Almagro et al. 2015) captures mitosis in the metaphase and is successful in treating intense leukemia (in youngsters), lymphocytic leukemia, Hodgkin's illness, Wilkins' tumor, neuroblastoma, and reticulum cell sarcoma (Aslam et al. 2010). A number of other indole alkaloids extricated from *C. roseus* have been found to display strong cytotoxic action against different malignant growth cell lines. Catharoseumine, another monoterpenoid indole alkaloid disengaged from the *C. roseus* entire plant was found to have an inhibitory impact on the human promyelocytic leukemia HL-60 cell line with an IC50 of 6.28 μM (Wang et al. 2012a). Also, three recently detached dimeric indole alkaloids including 140, 150- didehydrocyclovinblastine, 17-deacetoxycyclovinblastine, 17-deacetoxyvinamidine and five known mixtures (vinamidine, leurosine, catharine, cycloleurosine, and leurosidine) had in vitro restraint of cell practicality against the human

Table 2.1 List of therapeutic agents present in different plant parts of periwinkle and their applications

S. No	Part of plant	Therapeutic agents	Application	References
1	Ethanol extract of flower and leaves	Alkaloids, tannins, and tri-terpenoids	Diabetic wound healing activity	Mohammad et al. (2009), Retna and Ethalsha (2013)
2	Methanolic extracts of various parts	Flavonoids, coumarin, quinine, and phenolic compounds	Antioxidant activity	Mohammad et al. (2009), Retna and Ethalsha (2013)
3	Leaves extract	A Alkaloids and carbohydrates	Hypotensive effects	Mohammad et al. (2009), Retna and Ethalsha (2013)
		B. Vincristine and Vinblastine	chemotherapeutic agents, lymphoma, and childhood leukemia	Jonathan et al. (2010)
		C. Whole extract could be used as the prophylactic agent	Antimicrobial activity against the microorganisms like *Pseudomonas aeruginosa* NCIM 2036, *Salmonella typhimurium* NCIM 2501, *Staphylococcus aureus* NCIM 5021	Prajakta and Ghosh (2010)
		D. Vincamine and Vindoline alkaloids	Anti-ulcer property	Babulova et al. (2003)
		D. Vincamine	Cerebro-vasodilatory and neuroprotective activity	Sain and Sharma (2013)
4	Leaf juice of Catharanthus roseus (Linn.) G. Donn	Flavonoid, and probably, vinpocetine	Atherosclerotic activity	Patel et al. (2011)
5	Wax layer of leaves	Catharanthine	Antifungal activity against *Candida albicans*	Yordanov et al. (2005)
6	Ethyl acetate fractions of leaf extract	–	Potent Insecticidal	Deshmukhe et al. (2010)

bosom disease cell line MDA-MB-231 with an IC50 scope of 0.73–10.67 μM (Wang et al. 2012b). Significantly, cathachunine, another bisindole alkaloid from *C. roseus*, showed antitumor impacts on human leukemia cells, however at much lower cytotoxicity against typical human endothelial cells, demonstrating that its activity repressed specifically toward leukemia cells. Alongside the examinations on anticancer action of individual alkaloids from *C. roseus*, impact of the whole rough concentrate on different malignant growth cell lines was additionally explored. Late discoveries tracked down that *C. roseus* root and stem extract had solid in vitro cytotoxic action against a number of cancer cell lines (Pham et al. 2018, 2019). Also, the investigation of Fernández-Pérez et al. (Fernández-Pérez et al. 2013) affirmed that the incredible antitumor action of the indole alkaloid-improved concentrate got from *C. roseus* cells didn't result from the impact of a solitary compound, however, relied upon the joint activity of bioactive mixtures. These outcomes uncovered the synergistic impact and positive connection of the bioactive components present in *C. roseus* against malignant growth cells, which has likewise been found in other plant materials (Barth et al. 2007; Bhuyan et al. 2018)and considered as a technique for disease treatment (Jiang et al. 2010). This marvel is conceivable because of the help of mixtures with next to zero action to the primary dynamic mixtures through aiding the active principle to reach the target by improving bioavailability, diminishing absorption and discharge, supplementing systems of activity, inversion of obstruction, and adjusting results (Rasoanaivo et al. 2011).

The following sections will focus on *Vincristine* (VCR) and *Vinblastine* (VBL), well-characterized alkaloids which are used to cause great effects in cancer therapy. One of the major concerns regarding the extraction and availability of both the compounds is their particularly low yield in *Catharanthus*. VCR is obtained at a percent yield of less than 0.0003% with VBL having similar yields. Hence in future research greater focus needs to be placed on the production of *Catharanthus* lines that can yield greater amounts of the chemotherapeutic compounds.

2.3.1 Vincristine (VCR)

This purified alkaloid is obtained in the extract from periwinkle plants, primarily *Catharanthus rosesus*. *Vincristine* is classified as a vinca alkaloid, which is a class of compounds that express cytotoxic activity disrupting microtubule functionality. In addition to *Vincristine, Vinblastine* and *Vinorelbine* are also classified as Vinca Alkaloids and have similar functions. VCR has been used for the treatment of conditions like rhabdomyosarcoma, acute leukemia, Hodgkin's disease, Wilm's tumor, etc. A combination therapy of VCR and other oncolytic agents are used in the treatment process (Kumar 2016).

As mentioned above, VCR has well-characterized anti-mitotic and antitumor properties. Several studies exist regarding the activity of VCR against tumors in both animals and humans. It is effective in the treatment of Hodgkin's disease as well lymphomas and acute leukemias. This is primarily attributed to the Stathmokinetic nature of Vinca Alkaloids, which naturally include *Vincristine*. The term Stathmokinesis is derived from the Greek words *Stathmos* (arrest/stop) and *Kinesis* (movement), to mean "prevention of motion". In the biological field, the word is used to refer to the prevention of the mitotic cycle. Such agents which are able to arrest mitosis are called Stathmokinetic agents. These agents arrest the process of mitosis in the metaphase stage by various means (Gedda and Cardinali 1968).

Microtubules are important intracellular structures that play various roles. They are particularly essential for the progression of the mitotic cell cycles. Damaged microtubules not only prevent the furthering of the cell cycle but also trigger apoptosis. Stathmokinetic agents prevent the multiplication of the tumor cells by disrupting the process of mitotic spindle formation. The lack of mitotic spindles prevents the segregation of chromosomes, which triggers apoptosis resulting in cellular death. Based on which part of the formation process is disrupted, Stathmokinetic compounds can be categorized into two classes. One class interacts with and binds to the tubulin protein, which prevents their polymerization into mitotic spindles. The other class binds to the microtubules and stabilizes them. This prevents them from depolymerizing into the compound's units, which ultimately prevents mitotic spindle formation. Vinca Alkaloids fall into the first class. *Vincristine* and *Vinblastine* both specifically bind to tubulin at the β-subunit of the *Vinca* domain (Avendaño and Menendez 2015). Once the Vinca Alkaloids bind to the free tubulin, they cause conformational changes in the structure of tubulin. These changes stabilize the protein, resulting in the formation of crystal aggregates. The lack of free tubulin disrupts the assembly of microtubules into mitotic spindles, which ultimately results in the destruction of existing mitotic spindles and arresting the cell cycle. The lack of the mitotic spindles entails that the duplicated chromosomes are not segregated in the cell cycle. A checkpoint system exists to ensure that cells that do not have mitotic spindles or segregated nuclear material in the metaphase are restricted from entering the anaphase and are ultimately tagged for destruction. This basically outlines how both *Vincristine* and *Vinblastine* interact with and prevent the growth of cancerous cells (Avendaño and Menendez 2015).

As with all things, the use of the Vinca Alkaloids for chemotherapy has its own share of downfalls. The biggest problem with *Vincristine* and *Vinblastine* is that they have no specificity when it comes to tubulin binding. This translates into cancerous as well as non-target, normal cells that divide rapidly being affected by the compounds. *Vincristine's* neurotoxicity is a major hindrance to neurons due to its activating apoptotic pathways and altering their functioning. Another factor affecting the use of both *Vincristine* and *Vinblastine* is that cells are able to adapt to their presence and develop resistance towards them. P-glycoprotein a.k.a Multi-Drug Resistance Protein (MDR protein) is responsible for this aspect. Cells containing chemotherapeutic drugs (as well as other foreign compounds) can cause the over expression of the P-gp proteins which then effectively pump them out of the cell. Studies into P-gp inhibitors are being carried out to increase the efficacy of therapeutic drugs. However, most P-gp inhibitors are practically unusable due to their toxic side effects (Syed et al. 2021).

Interestingly, several marine organisms were discovered to produce substances that are functionally quite similar to our Vinca Alkaloids. Compounds like Halichondrin B, obtained from *Halichondria okada*, Dolastatin 10, obtained from *Dolabella auricularia,* and Vedotin, obtained from *Symploca cyanophyte* have been observed to disrupt microtubule activity in rapidly dividing cells by binding to the β-subunit of the Vinca domain in tubulin proteins. Several studies are currently being carried out to test their efficacy as well as increase their yield via alternate as well synthetic sources (Towley et al. 2001).

Vincristine is primarily metabolized within the liver. Cytochrome P450 3A plays a key role in clearing the drug from our systems. It is excreted from our body via bile and feces either in a non-metabolized form or a metabolite. Studies have also discovered a link between the presence of Vinca Alkaloids and the increase in concentration of cyclic AMP. This is even more noteworthy given the fact that microtubules express both adenylatecyclase activities as well as phosphodiesterase activity. The increased cAMP levels have further implications on enzymatic activity as well as increased rates of microtubule polymerization.

2.3.2 Vinblastine (VBL)

Similar to the previous vinca alkaloid, Vinblastine is obtained from *Catharanthus rosesus* extracts. They have potent cytotoxic activity and prevent the transition of cells from the metaphase to the anaphase. The mode of action of both the Vinca Alkaloids is quite similar and produces identical results on a cellular scale. However, their general usage differs based on the afflicting medical condition. Vinblastine is used in the treatment of Hodgkin's disease, different Lymphomas, Kaposi's Sarcoma, etc. Vinblastine is administered in combination with other anticancerous drugs to enhance the overall effect of the combination. A combination of drugs is used to ensure that the toxic side effects of the individual drugs do not occur while also resulting in therapeutic activity. Additionally, cancers that affect testicular germinal cells are shown to be especially sensitive to Vinblastine.

Hematologic Toxicity is the primary side effect associated with Vinblastine usage. It may also result in acute shortness of breath as well bronchospasm, which may be fatal in severe cases. Similar to Vincristine, Vinblastine attaches to the tubulin subunit and prevents microtubule formation, which ultimately results in the destruction of the affected cell. Upon binding to tubulin, Vinblastine crystallizes the protein and prevents its polymerization. In addition to their antineoplastic effects, Vinblastine is also considered to have certain immunosuppressant effects.

Like all Vinca Alkaloids, Vinblastine is composed of catharanthin and vindoline, which are ringed units. The cytochrome P450 3A is responsible for the metabolism of Vinblastine as well. And just like Vincristine, Vinblastine is eliminated from the body via bile and feces. (Kumar 2016).

2.3.3 Mode of Action of Vincristine and Vinblastine

A detailed description regarding the exact mode of action of the two Vinca Alkaloids is not possible currently, as the mechanism of action is not fully understood. What we do know is that these bioactive compounds interact with the Vinca domain adjacent to the GTP-binding site of the tubulin molecules. After a certain threshold concentration of tubulin binding, the microtubule activity within the cell is changed. Beyond this concentration, the tubulin binding reaches a critical stage at which pre-existing microtubules are depolymerized resulting in the absence of microtubules within the cell. At even higher concentrations, the Vinca Alkaloids cause the aggregation of the tubulins into para crystals. This aggregation, however, is not of much importance from a therapeutic point of view as the concentration of Vinca Alkaloids required for this happen is in the clinically toxic range (Alam et al. 2017).

When compared to the effects of taxanes on tubulin molecules (another class of stathmokinetic drugs), the interaction of Vinca Alkaloids with tubulin either results in the inactivation of the mitotic spindle or its depolymerization. Taxanes bind to tubulin molecules and cause conformational changes which prevent normal microtubule formation. Such microtubules are able to attach to chromosomes during the mitosis cycle, but are unable to separate the sister chromatids due to the changed microtubule structure (Kelling et al. 2003). When the sister chromatids do not separate, the Spindle Assembly Checkpoint (SAC) is activated which results in the mitotic arrest eventually resulting in cellular destruction. However, alternate paths also exist that the affected cell may occasionally take. The cell may divide to give rise to unequal daughter cells (aneuploid cells) or they may exit cell division without dividing (Gascoigne and Taylor 2009).

2.4 Regulatory Aspects of Herbal Anticancer Drugs

Herbal medicines have been in practice for many centuries now, with the written records going back to almost 5000 years ago (Swerdlow 2000). Till 1890, about 59% of the listing available on the US Pharmacopeia were related in one way or another to the herbal origins, while till recently, it has been found that one out of five adults in the USA claimed to be using herbal products (Barnes et al. 2004; Revathy et al. 2012). Botanical ingredient-based herbal food supplements came unscathed from global remission, counting for a massive 7.2 billion Euros in 2015, and were expected to grow and reach 7.9 billion Euros by 2020 (Gray 2015; Jeurissen et al. 2018). While the American Botanical Council reported sales of US$ 5.3 billion in 2011, that grew to US$ 6 billion by end of 2013 (Lindstrom et al. 2014). Besides these, a huge proportion of the African and Asian population depends upon traditional medicine and medicinal herbs for their primary healthcare. Some of these were taken forward by the ancient systems of medicinal knowledge like Ayurveda, Siddha, and Chinese medicine (WHO 2008; Pan et al. 2014).

Recently, an upsurge in the preference of herbal drugs together with a flexible regulatory system and associated toxicity has possessed many questions on the application of these drugs and supplements (Elvin-Lewis 2001; Cuzzolin et al. 2006; Sahoo et al. 2010). Despite Herbal medicine's presence for many centuries, there are no common ground regulations (Thakkar et al. 2020). While FDA (Food and Drug Administration) is the primary body regulating herbal supplements in the USA. They allow the herbal supplements to be sold as dietary supplements and not as the prescribed over-the-counter drugs. This provides the manufacturer enough opportunities to bypass the stringent process of demonstrating safety and efficacy as the law defines it broadly, and even though they are not allowed to make direct claims they can still make structural and functional claims on health benefits (Bent 2008).

The research and development plan for a new pharmaceutical drug is well laid out and straightforward, while the same is not true for herbal drugs. As plants rarely have one specific active pharmacological compound it makes it difficult to report the specific functions (Sahoo et al. 2010). The relative concentration or chemical profiles of the different compounds found in a plant could vary a lot and can significantly influence the results. Besides the above-mentioned heterogeneity, there are various other things to be taken under consideration, like adulteration of either synthetic or chemical ingredients that are common with the trade of medicinal plant is one (Chan 2003; Yee et al. 2005; Lam et al. 2008), while others are misidentification of plant based on morphology (Ramawat and Goyal 2008; Yap et al. 2008), intentional or accidental addition of toxic heavy metals (Rai et al. 2001; Dargan et al. 2008; Mazzanti et al. 2008; Zhang et al. 2012;), and pesticides (Rai et al. 2008; Xue et al. 2008a, 2008b) or microbial infection (Bugno et al. 2006) could potentially affect the therapeutic values and need to be regulated.

Phytomedicine development is a multi-step process, that begins with the collection of raw material, followed by identification, stabilization, extractions, quality assurance, isolation of active compound, purification, fractionation, and toxicity evaluation. Throughout the development, process standardization is a critical step that depends on many factors including but not limited to cultivation conditions, distribution, manufacturing, and marketing. Further, the biochemical profile of a plant and its secondary metabolites are also reported to be affected by genetic, moisture, climate, nutrient, photoperiod, storage, harvesting, extraction, and packaging. In order to keep the standard of drug efficacy and safety, chemical consistency should be regularly checked and kept in optimum range across the process (Sahoo et al. 2010).

Due to inherent complexities and possible variations at every step of the process, a multi-technique approach encompassing botanical technique, chemical marker, etc. needs to be taken together for quality control. Recently following techniques and steps are used for the identification and regulation of herbal drugs:

- Qualitative analysis is done through moisture content, ash value, extractive value, pesticides, heavy metal, essential oil, phytochemical profile (flavonoids, saponins, tannins, phenolics, steroids, etc.) and crude fiber (Holst 1973; Idu et al. 2008; Abdelhadi et al. 2015; Akram et al. 2015; Chen et al. 2015).
- Morphological identification and Macroscopic evaluation involve organoleptic study, micromorphology study of roots, bark, stems, rhizomes, woods, leaves, seeds, fruits, and aerial parts (Bauer 1998; Quettier-Deleu et al. 2000; Grubesic et al. 2005; Smillie and Khan 2010).
- Downstream evaluation is done for better identification of solvent, extract, polarity, pH, thermostability, phytochemical group, and evaluation of the drug residue post downstream extractions (Azmir et al. 2013; Varma 2016; Ngo et al. 2017).
- Chemical analysis such as thin-layer chromatography, High-performance thin-layer chromatography, and High-performance liquid chromatography is most frequently used for better identification (Lazarowych and Pekos 1998; Xie et al. 2006; Raclariu et al. 2017).
- Metabolite analyses are usually done through LC/MS, GC–MS, LC-NMR, protein analysis, and HPLC/MS (Alali and Tawaha 2009; Farag et al. 2012; Trivedi et al. 2017; Bendif et al. 2020).
- Bioassay and Herbal drugs evaluation could be done in vitro/in vivo. Additionally, one can also study different enzymes, receptors, and potentially can run various high-throughput screens for the drug evaluation and its parasiticidal, anti-viral, and antimicrobial effect (Chanda and Nagani 2013; Tiong et al. 2013; Naz et al. 2015).
- Safety study involves toxicity study, herb-drug interaction, induction, and inhibition of CYP450, and pharmacovigilance (Nasri 2013).
- Quality control is usually guided through the set of technical guidelines published by WHO

over the years. A few of them are "Quality control methods for medicinal plant material", "Guidelines of good agricultural and collection practices GACP" and "WHO guidelines for assessing the quality of herbal medicines with reference to contaminants and residues". Besides these many countries like Japan, China, India, and European associations have their own sets of rules governing the collection, manufacturing, storage, applications, and propagation of herbal medicinal plants (WHO 1998, 2003, 2007a; Liu et al. 2018).

For further information on regulatory norms development for herbal drugs across the different countries, the following reading could be a great insight: Mukherjee (2019) and Thakkar et al. (2020).

2.5 Modern Trends in Traditional Medicine Informatics and Opportunities for Anticancer Plant Products

The steps engaged with cancer development in the body incorporate metastasis, attack, angiogenesis, and aggravation. These cycles are managed by microRNAs and diverse signaling pathways (Wnt/β-catenin, Notch, Sonic hedgehog, COX-2, EGFR, MAPK-ERK, JAK-STAT, Akt/PI3K/mTOR, NF-κB, AP-1, and so on). Phytochemicals, for example, dietary polyphenols assume an effective part in lessening the speed of metastasis and different signs of malignancy as it constrict miRNAs related to different signaling pathways associated with cancer. Phytochemicals control epigenetic events (non-coding miRNAs, histone alteration, DNA methylation) and by implication influence various pathways. Photochemical apply inhibitory impact on progression of metastasis and has anticancer potential by inciting apoptosis and by restraining cell development, relocation, intrusion, and apoptosis (Appendino et al. 2014).

Vinca Alkaloids, Vincristine (VCR), and Vinblastine (VBL) were the first plant derivative products to be introduced for clinical use. These agents were sequestered from *Catharanthus roseus* G.Don (Apocynaceae). The action probably won't get affirmed by the research analysts since it was distinguished that the concentrates from the plants start the bone marrow dejection in mice rodents and furthermore gradually decline the number of white platelets. The concentrate from the plants likewise broadened life span of mice that bears transplantable leukemia in lymphocytes. The two powerful alkaloids vincristine and vinblastine were isolated by extra fractionation and extraction. The plants were mostly found in the specific spots in Madagascar yet the plant tests utilized for the innovation of vinblastine and vincristine were taken from the Jamaica and Philippines. As of late, it was tracked down that the semi-engineered vinca alkaloid is identical to vindesine (VDS) and vinorebline (VRL). To battle with different malignancy infections at first these drugs are utilized without any help or in combination with other chemotherapeutic medications. Vinblastine is chiefly utilized for the treatment of different malignancy illnesses like bosom disease, Kaposi's sarcoma, lymphomas, lung tumors, testicular malignancy, and leukemia. It showed excellent outcomes in the treatment of leukemia explicitly in the treatment of intense lymphocytic leukemia at a young age. Public disease foundation in July 2004 revealed that there are around 2069 clinical preliminaries for cancer and alongside these specialists, 160 mixes of medications were designed for the therapy of different malignancy illnesses (Cragg and Newman 2005).

Back in the 1950s, Artemisinin, the most well-known antimalarial drug, was likewise gotten through large scale screening from plant sources. From that point onward, a few FDA endorsed drugs were created from regular herbs or plant products through the *in-silico* approaches. Angiotensin converting catalyst inhibitor-Captopril was the primary example of the use of chemical inhibitor structure to create drugs in the 1970s (Tu 2011, 2016). Furthermore, the primary FDA affirmed HIV-1 chemical inhibitor–Saquinavir was additionally created by *in-silico* approaches (Zaman et al. 2002). These days, with the persistent headway of software engineering, fruitful instances of discovering drugs from characteristic items utilizing computer helped drug plan strategies have become all the more often, like Dozamide (Affirmed by FDA in 1995), Imatinib (Endorsed by FDA in 2001), Dasatinib (Endorsed by FDA in 2006), and Ponatinib (Endorsed by FDA in 2012) (Ghosh and Gemma 2015). With the persistent development of computer technology, the *in-silico* approach of using a computer science stage to compute the mixes of reenacted mixtures and targets has gotten progressively precise. Moreover, the improvement of network pharmacology advances has empowered the fast explanation of the intricate connections between compounds and their different action targets (Rubio-Perez et al. 2015). Traditional medicines (TM) include an exceptionally unpredictable arrangement of compound substances. Its intricacy isn't just reflected in the piece of the compound constituents yet in addition epitomized in the organization of connections between the remedy and the human body and the effort of pharmacological impacts through numerous channels, various targets, and the by and large administrative system (Zhang et al. 2016).

Concerning ramifications of traditional medication informatics, the use of current research technology to examine these should address information on a collection of qualities of the actual prescriptions, and the cooperation among traditional medical components and the human body (Ehrman et al. 2007). The meaning of herbal medication informatics is as per the following: an interdisciplinary science that applies data science hypothesis and computer science innovation to examine the guideline and interaction of traditional medicinal data stream and researches the perspectives on traditional medicine experts for the same (Yi et al. 2016).

The meaning of network pharmacology is like a discipline dependent on systems biology hypothesis, in which the network analysis of natural frameworks is utilized to choose the attributes of signal nodes for multi-target small molecular drug design (Liu et al. 2015). Subsequently, we can treat a solitary therapeutic plant with a similar intricacy as traditional medicine using specialized methods from Traditional Medicine (TM) informatics and network pharmacology to investigate the compound creation and the possible pharmacological establishment of the plant's belongings.

A wide-ranging complete specialized course uses a progression of in-silico ways to deal with so as to uncover the pharmacological premise of the impacts of therapeutic plants as depicted in Fig. 1, first to decide the object of study and afterward to acquire its characteristic constituents and build the compound data set. After pre-handling of the included mixtures, a three-venture virtual screening is performed. At last, network pharmacology instruments can be utilized to break down the objective set acquired from the screening; sketch out the organization topological guide of the therapeutic plants, regular constituents, natural targets, related pathways, related illnesses, and their associations; and expand on the outcomes.

The anticipated joint technique consolidates various *in-silico* approaches and, through the full utilization of a progression of virtual screening and network pharmacology machinery, can explain the substance premise and the system of therapeutic plants. The main portion of this procedure utilizes programming innovation, including active principle activity and virtual screening, all dependent on mass exploratory information, to give pinpoint exactness.

This approach is better than customary screening since it requires just primary data on compounds and can perform multi-target screening of countless mixtures in a moderately brief timeframe, which can radically diminish the expense and time of drug advancement. Additionally, joining this *in-silico* system with relevant pharmacological tests could essentially improve the proficiency of therapeutic plant research and new drug discovery.

The last part of the technique is guided by network pharmacology hypothesis and utilizations of various investigation devices and related data sets to acquire important data from large bio-data assets; construct an organization topological guide of therapeutic plants, substance constituents, qualities, proteins, pathways, and diseases; and perform information mining and examination (Gandhi et al. 2015; Tu 2016; Alamgir 2018; Chakraborty 2018; Uhlenbrock et al. 2018).

Figure 2.1 summarizes the steps involved in screening antimicrobial compounds from herbal plants and a similar approach can be adapted for identifying therapeutic compounds from the medicinal plants using in-silico and other modern advanced techniques and then later isolating those secondary metabolites or active principles to synthesize new anticancer drugs.

This in-silico technique can resolve the situation with therapeutic plants that are hard to concentrate from a down-to-earth perspective and can anticipate and explain the components of the principal component in medicinal plants. Despite the fact that computer-aided drug design helped drug plan and network pharmacology have been generally utilized and created, there still have lacks and impediments:

[1] The model development and computational accuracy of docking calculations should be additionally improved.

[2] Because of the structure based philosophy, a few mixtures are not appropriate for computer supported plans as a result of their unique design qualities.

[3] An enormous number of data sets can improve diverse data for the obtained potential targets, the advancement of the determination of these data sets and their compelling data explanation actually requires persistently functional exercises to advance.

With the advancement of software engineering, and the steady improvement of calculations, including the development of the protein model and through more useful investigations and advancement guides, the whole interaction of *in-silico* strategy can be upgraded. It is also

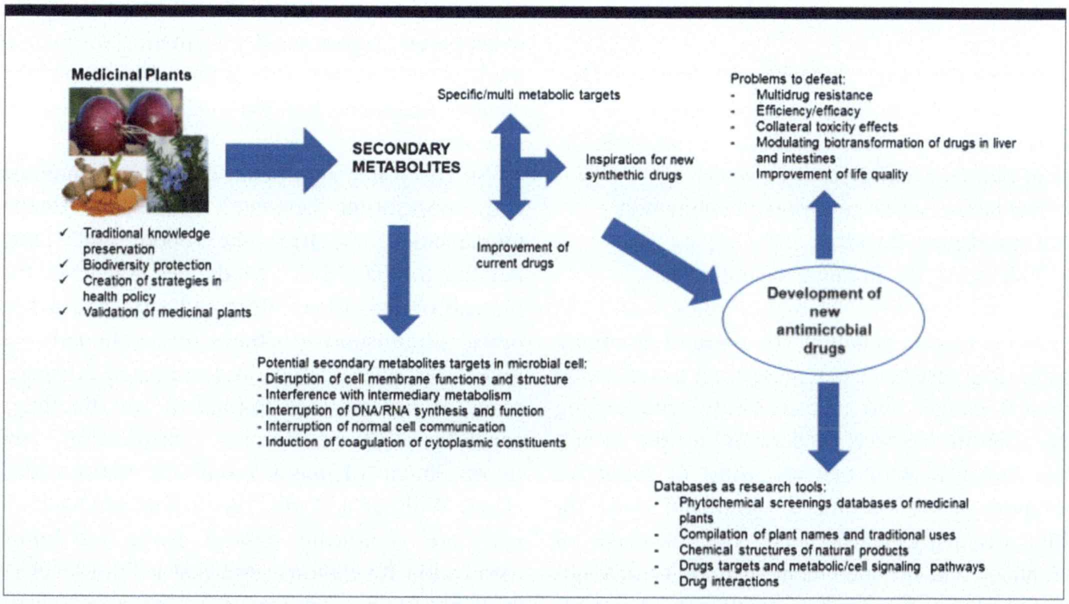

Fig. 2.1 Flow diagram summarizing the potential of natural products in the discovery of new antimicrobial agents (adapted from an open access article distributed under the terms and conditions of the Creative Commons Attribution (CC BY) license (http://creativecommons.org/licenses/by/4.0/) by Anand et al. (2019)

assumed that later on, this methodological cycle will empower the revelation of new medications all the more productively, precisely, and rapidly. This philosophy will be all the more broadly usable in future work on uncovering and foreseeing the premise of therapeutic materials (Gandhi et al. 2015; Tu 2016; Alamgir 2018; Chakraborty 2018; Uhlenbrock et al. 2018).

2.6 Conclusion

Present-day drugs have numerous side effects and limitations which adversely affect the quality of health. Additionally, the treatment cost is likewise higher than can be managed by a great many patients who are living in less or underdeveloped countries. Henceforth, there is no uncertainty that there is an urgent requirement for new viable and safe medications, and for promising new medications that beat the inconveniences of utilizing current prescriptions. Therefore, here comes the role of herbal medicines which act as bio-enhancers by.

- Expanding drug ADME (Absorption, Dispersion, Metabolism, and Excretion)
- Balancing biotransformation of medications in the liver and digestive organs
- Balancing active transport phenomenon
- Diminishing disposal
- Enhancing the immune system.

In the current situation, the issue of emerging multi-drug resistance is representing a worldwide clinical danger and is consistently challenging the scientific society. Understanding the principles engaged with the screening of bioactive compounds has become a significant issue for drug discovery researchers. The decrease of adequacy and the increment of toxicity of engineered drugs are further increasing the complexity of the problem. This has driven specialists to look towards native herbal medications for an answer, as they are presently known to assume a significant part in the advancement of powerful therapeutics. The example of overcoming adversity of artemisinin fills in as the best model for urging ethnobotanists to seek after more plant-determined medications to battle drug resistance. Biotechnology is the most integral asset and will start the assembling of new medications from plant sources at a lot quicker and controlled interactions. It supports the pipeline of drug discovery and advancement.

Taking everything into account, there is a dire need to proceed with research models to help the improvement of plant medicinal products to counter drug resistance, similarly administrative changes of clinical advancement programs. The utilization of plant drugs is speeding up and improving the channel of medication management. There are a few motivations to utilize botanical medicines, of which two may assume to play important roles. In the first place, natural treatment gives different components of activity, enveloping, as a rule, a solitary medication to treat a solitary infection. Second, the usage of exceptional conventional information on natural medication can possibly create biocompatible, cheaper treatments and will rush the disclosure of new drugs.

Surely, it is a significant call for coordination and cooperation between the World Health Organization (WHO), the Food and Drug Administration (FDA), Medicines Agencies, the biotech organizations, drug industry, and a few other administrative offices internationally to give clear rules to the disclosure and advancement of natural medications to use the huge capability of customary medication for improvement of medications for various diseases. Without a doubt, therapeutic phytochemicals are significant natural assets for future medication revelations, and just a little level of the phytochemical properties of medicinal plants have been explored. Full exertion should be given to investigate and assess possible characterization of the restorative components with the assistance of information bases and interdisciplinary collective endeavors. Eventually, the finding of more compelling and less poisonous medications will profit the worldwide populace.

Conflicts of Interest The authors declare no conflict of interest.

Author's Contribution Conceptualization: SK.; designed and planned the major original draft preparation: SK, NS, IS and RSA.; carried out, editing and proofreading the main manuscript text: SK.

Contribution	S. Kaushik	N. Sharma	I. Singh	RS. Ajee
1 Basic Introduction	*			
2 Brief overview of *Catharanthus*		*		
3 Alkaloids present in *Catharanthus*				*
4 Regulator aspects			*	
5 Modern trends	*			
6 Conclusion	*			
7 References	*	*	*	*

Funding The present study received no any extramural funding support from any organization.

References

Abdelhadi M, Meullemiestre A, Gelicus A, Hassani A, Rezzoug SA (2015) Intensification of *Hypericum perforatum* L. oil isolation by solvent-free microwave extraction. Chem Eng Res Design 93:621–631

Akram S, Najam R, Rizwani GH, Abbas SA (2015) Determination of heavy metal contents by atomic absorption spectroscopy (AAS) in some medicinal plants from Pakistani and Malaysian origin. Pak J Pharmaceut Sci 28(5):1781–1787

Alali FQ, TawahaK (2009) Dereplication of bioactive constituents of the genus hypericum using LC-(+,–)-ESI-MS and LC-PDA techniques: *Hypericum triquterifolium* as a case study. Saudi Pharmaceut J 17(4):269–274

Alam MM, Naeem M, Khan MMA, Uddin M (2017) Vincristine and vinblastine anticancer Catharanthus alkaloids: Pharmacological applications and strategies for yield improvement. In: Naeem M, Aftab T, Khan M, Masroor A (eds) *Catharanthus roseus*. Springer, Cham, Switzerland, pp 277–307

Alamgir ANM (2018) Biotechnology, in vitro production of natural bioactive compounds, herbal preparation, and disease management (treatment and prevention). In: Sayeed AM, Swamy MK, Sinniah UR (eds) Therapeutic use of medicinal plants and their extracts, Vol 2. Springer, Cham, Switzerland, pp 585–664

Almagro L, Fernández-Pérez F, Pedreño MA (2015) Indole alkaloids from Catharanthus roseus: bioproduction and their effect on human health. Molecules 20(2):2973–3000

Ansani NT, Ciliberto NC, Freedy T (2003) Hospital policies regarding herbal medicines. Amer J Health-Syst Pharm 60(4):367–370

Appendino G, Minassi A, Taglialatela-Scafati O (2014) Recreational drug discovery: natural products as lead structures for the synthesis of smart drugs. Nat Prod Rep 31(7):880–904

Aslam J, Khan SH, Siddiqui ZH, Fatima Z, Maqsood M, Bhat MA, Nasim SA, Ilah A, Ahmad IZ, Khan SA (2010) *Catharanthus roseus* (L.) G. Don. An important drug: it's applications and production. Pharmacie Globale (IJCP) 4(12):1–16

Avendaño C, Menendez JC (2015) Medicinal chemistry of anticancer drugs, 2nd edn. Elsevier

Azmir J, Zaidul ISM, Rahman MM, Sharif KM, Mohamed A, Sahena F, Sahena MHA, Jahurul K, Ghafoor NA, Norulaini N, Omar AKM (2013) Techniques for extraction of bioactive compounds from plant materials: a review. J Food Eng 117 (4):426–436

Bai J, Li Y, Zhang G (2017) Cell cycle regulation and anticancer drug discovery. Cancer Biol Med 14:348

Barnes PM, Powell-Griner E, McFann K, Nahin RL (2004) Complementary and alternative medicine use among adults: United States, 2002. In: Seminars in Integrative Medicine, vol 2, No 2. WB Saunders, pp 55–71

Barth SW, Faehndrich C, Bub A, Watzl B, Will F, Dietrich H, Rechkemmer G, Briviba K (2007) Cloudy apple juice is more effective than apple polyphenols and an apple juice derived cloud fraction in a rat model of colon carcinogenesis. J Agric Food Chem 55 (4):1181–1187

Bauer R (1998) Quality criteria and standardization of phytopharmaceuticals: Can acceptable drug standards be achieved? Drug Inform J 32(1):101–110

Bendif H, Peron G, Miara MD, Sut S, Dall–Acqua S, Flamini G, Maggi F (2020) Total phytochemical analysis of *Thymus munbyanus subsp. coloratus* from Algeria by HS-SPME-GC-MS, NMR and HPLC-MSn studies. J Pharmaceut Biomed Analys 186:113330

Bent S (2008) Herbal medicine in the United States: review of efficacy, safety, and regulation. J Gen Int Med 23(6):854–859

Berman HM, Westbrook J, Feng Z, Gilliland G, Bhat TN, Weissig H, Bourne PE (2000) The protein data bank. Nucleic Acids Res 28(1):235–242

Bhuyan DJ, Vuong QV, Bond DR, Chalmers AC, Bowyer MC, Scarlett CJ (2018) Eucalyptus microcorys leaf extract derived HPLC-fraction reduces the viability of MIA PaCa-2 cells by inducing apoptosis and arresting cell cycle. Biomed Pharmacoth 105:449–460

Bjering H, Ginige A, Maeder A, Bensoussan A, Zhu X, Lattuca C (2011) Electronic medical record information system for patient consultations in Chinese Medicine. Health informatics: The transformative power of innovation: Selected Papers from the 19th Australian National Health Informatics Conference (HIC 2011). IOS Press, Amsterdam, pp 10–15

Bugno A, Almodovar AAB, Pereira TC, Pinto TDJA, Sabino M (2006) Occurrence of toxigenic fungi in herbal drugs. Braz J Microbiol 37(1):47–51

Carvalho ACB, Ramalho LS, de Oliveira Marques RF, Perfeito JPS (2014) Regulation of herbal medicines in Brazil. J Ethnopharmacol 158:503–506

Chakraborty P (2018) Herbal genomics as tools for dissecting new metabolic pathways of unexplored medicinal plants and drug discovery. Biochim Open 6:9–16

Chan K (2003) Some aspects of toxic contaminants in herbal medicines. Chemosphere 52(9):1361–1371

Chanda S, Nagani K (2013) In vitro and in vivo methods for anticancer activity evaluation and some Indian medicinal plants possessing anticancer properties: an overview. J Pharmacog Phytochem 2(2):140–152

Chen M, Zhao Z, Lan X, Chen Y, Zhang L, Ji R, Wang L (2015) Determination of carbendazim and metiram pesticides residues in reapeseed and peanut oils by fluorescence spectrophotometry. Measurement 73:313–317

Collu G, Unver N, Peltenburg-Looman AM, van der Heijden R, Verpoorte R, Memelink J (2001) Geraniol 10-hydroxylase, a cytochrome P450 enzyme involved in terpenoid indole alkaloid biosynthesis. FEBS Lett 508(2):215–220

Cotto KC, Wagner AH, Feng YY, Kiwala S, Coffman AC, Spies G, Wollam A, Spies NC, Griffith OL, Griffith M (2018) DGIdb 3.0: a redesign and expansion of the drug—gene interaction database. Nucleic Acids Res 46(D1):D1068–D1073

Cragg GM, Newman DJ (2003) Plants as a source of anticancer and anti-HIV agents. Ann Appl Biol 143 (2):127–133

Cragg GM, Newman DJ (2005) Plants as a source of anticancer agents. J Ethnopharmacol 100(1–2):72–79

Cuzzolin L, Zaffani S, Benoni G (2006) Safety implications regarding use of phytomedicines. Eur J Clin Pharmacol 62(1):37–42

Dargan PI, Gawarammana IB, Archer JR, House IM, Shaw D, Wood D (2008) Heavy metal poisoning from Ayurvedic traditional medicines: an emerging problem? Intl J Environ Health 2(3–4):463–474

Das S, Sharangi AB (2017) Madagascar periwinkle (*Catharanthus roseus* L.): Diverse medicinal and therapeutic benefits to humankind. J Pharmacog Phytochem 6(5):1695–1701

Deshmukhe PV, Hooli AA, Holihosur SN (2010) Bioinsecticidal potential of Vinca rosea against the tobacco caterpillar, *Spodoptera litura Fabricius* (Lepidoptera: Noctuidae). Recent Res Sci Technol 2(2):1–5

Ehrman TM, Barlow DJ, Hylands PJ (2007) Phytochemical informatics of traditional Chinese medicine and therapeutic relevance. J Chem Inform Model 47 (6):2316–2334

Elvin-Lewis M (2001) Should we be concerned about herbal remedies. J Ethnopharmacol 75(2–3):141–164

Erdogrul DT (2002) Antibacterial activities of some plant extract used in folk medicine. Pharm Biol 40:269–273

Evans S (2008) Changing the knowledge base in Western herbal medicine. Soc Sci Med 67(12):2098–2106

Fang X, Shao L, Zhang H, Wang S (2005) CHMIS-C: a comprehensive herbal medicine information system for cancer. J Med Chem 48(5):1481–1488

Farag MA, Porzel A, Wessjohann LA (2012) Comparative metabolite profiling and fingerprinting of medicinal licorice roots using a multiplex approach of GC–MS, LC–MS and 1D NMR techniques. Phytochemistry 76:60–72

Fernández-Pérez F, Almagro L, Pedreño MA, Gomez Ros LV (2013) Synergistic and cytotoxic action of indole alkaloids produced from elicited cell cultures of *Catharanthus roseus*. Pharmaceut Biol 51(3):304–310

Ferlay J, Laversanne M, Ervik M, Lam F, Colombet M, Mery L, Piñeros M, Znaor A, Soerjomataram I, Bray F (2020) Global Cancer Observatory: Cancer Tomorrow. Lyon, France. Int Agency for Res on Cancer. Available from: https://gco.iarc.fr/tomorrow

Fridlender M, Kapulnik Y, Koltai H (2015) Plant derived substances with anti-cancer activity: from folklore to practice. Front Plant Sci 6:799

Gandhi SG, Mahajan V, Bedi YS (2015) Changing trends in biotechnology of secondary metabolism in medicinal and aromatic plants. Planta 241(2):303–317

Gascoigne KE, Taylor SS (2009) How do anti-mitotic drugs kill cancer cells? J Cell Sci 122(15):2579–2585

Ghosh AK, Gemma S (2015) Structure-based design of drugs and other bioactive molecules. Hoboken: John Wiley & Sons. p. 397–409

Gray N (2015) Number cruncher: Europe's key supplement markets, brands and opportunities revealed. Article published on https://www.nutraingredients.com/Article/2015/09/22

Grubesic RJ, Vukovic J, Kremer D (2005) Spectrophotometric method for polyphenols analysis: prevalidation and application on *Plantago* L. species. J Pharmaceut Biomedanalys 39(3–4):837–842

Gupta SC, Sung B, Prasad S, Webb LJ, Aggarwal BB (2013) Cancer drug discovery by repurposing:

teaching new tricks to old dogs. Trends Pharmacol Sci 34(9):508–517
Hasanpour-Dehkordi A, Dehghani A, Solati K (2017) A comparison of the effects of pilates and McKenzie training on pain and general health in men with chronic low back pain: A randomized trial. Indian J Palliat Care 23(1):36–40
Holst DO (1973) Holst filtration apparatus for Van Soest detergent fiber analyses. J Assoc Official Analyt Chem 56(6):1352–1356
Idu M, Omonigho SE, Igeleke CL, Oronsaye FE, Orhue ES (2008) Microbial load on medicinal plants sold in Bini markets. Nigeria Ind J Trad Know 7 (4):669–672
Iqbal J, Abbasi BA, Mahmood T, Kanwal S, Ali B, Shah SA, Khalil AT (2017) Plant-derived anticancer agents: a green anticancer approach. Asian Pac J Trop Biomed 7(12):1129–1150
Janamian T, Myers SP, O'Rourke P, Eastwood H (2011) Responding to GPs' information resource needs: implementation and evaluation of a complementary medicines information resource in Queensland general practice. BMC J Complement Altern Med 11(1):77
Jeurissen SM, Buurma-Rethans EJ, Beukers MH, Jansen-van der Vliet M, van Rossum CT, Sprong RC (2018) Consumption of plant food supplements in the Netherlands. Food Funct 9(1):179–190
Jiang H, Shang X, Wu H, Huang G, Wang Y, Al-Holou S, Gautam SC, Chopp M (2010) Combination treatment with resveratrol and sulforaphane induces apoptosis in human U251 glioma cells. Neurochem Res 35(1):152–161
Jivad N, Bahmani M, Asadi-Samani M (2016) A review of the most important medicinal plants effective on wound healing on ethnobotany evidence of Iran. Der Pharm Lett 8(2):353–357
Kelling J, Sullivan K, Wilson L, Jordan MA (2003) Suppression of centromere dynamics by Taxol® in living osteosarcoma cells. Cancer Res 63(11):2794–2801
Kim EJ, ChenY HJQ, Li KM, Razmovski-Naumovski V, Poon J, Li GQ (2013) Evidence-based toxicity evaluation and scheduling of Chinese herbal medicines. J Ethnopharmacol 146(1):40–61
Krishnamurthi K (2007) 17-screening of natural products for anticancer and antidiabetic properties. Cancer 3:4
Lam YH, Poon WT, Lai CK, Chan AYW, Mak TWL (2008) Identification of a novel vardenafil analogue in herbal product. J Pharmaceut Biomed Analys 46 (4):804–807
Lata B (2007) Cultivation, mineral nutrition and seed production of Catharanthus roseus (L.) G. Don in the temperate climate zone. Phytochem Rev 6(2–3):403–411
Lazarowych NJ, Pekos P (1998) Use of fingerprinting and marker compounds for identification and standardization of botanical drugs: strategies for applying pharmaceutical HPLC analysis to herbal products. Drug Inform J 32(2):497–512
Lindstrom A, Ooyen C, Lynch ME, Blumenthal M, Kawa K (2014) Sales of herbal dietary supplements increase by 7.9% in 2013, marking a decade of rising sales: turmeric supplements climb to top ranking in natural channel. Herbal Gram 103:52–56
Liu C, Guo DA, Liu L (2018) Quality transitivity and traceability system of herbal medicine products based on quality markers. Phytomedicine 44:247–257
Mazzanti G, Battinelli L, Daniele C, Costantini S, Ciaralli L, Evandri MG (2008) Purity control of some Chinese crude herbal drugs marketed in Italy. Food Chem Toxicol 46(9):3043–3047
Muhammad LRN, Muhammad AT, Bazir SN (2009) Antimicrobial activity of different extracts of *Cathranthus roseus*. Clin Exp Med J 3:81–85
Mukherjee PK (2019) Quality control and evaluation of herbal drugs: evaluating natural products and traditional medicine, 1st edn. Elsevier
Mukherjee PK, Bahadur S, Chaudhary SK, Kar A, Mukherjee K (2015) Quality related safety issue-evidence-based validation of herbal medicine farm to pharma. In: Evidence-based validation of herbal medicine. Elsevier, pp 1–28
Nasri H (2013) Toxicity and safety of medicinal plants. J HerbMed Pharmacol 2:21–22
Naz S, Haq R, Aslam F, Ilyas S (2015) Evaluation of antimicrobial activity of extracts of in vivo and in vitro grown *Vinca rosea* L.(*Catharanthus roseus*) against pathogens. Pak J Pharm Sci 28(3):849–853
Ngo TV, Scarlett CJ, Bowyer MC, Ngo PD, Vuong QV (2017) Impact of different extraction solvents on bioactive compounds and antioxidant capacity from the root of *Salacia chinensis* L. J Food Qual 2017, Article ID 9305047, 8 pages
Niemeyer K, Bell IR, Koithan M (2013) Traditional knowledge of Western herbal medicine and complex systems science. J Herb Med 3(3):112–119
Noraziah A, Abdella AN, Hamid RA, Sidek RM, Omardin MA (2011) Empirical study on medicinal herbs information system (MHIS) in Malaysia. Afr J Bus Manag 5(13):5292–5295
Ogirima SA, Olabiyisi SO, Omidiora EO, Oke AO (2015) Web based decision support system for prescription in herbal medicine. J Emerg Trends Eng Appl Sci 6 (7):245–254
Pan SY, Litscher G, Gao SH, Zhou SF, Yu ZL, Chen HQ, Zhang SF, Tang MK, Sun JN, Ko KM (2014) Historical perspective of traditional indigenous medical practices: the current renaissance and conservation of herbal resources. Evid Based Complement Alternat Med 2014:525340
Parsaei P, Bahmani M, Naghdi N, Asadi-Samani M, Rafieian-Kopaei M, Tajeddini P, Sepehri-Boroujeni M

(2016) Identification of medicinal plants effective on common cold: an ethnobotanical study of Shiraz. South Iran. Der Pharm Lett 8(2):90–97

Patel Y, Vadgama V, Baxi S, Tripathi CB (2011) Evaluation of hypolipidemic activity of leaf juice of *Catharanthus roseus* (Linn.) G. Donn. in guinea pigs. Acta Pol Pharm 68(6):927–935

Pham HNT, Sakoff JA, Van Vuong Q, Bowyer MC, Scarlett CJ (2018) Screening phytochemical content, antioxidant, antimicrobial and cytotoxic activities of *Catharanthus roseus* (L.) G. Don stem extract and its fractions. Biocatalys Agric Biotechnol 16:405–411

Pham HNT, Sakoff JA, Van Vuong Q, Bowyer MC, Scarlett CJ (2019) Phytochemical, antioxidant, anti-proliferative and antimicrobial properties of *Catharanthus roseus* root extract, saponin-enriched and aqueous fractions. Mol Biol Rep 46(3):3265–3273

Pham HNT, Vuong Q, Bowyer M, Scarlett CJ (2017) Effect of extraction solvents and thermal drying methods on bioactive compounds and antioxidant properties of Catharanthus roseus (L.) G. Don (Patricia White cultivar): effect of solvents and drying on *C. roseus*. J Food Process Preserv 41(5):e13199

Prajakta JP, Ghosh JS (2010) Antimicrobial activity of catharanthus roseus. A detailed study. British J Pharmacool Toxicol 1(1):40–44

Quettier-Deleu C, Gressier B, Vasseur J, Dine T, Brunet C, Luyckx M, Trotin F (2000) Phenolic compounds and antioxidant activities of buckwheat (Fagopyrum esculentum Moench) hulls and flour. J Ethnopharmacol 72(1–2):35–42

Raclariu AC, Paltinean R, Vlase L, Labarre A, Manzanilla V, Ichim MC, de Boer H (2017) Comparative authentication of Hypericum perforatum herbal products using DNA metabarcoding. TLC and HPLC-MS Sci Rep 7(1):1–12

Rai V, Kakkar P, Khatoon S, Rawat AKS, Mehrotra S (2001) Heavy metal accumulation in some herbal drugs. Pharmaceut Biol 39(5):384–387

Rai V, Kakkar P, Singh J, Misra C, Kumar S, Mehrotra S (2008) Toxic metals and organochlorine pesticides residue in single herbal drugs used in important ayurvedic formulation—'Dashmoola.' Environ Monitor Assess 143(1):273–277

Ramawat KG, Goyal S (2008) The Indian herbal drugs scenario in global perspectives. Bioactive Mol Med Plants 325–347

Rasoanaivo P, Wright CW, Willcox ML, Gilbert B (2011) Whole plant extracts versus single compounds for the treatment of malaria: synergy and positive interactions. Malaria J 10(1):1–12

Rayan A, Raiyn J, Falah M (2017) Nature is the best source of anticancer drugs: Indexing natural products for their anticancer bioactivity. PLoS One 12(11): e0187925

Retna AM, Ethalsha P (2013) A review of the taxonomy, ethnobotany, chemistry and pharmacology of Catharanthus roseus (Apocyanaceae). Int J Eng Res Technol 2(10):3899–3912

Revathy SS, Rathinamala R, Murugesan M (2012) Authentication methods for drugs used in Ayurveda, Siddha and Unani systems of medicine: an overview. Int J Pharmaceut Sci Res 3(8):2352

Rouhi-Boroujeni H, Asadi-Samani M, Moradi MT (2016) A review of the medicinal plants effective on headache based on the ethnobotanical documents of Iran. Der Pharm Lett 8(3):37–42

Roepke J, Salim V, Wu M, Thamm AMK, Murata J, Ploss K, Boland W, De Luca V (2010) Vinca drug components accumulate exclusively in leaf exudates of Madagascar periwinkle. In: Proceedings of the National Academy of Sciences Aug 2010, 107(34): 15287–15292

Rubio-Perez C, Tamborero D, Schroeder MP, Antolín AA, Deu-Pons J, Perez-Llamas C, Lopez-Bigas N (2015) In silico prescription of anticancer drugs to cohorts of 28 tumor types reveals targeting opportunities. Cancer Cell 27(3):382–396

Sahoo N, Manchikanti P, Dey S (2010) Herbal drugs: standards and regulation. Fitoterapia 81(6):462–471

Sain M, Sharma V (2013) Catharanthus roseus (an anti-cancerous drug yielding plant)-a review of potential therapeutic properties. Intl J Pure Appl Biosci 1 (6):139–142

Sen S, Chakraborty R (2017) Revival, modernization and integration of Indian traditional herbal medicine in clinical practice: importance, challenges and future. J Tradition Complement Med 7(2):234-244S

Shahriyari L, Komarova NL (2013) Symmetric vs. asymmetric stem cell divisions: an adaptation against cancer? PLoS One 8(10):e76195

Shim JS, Liu JO (2014) Recent advances in drug repositioning for the discovery of new anticancer drugs. Intl J Biol Sci 10(7):654

Smillie TJ, Khan IA (2010) A comprehensive approach to identifying and authenticating botanical products. Clin Pharmacol Therapeut 87(2):175–186

Smith A, Jogalekar S, Gibson A (2014) Regulation of natural health products in Canada. J Ethnopharmacol 158:507–510

Snow J (2016) Context effects in western herbal medicine: fundamental to effectiveness? Explore 12(1):55–62

Solati K (2016) Effectiveness of cognitive-behavior group therapy, psycho-education family, and drug therapy in reducing and preventing recurrence of symptoms in patients with major depressive disorder. J Chem Pharm Sci 9(4):3414–3418

Stark C, Breitkreutz BJ, Reguly T, Boucher L, Breitkreutz A, Tyers M (2006) BioGRID: a general repository for interaction datasets. Nucleic Acids Res 34(suppl1):D535–D539

Swerdlow JL (2000) Nature's medicine: Plants that heal: A chronicle of mankind's search for healing plants through the ages. National Geographic

Syed SB, Lin S, Arya H, Fu I, Yeh T, Charles MRC, Periyasami L, Hsieh H, Coumar MS (2021) Overcoming vincristine resistance in cancer: Computational design and discovery of piperine-inspired P-glycoprotein inhibitors. Chem Biol Drug Design 97 (1):51–66

Tay KC, Tan LTH, Chan CK, Hong SL, Chan KG, Yap WH, Goh BH (2019) Formononetin: a review of its anticancer potentials and mechanisms. Front Pharmacol 10:820

Thakkar S, Anklam E, Xu A, Ulberth F, Li J, Li B, Tong W (2020) Regulatory landscape of dietary supplements and herbal medicines from a global perspective. Regul Toxicol Pharmacol 114:104647

Tiong SH, Looi CY, Arya A, Wong WF, Hazni H, Mustafa MR, Awang K (2015) Vindogentianine, a hypoglycemic alkaloid from *Catharanthus roseus* (L.) G. Don (Apocynaceae). Fitoterapia 102:182–188

Tiong SH, Looi CY, Hazni H, Arya A, Paydar M, Wong WF, Awang K (2013) Antidiabetic and antioxidant properties of alkaloids from *Catharanthus roseus* (L.) G. Don. Molecules 18(8):9770–9784

Trivedi MK, Panda P, Sethi KK, Jana S (2017) Metabolite profiling in *Withania somnifera* roots hydroalcoholic extract using LC/MS, GC/MS and NMR spectroscopy. Chem Biodivers 14(3):e1600280

Tu Y (2011) The discovery of artemisinin (qinghaosu) and gifts from Chinese medicine. Nat Med 17 (10):1217–1220

Tu Y (2016) Artemisinin'a gift from traditional Chinese medicine to the world (Nobel lecture). Angew Chem Int Ed 55(35):10210–10226

Uhlenbrock L, Sixt M, Tegtmeier M, Schulz H, Hagels H, Ditz R, Strube J (2018) Natural products extraction of the future–Sustainable manufacturing solutions for societal needs. Processes 6(10):177

Varma N (2016) Phytoconstituents and their mode of extractions: an overview. Res J Chem Environ Sci 4 (2):8–15

Wachtel-Galor S, Yuen J, Buswell JA, Benzie IF (2011) *Ganoderma lucidum* (Lingzhi or Reishi). In: Benzie IFF, Wachtel-Galor S (eds) Herbal medicine: biomolecular and clinical aspects. 2nd edn. CRC Press/Taylor & Francis, Boca Raton, FL

Walker DR (2015) Report on the regulation of herbal medicines and practitioners. Department of Health and the Medicines (UK), Healthcare Products Regulatory Agency

Wang CH, Wang GC, Wang Y, Zhang XQ, Huang XJ, Zhang DM, Ye WC (2012a) Cytotoxic dimeric indole alkaloids from *Catharanthus roseus*. Fitoterapia 83 (4):765–769

Wang L, He HP, Di YT, Zhang Y, Hao XJ (2012b) Catharoseumine, a new monoterpenoid indole alkaloid possessing a peroxy bridge from *Catharanthus roseus*. Tetrahedron Lett 53(13):1576–1578

Wiesner J, Knöss W (2014) Future visions for traditional and herbal medicinal products—a global practice for evaluation and regulation? J Ethnopharmacol 158:516–518

Wishart DS, Feunang YD, Guo AD et al (2018) DrugBank 5.0: a major update to the DrugBank database for 2018, Nucleic Acids Res 46: 1074–1082

World Health Organizatio (2013) WHO Traditional Medicine Strategy, pp 2014–2023

World Health Organization (1998) Quality Control Methods for Medicinal Plant Materials. World Health Organization

World Health Organization (2004) WHO guidelines on safety monitoring of herbal medicines in pharmacovigilance systems. World Health Organization

World Health Organization (2007) WHO international standard terminologies on traditional medicine in the western pacific region

World Health Organization (2008) Traditional medicine http://www.who.int/mediacentre/factsheets/fs134/en/

Xie P, Chen S, Liang YZ, Wang X, Tian R, Upton R (2006) Chromatographic fingerprint analysis—a rational approach for quality assessment of traditional Chinese herbal medicine. J Chromatograph A 1112(1–2):171–180

Xue J, Hao L, Peng F (2008a) Residues of 18 organochlorine pesticides in 30 traditional Chinese medicines. Chemosphere 71(6):1051–1055

Xue J, Liu D, Chen S, Liao Y, Zou Z (2008b) Overview on external contamination sources in traditional Chinese medicines. World Sci Technol 10(1):91–96

Yap KYL, Chan SY, Lim CS (2008) The reliability of traditional authentication—a case of ginseng misfit. Food Chem 107(1):570–575

Yee SK, Chu SS, Xu YM, Choo PL (2005) Regulatory control of Chinese proprietary medicines in Singapore. Health Policy 71(2):133–149

Yi F, Tan XL, Yan X, Liu HB (2016) *In silico* profiling for secondary metabolites from *Lepidium meyenii* (maca) by the pharmacophore and ligand-shape-based joint approach. Chin Med 11(1):1–17

Yordanov M, Dimitrova P, Patkar S, Falcocchio S, Xoxi E, Saso L, Ivanovska N (2005) Ibogaine reduces organ colonization in murine systemic and gastrointestinal *Candida albicans* infections. J Med Microbiol 54(7):647–653

Zaman MA, Oparil S, Calhoun DA (2002) Drugs targeting the renin—angiotensin—aldosterone system. Nat Rev Drug Discov 1(8):621–636

Zhang J, Wider B, Shang H, Li X, Ernst E (2012) Quality of herbal medicines: challenges and solutions. Complement Therap Med 20(1–2):100–106

Zhang Y, Mao X, Guo Q, Bai M, Zhang B, Liu C, Lin N (2016) Pathway of PPAR-gamma coactivators in thermogenesis: a pivotal traditional Chinese medicine-associated target for individualized treatment of rheumatoid arthritis. Oncotarget 7(13):15885

Zhou X, Chen S, Liu B, Zhang R, Wang Y, Li P, Guo Y, Zhang H, Gao Z, Yan X (2010) Development of traditional Chinese medicine clinical data warehouse for medical knowledge discovery and decision support. J Artif Intell Med 48(2):139–152

Classical Breeding and Trait Genetics in Catharanthus

3

Tripta Jhang, Shweta Dwivedi, and Sarika Sharma

Abstract

Catharanthus roseus (L) G. Don (Madagascar periwinkle) is one of the most extensively investigated ornamental and medicinal plants due to the high demand for its monoterpene indole alkaloid (MTIAs), including vinblastine, vincristine, and its monomers. Since vincristine's discovery in 1961, it maintains a steady, strong foothold in clinical demand and drug production. These dimeric alkaloids are biosynthesized in very low levels *in planta* due to their high cytotoxicity, restricted by spatial and temporal tissue-specificity, essentially for the plant's own survival. Commercial production of vincristine and vinblastine by large-scale upscaling of catharanthine, vindoline by bio-fermenters, and their chemical semi-synthesis are yet not economic leaving only viable option through *in planta* cultivation. They are still harvested and purified from the leaves and roots of field-grown plants. Its age-old cultivation since 1658, propelled by profitable horticultural commerce of its attractive corolla and ease of culture as bedded plant, pushed the public sector landscape industry to breed and release a series of more than 334 cultivars to date of which patents and IPR protected many. Catharanthus breeders focused on breeding for dwarf or semidwarf, contrasting colored corolla with its eye color, floriferous, high biomass, resistance to aerial blight and dieback diseases, heat and drought tolerance as the prime targets to develop new cultivars for wider horticultural adoption. Systematic breeding and genetic improvement for the higher content of monomers MTIAs ajmalicine, vindoline, and catharanthine are sustained by just 6% of the total research efforts on this plant, evident by a total of 3,861 publications produced till 2020. Catharanthus breeding started with the initiative of commercial horticultural seed companies in1980s. Catharanthus new strains were bred through interspecific and intervarietal al. hybridization and selections for corolla and eye color recombination, enhancement in leaf size, and shortening the plant height. Hybrid cultivars are being generated by the development of artificially induced, stable male-sterile lines. In limited initiatives of research organizations, total alkaloid content could be improved from 0.0003% in the wild to 6.2% in hybrids. A valuable genetic screens resource is now available for Catharanthus genomic research using both natural and induced mutations. Since most of the MTIAs pathways in Catharanthus are now deciphered, they can serve as a valuable tool for mining for allelic polymorphism, SNPs in germplasm, assessing

T. Jhang (✉) · S. Dwivedi · S. Sharma
Division of Plant Breeding and Genetic Resource Conservation, CSIR- Central Institute of Medicinal Plants (CIMAP), Lucknow 226015, India

© The Author(s), under exclusive license to Springer Nature Switzerland AG 2022
C. Kole (ed.), *The Catharanthus Genome*, Compendium of Plant Genomes,
https://doi.org/10.1007/978-3-030-89269-2_3

heterozygosity, and uncovering variants for MTIA content by detecting natural variants for MTIA targeted breeding. Genetic linkage maps have been generated, paving the way for high throughput QTL mapping. Whole-genome sequencing of 'Sunstorm Apricot' and chloroplast sequence of 'Pacifica Punch Halo' together facilitates the fine dissection of gene and QTLs towards MTIA targeted breeding by breaking the bottlenecks of MTIA biosynthesis in plants and in heterologous systems to meet the sharpshooting demand from the neoplasmic pharmaceuticals. Here we have tried to compile the work carried out on various aspects of Catharanthus genetics and breeding to help understand the plant better with the aim that the generated information will help overcome the major obstacles in its genetic improvement program.

3.1 Introduction

Catharanthus roseus (L.) G. Don has evolved as a model medicinal plant for its commercialized and clinically used bis-indole mono terpenoid alkaloids (MTIA). *C. roseus* is fondly popularized as 'Madagascar periwinkle, and the plant has spread worldwide more as a cultivated ornamental than a medicinal plant. It is a widely adaptable, ever-blooming perennial herb, now globally found naturalized in almost every tropical and subtropical region, occurring on every continent except Antarctica. Madagascar forest, its site of origin, today largely stands deforested due to the unmindful collection and uncontrolled uprooting that only 10% of the native forest is left with its ecological biodiversity. Hence, at its site of origin, it falls under the endangered biodiversity category (Consiglio et al. 2006).

Though the medicinal properties of *C. roseus* were recorded as early as 50 BC in traditional medicine literature, it was systematically recognized by western physicians with proven anticancer activity in 1958 (Noble et al. 1958; Husain et al. 1993). The American National Cancer Institute screened several thousand (about 40,000) plants for their therapeutic properties (Tyler 1988), of which *C. roseus* was the 40th priority plant tested (Svoboda et al.1962). As a part of this screening program, the hypoglycemic property of *C. roseus* mentioned in folklore was investigated by two independent research groups resulting in breakthrough discoveries but away from anticipation. Peripheral granulocytopenia and bone marrow depression in rats induced by *C. roseus* leaf fractions led to the isolation of vinleucoblastine and leurosine as the first antitumor compounds by Noble, Beer, and Cutts's team at Collip Laboratories, University of Western Ontario in 1958. Antileukemic activity in mice resulted in the isolation of vincristine from leaves of *C. roseus* in 1961 by the Svoboda, Johnson, Neuss, and Gormanteam from Eli Lilly Research Laboratories (Johnson et al. 1963). Eli Lilly was the first company to fund the extraction and isolation and commercialization of the natural antitumor drug vinblastine as Velban® in 1961 for the treatment of Hodgkin's disease, lymphosarcoma, neuro-blastoma, and choriocarcinoma. It was modified semi synthetically by Pierre Potier (1989) at Institute de Chimie des chemical CSN in Gif-sur-Yvette and commercialized in 1989 as Navelbine by Pierre Fabre Laboratories (Roux and Gueritte 2017). Vincristine is a highly active, cell cycle-dependent anticancer drug that binds to tubulin, causing microtubule inhibition, metaphase arrest, and apoptotic death of cells undergoing mitosis. Vincristine sulphate, commercially known as Oncovin® or Vincovin® is used to treat leukemia in children and reticulum cell sarcoma (Svoboda and Blake 1975). Two semisynthetic bisindole alkaloids, vinorelbine, marketed as Navelbine®, and vindesine marketed as Eldisine®, and Fildesin® are used in the treatment of breast cancer, bronchial cancer, acute lymphoblastic leukemia, and refractory lymphoma, respectively (Bruneton 1995; Pezzuto 1997). Ajmalicine from the *C. roseus* roots imported from India was initially taken as an alternative valuable source adulterant in *Rauvolfia* roots and was readily accepted. Ajmalicine is reported in 20 species of the genus *Rauvolfia* (Tylor and Farnsworth 1975) and four species of *Catharanthus* (Singh 1996). It is used in the treatment of circulatory diseases, especially hypertension. Its chemotherapeutical pharmaceutical properties (vincristine and vinblastine) and hypotensive (serpentine and ajmalicine) soon caught the interest of the scientific world.

C. roseus is reported to biosynthesize around 150 terpenoid indole alkaloids, including major MTIA vincristine, vinblastine, ajmalicine, serpentine, yohimbine, catharanthine, tabersonine, vindoline, strictosidine, secologanin. It also produces a broad spectrum of phenolic compounds such as flavonoids, caffeic acid, cinnamic acid, and derivatives (van der Heijden et al. 2004; O'Connor and Maresh 2006; Ferreres et al. 2008; Verma et al. 2012; Dugé de Bernonville et al. 2015). These MTIAs being highly toxic to the plant, are generated in temporally and spatially separated compartments for the sake of plant's own survival, at very low levels. Huge quantities of plant materials were harvested, in order to obtain vinblastine (1 g/t) and vincristine (20 mg/t) (Tyler 1988) which has triggered continued high intensive research on this plant.

A literature mining analysis with keywords *C. roseus* was carried out from 1963 to 2020 and nearly 3,861 publications were recorded under the major classical subject fields with an average of about 68 publications per year (Fig. 3.1). The impetus comes from 1985 onwards with the discoveries of genomic tools adding about 3,310 more publications with an average of 95 publications per year, demarcating the most productive era of prime work. Thus, it is evident that the interest in this plant is increasing continuously over the last seven decades. In the 1950s initial interest in this plant was primarily due to its pharmacological properties and till 1968 research was focused only on its phytochemistry and pharmacological value. With established medicinal and economic value, research focused on enhancing its alkaloids production through the cell, tissue, and hairy root cultures. The decade 1980's was dedicated to understanding the Catharanthus plant physiology and alkaloid biochemistry. From 1994 workers started working out the molecular biology of the plant to overcome the bottlenecks of alkaloid production. The low content of medicinally important alkaloids (VLB and VCR) in this plant has led to considerable research on the production of these alkaloids through synthesis, semi synthesis, cell and tissue cultures (Moreno et al. 1995). An overview of the number of publications indicates extensive work has appeared in the area of physiology and biochemistry (18.47%), cell and tissue cultures (18.15%)followed by molecular biology and genetic engineering (14.09%) and phytochemistry (17.36%). Only 6.3% of publications were reported in agronomy and 6.2% in genetics and breeding (Fig. 3.1). A sudden increase of interest in Phytoplasma (pathology) in the last decade was also noticed (Chang 1998; Choi et al. 2004; Favali et al. 2004). Many of the workers felt that majority of the studies in Catharanthus were on chemical and pharmacological aspects and only a few with agronomy and genetics of the plant (Talha et al. 1975; Dovrat and Goldschmidt 1978; Roller 1978; Levy et al. 1983; Kulkarni et al. 2016).

C. roseus attracted commercial interest both as a medicinal and horticultural plant, as evident from 3,222 patents granted on its varied aspects during 1960–2020.Vincaleucoblastine was the first alkaloid to be patented in 1961 (GB870723 19,610,621), setting a patent surge. The aspects on some of these patents reviewed here cite numbers of patents granted in the bracket against the novel applications. Patents were granted on a novel method of preparation of MTIA alkaloids from *C. roseus* like pressure column chromatography, negative pressure evaporation crystallization, adsorption resin with hydrophobicity, dipole–dipole interaction, cation exchange resin, photochemical method, enzyme induction methods (359), synthetic processes for its alkaloids (7) and their derivatives (30). Subsequently, patents were granted for a wide variety of inventions such as usage of MTIA alkaloids as neoplasm inhibitors (58), pharmaceutical and medical applications such as injectable solutions of MTIA dimer salts, as a drug carrier, enhanced biological activity against MDR cancer cells (78), usage of the various herbal composition of *C.roseus* as traditional Chinese medicine in the treatment of lymphoma, hypertension, diabetes and for improving general health (64). Cancer

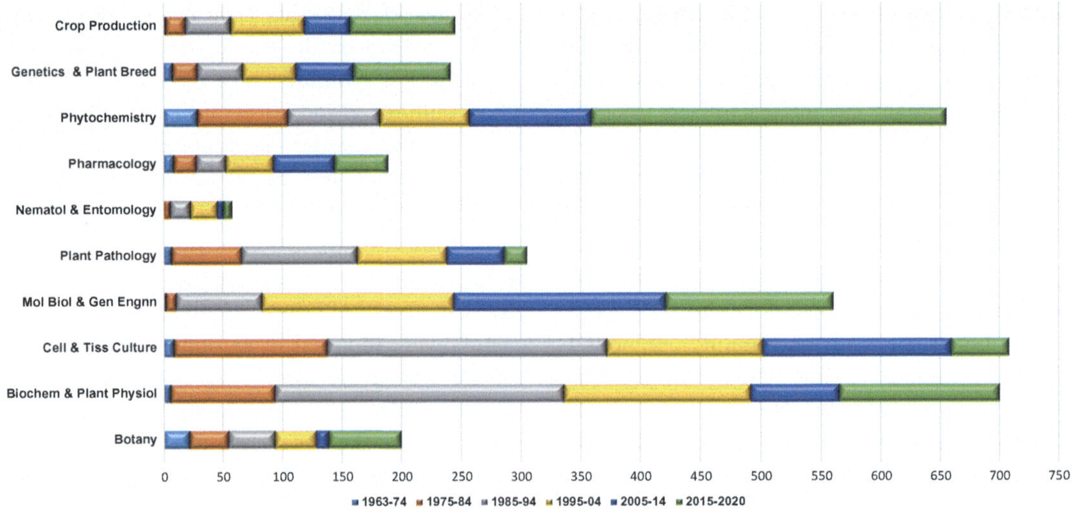

Fig. 3.1 Literature mining on *Catharanthus roseus* G Don in different subjects areas during the period (1963–2020)

diagnostics technologies based on gene mutation, gene expression, and protein profiling techniques such as single nucleotide polymorphism (SNP) based detection and protein profiling of aminoacyl-tRNA synthetases (regulators), nucleic acid aptamers, biomarkers, and microvesicle conjugated antigen-based detection (61), engineered antibody in conjugation with antitumor agents (82), administration of liposome-encapsulated MTIA alkaloids or derivatives (53), nanotechnology ways in targeted drug delivery system (30) fetched many patents. More than 160 patents were granted on various plant genes, encoded polypeptides, homologs, and their uses for increasing plant yield or agricultural characteristics such as enhanced water use efficiency, cold tolerance, salt tolerance, drought tolerance, nitrogen use efficiency, seed protein, and seed oil, growth rate, vigor, oil content, lignin, protein, carbon, carotenoid or triterpenoid levels by transgenic protein expression, salicylate biosynthesis for the induction of disease resistance, plant resistance to microorganism restricted to phloem, methods for increasing plant biomass and plant seed yield through over-expression of polyhydroxy butyrate (PHB) gene, metallothionein type II (MT-II) gene from *C. roseus* and its uses. Modulation of glutathione S-transferase (GST)/ lipid transfer protein LTP/ group three late embryogenesis abundant (LEA) regulatory proteins associated with cold tolerance from *C.roseus*, genetic engineering of Catharanthus by transferring glycine-rich RNA binding protein from *Camptotheca acuminata* for inducing drought resistance, Phytophthora resistance gene (3) for improved disease resistance and increased alkaloid levels, modulating plant root architecture using RAR (root architecture regulator) peptides, increasing nematode resistance in plants, disease treatment by identifying microRNA (miR) modulators, microbial isoprenoid production using a heterologous DXP pathway and by recombinant expression of modified HMG-CoA reductase with no regulation by protein kinase phosphorylation, induction of male sterility (2) with heterologous genes expressed from tissue-specific promoters, biomass-modulating proteins with various domains from plants, methods of altering biomass in a using the *STF* gene which affects lamina expansion in plants, modifying water user efficiency by regulatory region-regulatory protein associations with aquaporin-like polypeptides as reporter, stress-responsive transcription factors and alteration of patterns of gene expression to modulate metabolite, genes encoding strictosidine synthase (*str*) and strictosidine glucosidase (*sgd*) expressed in

Saccharomyces cerevisiae and fed with the precursors tryptamine and secologanin, transgenic plants with altered internode lengths SHI (short internode) genes, production of ribosome-inactivating protein (RIP) through callus cultures. Patents were granted on modulation of primary and secondary metabolites from *C. roseus* through genetic engineering and plant tissue culture methods such as crown gall cell cultures, hairy root cell cultures, pathological or physiological conditions like blue light irradiation forming indole alkaloids by a large-scale culture of the adaptive cell of *C. roseus*, enhancing the yield of terpenes via the expression of fusion proteins comprising various combinations of geranyl diphosphate synthase large and small subunits and limonene synthases, cultivation of plant non-tissue multilayer cell pack for use in the generation of heterologous proteins and secondary metabolites. Seven patents were granted for flower color modulation with either molybdenum or tungsten application, by control of vacuolar pH, cloning chalcone 3-hydroxylase transgene, or engineering flavonoid 3′ 5' hydroxylase gene and dihydroflavonol-4-reductase gene, or production of flavonoids by recombinant microorganisms expressing plant metabolic pathway genes. Novel in vivo techniques for improving alkaloid content of *C. roseus,* e.g., use of plant growth regulators, harvesting pre-treatment method, combined use of cyclodextrins and methyl jasmonate, irradiation culture method of stem and leaf using a fluorescent lamp, with plant growth regulator such as benzyladenine, the standard cut method of the *V. rosea*, and its method, cultivating technology for increasing vinblastine content, increasing the level of nutrients in plants by irradiating with visible light, radioactive decontamination of soil fungal extract for eliciting immunity and pattern recognition receptor in plants got 15 patents. Patents on improved plant varieties and hybrids (29) are discussed in relevant sections in this chapter. Rest 1448 patents included Catharanthus in concept or its various aspects beyond the scope of this review.

The *in planta* low concentration of both vinblastine and vincristine in the plant materials (0.0005%) ushered extensive research in the past seven decades since the first report of callus suspension culture of Catharanthus (Carew 1966). Rigorous efforts to genetically enhance the MTIA concertation followed mainly through phytochemical, metabolomics or through genetic engineering and in yesteryears by applying bio-fermentors (Moreno et al. 1993; De Luca and St-Pierre 2000; Whitmer et al. 2002a; Whitmer et al. 2002b; El-Sayed and Verpoorte 2002; Zhao and Verpoorte 2007). These efforts demonstrated that breeding for high alkaloid content Catharanthus varieties and hybrids is the only sustainable, viable, and most economical option to increase the alkaloid content to meet the shooting demand of anti-cancerous alkaloids. This was proved by the reduction of publications under Catharanthus tissue culture and genetic engineering in the last two decades. Only 6% of publications account for unraveling the genetics of agronomic traits that constitute the biomass yield and have accomplished the breeding of MTIA enriched disease resistance and stress-tolerant varieties that now stands protected with variety patents or intellectual property rights (IPR). Private seed companies, mainly from Europe, America, and Japan, have remarkably generated 324 Catharanthus cultivars with attractive corolla colors with contrast eyes and shapes with an upright habit, tolerance to temperature, water use efficiency, and aerial phytoplasma capturing the landscape bedded plants and pot plants ornamental commercial sector. Plant breeding assisted with omics technology such as single-cell metabolomics, imaging mass spectrometry (IMS) Smart cell consortium, Phytometa syn, Medicinal Plant Genomic Resource, and CathaCyc contributed to the creation of metabolome, proteome, and transcriptome databases and dedicated libraries of *C. roseus* (O'Connor and Maresh 2006; Yamamoto et al. 2016, 2019; Rai et al. 2017). The resources have triggered the pace of genetic selection of the right genotypes for generation advancement for inclusion in the breeding program. Recently quick chemotype profiling assay in Catharanthus such as high throughput extraction technologies, high spatial resolution quantification techniques with high sensitivity and specificity of detection such as surface-assisted laser desorption/ionization imaging mass spectroscopy which is much sensitive than matrix-assisted laser desorption/ ionization mass spectrometry (MALDI-MAS) (Yamamoto et al. 2016, 2019; Dutkiewicz et al. 2021).

Allelix Inc in Canada led by Misawa, in collaboration with Kurz (National Research Council, Canada) and Kutney (University of British Columbia), established an economically viable way to produce catharanthine by using bioreactors and enzymatic coupling artificial reactions. They observed the concentration of vindoline in the intact *C. roseus* plant is approximately 0.2% as a dry weight basis, cathranthine, and vinblastine were expensive isolation from *in planta*. The technology was transferred from the Canadian company to a Japanese company, Mitsui Petrochemicals Industry, for further development. Hara and his team from Mitsui Petrochemical could increase the yield of catharanthine up to 150 mg/L in the MS medium supplemented with specified phytohormones, and the yield of vinblastine by the chemical coupling reaction was improved by the addition of ferric chloride, oxalate, malate, and sodium borohydride from anhydrovinblastine up to 50%. Although chemical synthesis presents an alternative for plant-based production, the structural complexity of these MTIA obstructs their chemical synthesis at commercial levels and has not reached a commercially viable production scale (Ehrenworth and Peralta-Yahya 2017). Catharanthus alkaloids are also reported for insect sterilants properties asn-hexane fraction of acetone extract from *C. roseus* leaf contains a-amyrinacetate and oleanolic acid against tobacco caterpillar and gram pod borer (Singh et al. 2003).

The production of MTIA alkaloids now relies on the cultivation of plants. It is currently cultivated in China, India, Madagascar, Israel, the United States, and central and southern Europe. The current production strategies are limited due to dependence on industrial-scale agricultural production due to the low yields of desired products in plants, extensive downstream processing, and inefficient chemical semi synthesis of desired products from plant-derived precursors, and crop loss due to environmental conditions (climate change, pests, and diseases). The onus is now on the Catharanthus breeders to come up with high content cultivars which produce high monomeric MTIA like vindoline and catharanthine (fda.gov/media/134349).

3.1.1 History of Catharanthus Cultivation

The genus *Catharanthus* has been revisited many times as its parent family Apocynaceae (*Christened* by Al de Jussieu 1789) underwent a series of revisions (Van Bergen and Snoeijer 1996). Genus *Catharnathus* has been variously designated by taxonomists as *Vinca rosea* L., *Lochnera rosea* Reichb., and *Ammocallis rosea* small in the past. The plant of Catharanthus was first noticed by Etienne de Flacourt, French Governor of East India Company, in Madagascar island in 1658 (Curry 2012), who brought it to the Royal Garden in Versailles near Paris, France in 1715, the sample of *C.roseus* is preserved in herbarium museum of Paris. Mr. Richard, the gardener of King Louis XV, grew the introduction, and later its seeds were sent to Chelsea Botanic Garden of London in 1757. It was raised by Philip Miller as a bedded garden plant; from here, it spread to the whole world initially as an ornamental naturalized plant. Till then, only purple, white, and white with red-eye corolla *Catharanthus roseus* genotypes were known. Carl von Linne, Swedish Naturalist, in 1753 (Plaizier AC1981) established the genus *Vinca* with two species *V. minor* and *V. major*. In 1759, tropical *V. rosea* was added into the temperate genus *Vinca* although it did not fit into its generic description with regard to stamens. Heinrich Gottlieb Ludwig Reichenbach pointed out the taxonomic difference between *Vinca rosea* from *V. major* and *V. minor*, first to recognize *V. rosea* as generically different from *Vinca* (Stearn 1975). British botanist George Don while describing these genera in 1835, retained the name *Vinca* for the genus containing *Vinca minor*, *Vinca major*, and *Vinca herbaceae* and established the genus *Catharanthus* typified by *Vinca rosea* renamed as *Catharanthus roseus* (from the Greek kathospure and Anthos flower) in 1838 and till date, the horticulture literature relates it with its old name *Vinca rosea* the *C. roseus*. The plant spread to almost every tropical and subtropical part of the two hemispheres from Africa, Southeast Asia, Latin America and the Caribbean, US pacific and Indian Ocean islands. A founding population of this plant was introduced into southern parts of India via Goa by Portuguese missionaries in 1794 (Morton 1977), augmented with several new introductions during the course of time. On account of its perpetual flowering, it was initially grown at cemeteries (Mishra et al. 2001a). These widespread introductions were primarily due to two reasons, originally, it was used as an appetite suppressant in the long journey of sailors and because of its ornamental large flowers and ease of cultivation supported by the traditional hypoglycemic properties (Roux and Gueritte 2017). The earliest herbarium voucher specimen is from Java (1828) and Brazil (1833) at the National Herbarium of the Netherlands. The two cultivars, '*rosea*' and '*alba*,' were first cataloged by Gordon, Dermer, and Thompson Seed Nursery of London in 1783. Eli Lily Company initiated contractual farming by wild gathering and cultivation at many tropical locations for the constant uniform plant raw material supply. Southern coastal Madagascar genotypes yielded higher bioactive phytochemicals, probably due to genotype x environment interaction effects (Curry 2012). The threatened Madagascar forest has been a conspicuous supply of *Catharanthus roseus* despite its worldwide spread. Though the plant does not need conservation efforts at its center of origin, Madagascar forests are now endangered due to unmindful uprooting. It is currently cultivated in China, India, Madagascar, Israel, the United States, and in all countries of central and southern Europe.

3.1.2 Global Trade of Catharanthus and Economics of MTIA Production

The four FDA-approved MTIA-based pharmaceuticals on the market comprise the antihypertensive agent reserpine and the anticancer drugs vinorelbine, vinblastine, and vincristine. Reserpine is a natural product and is extracted from *Rauwolfia vomitoria* for commercial production. Vinorelbine is a semisynthetic drug made in two steps from anhydrovinblastine obtained chemically or enzymatically from plant-extracted catharanthine and vindoline (Ehrenworth and Peralta-Yahya 2017). Vincristine was the first plant-derived anticancer drug approved by FDA in 1963 (Seca and Pinto 2018). Since then, Catharanthus derived three major drugs approved by US FDA in clinically prescribed medications are Vinblastine (1965, Velban®), Vincristine (1963, Oncovin®), Vinorelbine (NavelBine®) and Vinflunine (Eldisine®) is approved by European Medicine's Agency (Rowinsky 2003; Schutz et al. 2011; Almagro et al. 2015; Seca and Pinto2018; Ozber et al. 2020). This long history of Catharanthusalkaloids as anti-cancerous potent drugs has constantly pushed the demand for Catharanthus leaves and roots raw material, reported carrying an estimated total market value of ~$60 billion (Ozber et al. 2020). Vincristine sells for around US$ 200,000 per Kg and requires about 530 tonnes of plant material for its production. The worldwide trade in vincristine is worth about US$ 50 million per annum.

The Catharanthus plant produces these alkaloids in a very low concentration 0.0003%–0.3% by weight. It takes 530 kg of dry herb to produce 1 g vinblastine and 20 mg vincristine (Tyler 1988). Limited by the poor supply, the market cost of Catharanthus leaf-derived drugs one Kg each of vinblastine, vindoline, vincristine, and catharanthine is 5US$M, 1.5US$M, 69US$M, and 2US$M respectively (ChemFaces 2021). Catharanthus root-derived drugs one Kg each of ajmalicine and serpentine is 13.3US$M, 30US $M (ChemFaces 2021). Ajmalicine picked up high demand in 1970–80 as a potential drug for cerebral blood flow and hypersensitivity, and 3600 kg was annually isolated by the pharmaceutical sector from either Catharanthus or Rauwolfia (Arens et al. 1978; Moreno et al. 1995). Ajmalicine is exclusively found in Catharanthus roots, whereas serpentine is also found in aerial parts (Arens et al. 1978). Recently ajmalicine is reported to have neuroprotective activity besides its well-known usage as antihypertensive and cardiovascular disorders reducing activity. It exhibits antioxidant activity, increases the blood flow to the cerebrum, and increases oxygen bioavailability, hence a potential candidate drug for Alzheimer's treatment (Ambrin et al. 2020). Economics of ajmalicine production using large scale bio-fermenter with roots imported from India with 0.3% ajmalicine concentration with the purchase cost US$619/Kg whereas bio-fermenter grown cell with the improved technology could produce 6% ajmalicine after 6 days of accumulation cost US$608/Kg costing almost same (Drapeau et al. 1987). Cell culture produces 0.20 mg/g a day while the plants in the cultivation field generated 0.0082 mg/g a day. They observed, to make his technology at par with the net amount being produced through field cultivation, the process needs modification of a 40X increase rate of accumulation of alkaloid by cell cultures (Drapeau et al. 1987). Recently ajmalicine production in 100 L bioreactor could achieve 36 kg ajmalicine biomass in 20 days with a concentration of 73.18 mg/l (Fulzele and Namdeo 2018); thus, with technology improvement and using elite genotypes, the cost could be reduced. Market survey reports of 2021 indicate catharanthine is picking demand from the nutraceutical sector besides its use in making anti-cancerous drugs.

France, Belgium and Hungary has been the topmost importer of Catharanthus leaves and root from India (HS12119060, Ministry of Commerce, dated 12 June 2021) for the record maintained over the last seven decades (Fig. 3.2). The demand for these drugs is sharpshooting. US market is solely supplied by only two companies Pfizer and Teva Pharmaceutical. Recently in 2019, there was an acute shortage of vincristine sulphate, a pediatric cancer chemotherapy drug, which left the oncology community struggling for an alternate solution, as Teva Pharmaceutical briefly ceased its vincristine production, putting enormous pressure on Pfizer, the sole supplier to the US market (Palmer 2019).

3.2 Basic Information Required for Planning of Breeding Strategies

3.2.1 Origin, Ecological Distribution, and Wild Relatives

The genus *Catharanthus* comprises nine species with the recent addition of a new species (Allorge et al. 2015). Eight of them are indigenous to Madagascar, namely *C. coriaceus* Markgraf, *C. lanceus* (Bojer ex A.DC.) Pichon, *C. longifolius* (Pichon) Pichon, *C. ovalis* Markgraf, *C. scitulus* (Pichon) Pichon, *C. makayensis* L Allorge, *C. trichophyllus* (Baker) Pichon. *Catharanthus pusillus* (Murray) G. Don is endemic to India and Sri Lanka. While the first four species are confined to upland area of Madagascar, the next two are localized to sub arid south of Madagascar on sandy soils (Allorge et al. 2015). The recent addition of *Catharanthus makayensis* L Allorge is typified by pendulous fruits, long narrow filiform sepals, hexagonal cross-section of stems, white or light pink corolla with magenta eye. Within Madagascar, these species cohabitate in the same vicinity but still can be differentiated from their core regions *C. trichophyllus* (Baker) Pichon was described to be from northern, western Madagascar, *C.lanceus* (Bojer ex DC) Pichon) from Antananarivo central Madagascar, *C. coriaceus* Markgraf from Itremo mountains, *C. ovalis* Markgraf around Parc Isalo and *C. longifolius* (Pichon) Pichon around Ambalavao, *C. scitulus* (Pichon) Pichon south-central part of Madagascar and *C. roseus* (L) G.Don around Fort Dauphin. Only *C. roseus*, *C. trichophyllus*, and *C. pusillus* are has grown commercially (Bergen and Snoeijer 1996). As per most documented reports *C. roseus* primary center of origin is Madagascar and nearby islands

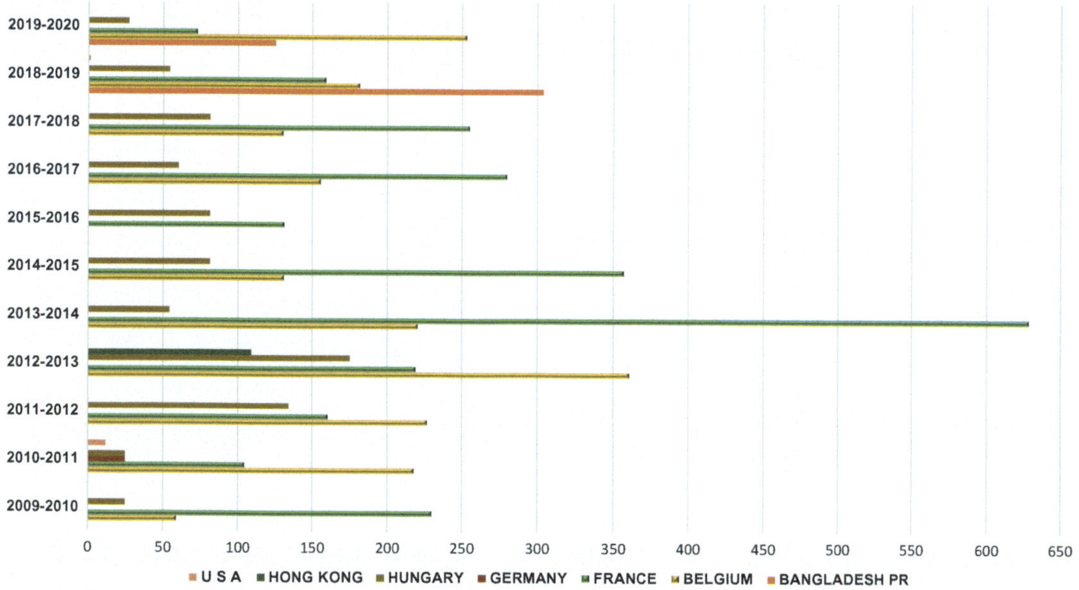

Fig. 3.2 Export of *Catharanthus roseus* plant parts (HS12119060) in tons to its major buyers over the last decade *Source* Ministry of Commerce and Industry dated 1.07.2021

(Veyret 1974; Plaizier 1981; Bergen and Snoeijer 1996). *C. roseus* at its center of origin Madagascar island remained unseparated by any geographical barrier and grew with other species alongside. Taxonomic rank varieties as documented are *Catharanthus roseus* var *albus* G. Don., *Catharanthus roseus* var. *angustus* (Steenis) Bakh. f., *Catharanthus roseus* var. *nanus* Markgraf, *Catharanthus roseus* var. *roseus* (Plaizier 1981). Catharanthus, though introduced as an ornamental species, have now naturalized all over tropical and subtropical regions of both the hemisphere somewhere, even as invasive species. Catharanthus is a hardy plant (USDA hardiness zone: 9–12). It grows well in most types of soils such as in red laterites soil, black soil, clay, rockeries, coastal sandy beaches, limestone, rocky terrains, wasteland dunes in savannah, open forest, woodland or grasslands, high altitude (0–900 m) except for high alkaline and waterlogged soils. Catharanthus performs best in a tropical and subtropical climate where it reaches 30–120 inch height. It has spread to the cooler temperate zone and can adapt well if it escapes the low winter temperatures. Catharanthus is normally heat, drought, and salt tolerant (up to 2000 ppm). The extent of tolerance varies as per genotypes adaptability in agro-geographical zones. The plant leaf and root yields often drop in extreme conditions (Mishra and Kumar, 2000).In India, it is cultivated in Tamil Nadu, Karnataka, Gujarat, Madhya Pradesh, Assam, and Tamil Nadu alone 435 mt is produced from 272 hectares with an average of 1.6 MT/ha productivity (tn.data.gov.in). As per Global Bioresource Information Facility total tabulated occurrences of *C.roseus* G Don are 23,326 with hotspots localized in India (5447), USA (2419), Italy (1430), Mexico (1231), Indonesia (1048), Australia (972), and Philippines (860) (Fig. 3.3) (www.gbif.org).

3.2.2 Cytogenetics and Genome Size

C. roseus is a diploid plant species with a chromosome number of $2n = 16$ (Simmonds 1960; Dnyansagar and Sudhakaran 1966; Pandeya and Jose 1970; De Padua et al. 1992). The karyotype formula in *C. roseus* was reconfirmed $2n = 2x = 16 = 2\ m = 12sm = 2\ T$ (Hong et al. 2008). Dnyansagar and Sudhakaran (1977)

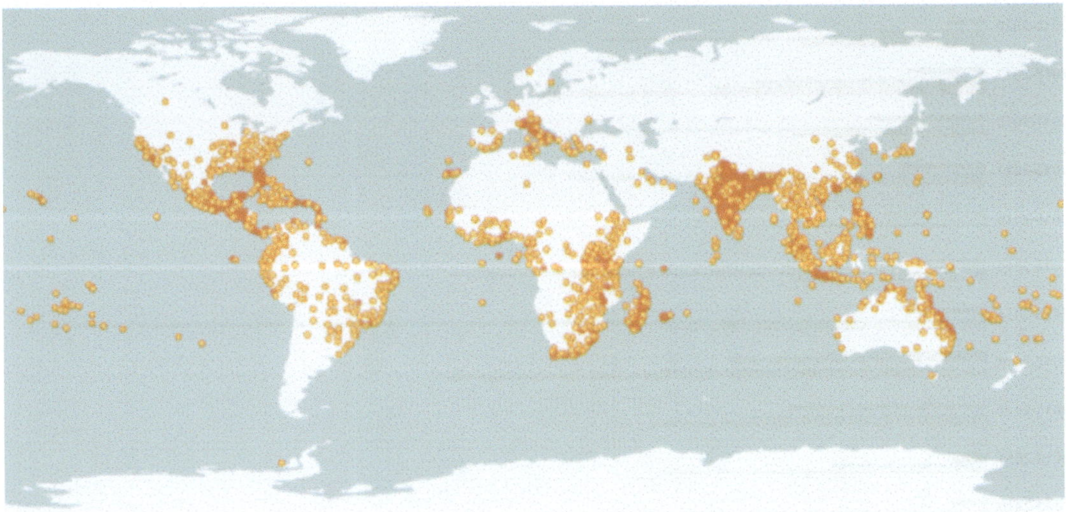

Fig. 3.3 Occurrences of *Catharanthus roseus* G Don across the globe along its hotspots *Source* The Global Biodiversity Information Facility

studied chiasma frequency, type of bivalents at diakinesis, and metaphase I of meiosis in pink and white varieties of *C. roseus.* The ring bivalents were found to be 73% at diakinesis and 72% at metaphase I. The chiasmata per bivalent varied from 2.8 to 1.7% from late diplotene to metaphase I. The mean number of chiasmata per cell at early metaphase I was 13.7 and did not vary between varieties. All Catharanthus species are exclusively diploid. The relative mean lengths of the chromosomes ranged from 0.042 to 0.076. The mitotic index was 22%. Few abnormal cells with $2n = 12$ and $2n = 14$ chromosomes were also observed (De Padua et al. 1992). The organization of *C.roseus* chromosomes was characterized using several chromatin staining techniques such as fluorescent in situ hybridization (FISH) of rDNA. Genome size was investigated by flow cytometry; the full chromosome complement of $2n = 16$ includes two metacentric, four sub telocentric, and two telocentric chromosome pairs, with the presence of a single nucleolus organizer region in chromosome 6. The C-value of this species was estimated to be 1C = 0.76 pg, corresponding to 738 Mbp (De Padua et al. 1992; Guimaraes et al. 2012). The complete plastid genome of *C. roseus* is 154.950 Kbp, with considerable variations in gene content and plastome organization within Apocynaceae, including the presence/absence of three essential genes (i.e., *accD*, *clpP*, and *ycf1*) and large size changes in non-coding regions (e.g., *rps2*-*rpoC2* and IRb-*ndhF*). The complete plastome was derived from the midrib of mature leaves of Pacifica Punch Halo infected with Phytophthora and was sequenced by Illumina Hiseq 2000 to generate 155 Kb, which includes86protein-coding, eight rRNA, and 37 tRNA genes (Ku et al. 2013). The whole genome of cultivar Sunstorm Apricot using 50xIlluminaHiSeqsequencingby Michigan State University has generated 523 Mb genome size consisting of 79,302 scaffolds with scaffold N5026249 bp, but this assembly still has 68,690 gaps (Kellner et al. 2015). The plastome and whole-genome sequence will pave the way for targeted breeding for MTIA content in Catharanthus.

3.2.3 Floral Biology, Pollination Mechanism, and Breeding Behavior

Knowledge of the morphology of the flower, its development, anthesis, pollination mechanism, mode of pollination, presence or absence of incompatibility, fruit and seed set of the plant is an essential prerequisite for understanding the breeding behavior of a plant species. Information on these aspects is necessary for artificial selfing and hybridization techniques, maintenance of varietal purity, selection, and execution of appropriate breeding methodology.

The vegetative phase in *C. roseus* is very short. Flowering begins when plants are only 4–6 inches tall or about ten weeks old and continue to flower as long as plants live. The flowers appear in pairs in the alternate leaf axils. These two flowers are never in the same stage of development, with one flower opening approximately 2 to 3 days before the other. In the early stages of development, the flower is polypetalous since pet al.s and stamens are at the same level as the floral receptacle. The corolla tube begins to form shortly after carpel initiation. The region below the stamen continues to elongate until the flower opens (Boke 1948). Stamen primordia appear soon after petal initiation. After a sequence of cell divisions, the stamen appears to consist of sessile anthers, and later the filaments are formed. Following the stamen initiation, the floral apex flattens. The central region grows at a slow rate and later develops into a carpel. The carpels are sessile and are not united to each other in the ovary region. The style remains short until the flower is nearly mature (Boke 1949). The corolla tube grows faster up to the day of flowering (Simmonds 1960). The elongation of style corresponds to the elongation of the corolla tube so that the relative position of the stigma and stamens remain almost constant. At anthesis, the stigma forms secretory cells on the sides of the stigma and a curious 'skirt' at its base. Based on these observations on the development of stamens and carpels, Boke (1949) concluded that the flower is well adapted for self-pollination. Anthesis generally starts from 15. 00 to 16. 00 h and continues until the next day. Anther dehiscence occurs just before anthesis, and the pollen is shed as a sticky mass. The stigma was found to be receptive between 6. 00 and 10 00 h and before 15 00 and 17 00 h (Rajasekaran et al. 1981). A characteristic feature of Apocynaceae is the disc-like or otherwise-shaped enlargement of the stigma head with a sticky secretion and a brush of hairs on which pollen collects as it is shed. The receptive portion of the stigma is at the base of the stigma head, and owing to the position of the anthers, self-pollination is rendered almost impossible, and insect visits are necessitated (Rendle 1971). Knuth (1909), in the book "Handbook of Flower Pollinations," stated that Darwin and Delpino had independently described the mechanism of pollination, which is brought about by nectar-seeking insects, which affect pollination by depositing pollen collected from the flowers during their previous visits. Automatic self-pollination in *C. roseus* is thus excluded. Many pollinators *Bombylius discolor*, *B. major*, *Anthophora pilipes*, *Apis mellifica*, *Bombus agrorum*, *B. hortorum*, *B. hypnorum*, *B. pratorum*, *B. terrester*, *B. vestalis*, *Osmia fusca*, *O. rufa*, thrips have been observed in the genus *Catharanthus* (Knuth 1909). Recently, two butterfly pollinators *Pachliopta hector* and *Catopsilia pyranthae* were observed in *C. roseus* (Kulkarni 1999). The corolla remains persistent generally for two days and sometimes for three days. Fruit (follicle) matures in four to five weeks. The follicles are dry dehiscent and shed seeds as they dehisce by the ventral suture.

C. roseus has also been considered as a self-pollinating species in several studies (Knuth 1909; Flory 1944; Boke 1949; Simmonds 1960; Krishnan et al. 1979; Levy et al. 1983; Milo et al. 1985). As *C.roseus* plants were generally found to be true-breeding for their flower color, Flory (1944) and Simmonds (1960) considered that *C. roseus* is a self-pollinating species. Darwin and Delpino reported that selfing within individual *C. roseus* flowers is not automatic and that pollination typically occurs through nectar-seeking insects (Knuth 1909; Rendle 1971). Krishnan et al. (1979) also found that *C. roseus* is a predominantly self-pollinating species with an outcrossing rate ranging from 1.7 to 13.8%. They determined outcrossing using flower color as a marker trait. Geitonogamy and phenotypic assortative mating for flower color brought about by pollinating butterflies give a false impression that *C. roseus* is autogamous when flower color alone is used as a marker trait to determine the breeding system (Kulkarni 1999; Sreevalli et al. 2000; Kulkarni et al 2001). Kulkarni (1999) reported an outcrossing rate of 79% when marker traits other than flower color were evaluated.

In normal allogamous strains, the stigma remained below the base of the anthers until flower drop, whereas in autogamous strains, self-pollination occurred due to an increase in the length of the gynoecium brought about by either an increase in the length of the style or of the ovary. Both self-pollinating strains *VI*, *MJ*, and *OR* possessing violet, magenta, and orange-red flowers with purple stems and non-self-pollinating strains, i.e., wavy leaf margin mutant and dwarf *Dw* mutant, is reported in *Catharanthus*. The F_1plants of *Wy* x *VI*, *Wy* x *MJ*, and *Dw* x *OR* and their reciprocals were normal for plant height and leaf margin but did not set any fruit or seed. The occurrences of intraflower self-pollination in these strains *MJ*, *VI*, and *OR* was governed by duplicate recessive genes, which were inherited independently of genes governing stem color, dwarfness and wavy leaf margin traits. Three alleles at two loci determine the occurrence or nonoccurrence of intraflower self-pollination in *C. roseus* (Kulkarni et al. 2005a). The self-pollinating plant of *C. roseus* has, however, not become common in nature, despite their reproductive advantage, due to non-dependence on pollinators as compared with normal cross-pollinating plants.

Heteromorphy in gynoecium length thus led to pin and thrum type of flowers in Catharanthus. A study on pollen less male-sterile *OR-EA* mutant with stigma position below the anthers 'thrum' flower (as in normal allogamous strain and EMS13-2 with thick leaves and long-styled mutant with stigma position above the anthers pin flower when crossed with their parental stains *OR* and 'Dhawal' respectively suggested that the production of pin (mutant) and thrum(normal) was under the control of inhibitory epistatic interaction between two independently inherited genes (Kulkarni et al. 2008). Earlier, a short style mutant conditioned by a single recessive gene *hsf* with pleiotropic effects on a number of other vegetative and reproductive characters has been reported (Mishra and Kumar 2003a; 2003b).

3.2.4 *Targets of Catharanthus Breeding*

The cultivation of Catharanthus has survived exactly three centuries and six decades since its first selection by Etienne de Flacourt in 1658 by the constant efforts of observant gardeners and commercial breeders funded by the private horticultural landscape seed companies and scarce efforts by the public sector for medicinal important MTIA targets.

Catharanthus breeding in the 1970s was more focused on its ornamental perspectives and natural selections in corolla color. *C. roseus* cv. *rosea* (purple), *C. roseus* cv. *alba* (white), and white corolla with purple eye color were prevalent as mixtures exhibiting a large range of variation in alkaloid content. Till the 1980s, there were no systematic breeding efforts and lacked pure lines in *C. roseus* (Kulkarni et al. 1984a). In the 1980s, Catharanthus breeders (mostly from commercial horticultural seed companies) focused their efforts to breed new recombinant strains of corolla and eye color with enhancement in leaf shape, size, color, and shortening the plant height based on inter cultivar hybridization and selections (Gogitidze and Maisuradze 1980).

In the 1990s, seed companies put intensive breeding efforts to produce frontrunner Catharanthus cultivars for summer bedding and landscape use like Thompson and Morgan, Park seed, Ball Horticultural, and Pan America, Goldsmith Incorporation, Syngenta, etc. (Bergen and Snoeijer 1996). Pushed by horticultural commerce dwarf, semidwarf, colored corolla, a high number of flowers with dense leaves and biomass was the Catharanthus breeders' targets to develop it as a bedding plant. These cultivars were planted in dense with the low plant to plant spacing, in long orderly rows with striking alternate corolla color and arranged in geometric shape or patterns to provide catchy beauty looks to the garden. The Catharanthus plant, in due course of time, evolved as a perennial plant. The plant remained under continuous marketing as an ornamental plant by the leading European seed companies for commerce. The International Registers of Catharanthus cultivars was published in 1998, with the first supplement in 2001 cataloged 50 cultivars (Hejidan et al. 2004). Currently, there are 364 varieties/cultivars from the public and private sector, where only a few are specifically bred for high MTIA (Cumulative seed catalog, of various horticultural companies).

The quest for new cultivars is pushed by the demand for greater productivity, higher alkaloid content, and tolerance to biotic and abiotic stresses. Cultivar development should incorporate an array of alternative breeding methods to develop a cultivar that is physiologically most fit to utilize the water and nutrient resource and tolerance to abiotic stresses, pests, and diseases. (Fehr 1987). New cultivars have resulted in higher yield per unit area, hence the adoption of such cultivar in a larger area supported by enhanced tolerance to losses to pests and environmental cues. Both self and cross-pollinated strains are reported in *C. roseus*. Hence initial cultivars were developed by pure line selection, having enhanced MTIA content. The breeding objective of Catharanthus genetic improvement decides the base plant material to induce variability, selection method, and resource allocation (Fehr 1987). MTIA content and quality stability are always a target for medicinally and ornamentally important Catharanthus breeding. In the current scenario, uniform and defined proportion of secondary metabolite along with their corolla and eye colors, compact canopy upright habit, early maturity, short internodal distance, biomass, disease resistance, and tolerance to heat, water, and light are primarily categorized the Catharanthus cultivars. The time of flowering of a cultivar is mainly the genotype response to the photoperiod and temperature of the geographical zone for which the cultivar is bred. Selection for traits like plant habit, early maturity, higher alkaloid content is exercised in early generations. The crop spacing and agronomic practices largely influence the productivity of a cultivar. Its ease of cultivation, harvest, and productivity is attributed to plant habit, strong stem, shorter internodal distance, compact canopy, indeterminate branching, high herbage yield, resistance or tolerance to major biotic stresses of the geographical and climatic zone.

3.3 Classical Breeding in Catharanthus

3.3.1 Pre-Breeding: Natural Variation and Its Assessment

Catharanthus flower structure, mating system, available natural variation assessment of the extent of genetic variability in the existing germplasm for targeted traits, their heritability, and inter trait correlations is the basic prerequisite for genetic improvement in Catharanthus as in other crops. The accurate evaluation of germplasm performance and targeted trait genetics is of great importance for the selection of parental lines and the development of new breeding populations. Having gene information for target loci (deduced from markers) can be extremely useful for breeders to efficiently use germplasm. Germplasm resources form the starting point of any breeding program. However, minimal information is available on the size of the available germplasm of *C. roseus*. There are 22 accessions at the ICAR-National Bureau of Plant Genetic Resources, India; 7 in Kew databases, 72 in National Plant Germplasm System, the USA and Global Biodiversity information facility documents 23,326 occurrences of *C. roseus* across the globe, of which unique accessions are not documented. The genetic diversity of *C. roseus* accessions has been characterized by different morphological, biochemical, chemical, metabolic, and molecular markers.

Information on inter trait correlations is essential to know the effect of selection for one trait of interest on other unselected traits and to know the possibility of carrying out indirect selection for characters of interest that are difficult or time-consuming to measure or are less heritable. In *C. roseus*, estimation of contents of total alkaloids and specific alkaloids viz., vinblastine, vincristine, vindoline, catharanthine, and ajmalicine is time-consuming and limits the plant population size that can be evaluated in a breeding program. Any trait with high heritability and a strong correlation with the contents of these alkaloids could be useful for preliminary screening for the content of alkaloids as well as for indirect selection for these important traits. Natural genetic diversity in Catharanthus was initially assessed using MahalanobisD^2 statistics. Two genotypes with high yields of leaf and root alkaloids were identified for direct cultivation after two years of evaluation of 17 genotypes collected from Poland, the then USSR, and different parts of India Virk et al. 1988). Genetic variability, heritability, genetic advance, and genetic diversity were assessed among selected 20 M_8 lines and six other selected lines (Dwivedi et al. 1999, 2000). Wide variation was observed for plant height, the number of branches per plant, leaf yield, and for the contents of total alkaloids, vindoline, catharanthine, vincristine, and vinblastine. High broad-sense heritability (83–91%) was observed for leaf yield and for the total alkaloids contents, vindoline, catharanthine, vincristine, and vinblastine, suggesting that the genetic differences between the studied genotypes were mainly due to additive gene action and could be exploited as pure lines and heterotic hybrid combinations. The 26 entries evaluated by Dwivedi et al. (2000) discriminated into nine clusters, suggesting hybridization between entries belonging to these clusters for realizing the greater magnitude of heterosis and wide genetic variability in the segregating generations (Dwivedi et al. 2000). Sharma et al. (2012) observed moderate heritability for leaf yield, root yield, and for contents of vindoline, vinblastine, catharanthine in leaves (40–79%) and low heritability for contents of total alkaloids in leaves and roots (5 and 13%). Heritability is observed to be higher for secondary metabolites as compared to primary metabolites; hence high genetic advance was observed for these MTIA (Dwivedi et al. 1999). The broad-sense heritability was observed to be higher for plant height and moderate for root and shoot weight, which can be used as selection indices for population improvement in Catharanthus. UPGMA and multivariate principle component analysis differentiated 32 accessions of *C. roseus* from geographical and climatically different areas of the Indian subcontinent based on 53 morphological, agronomic, and biochemical traits (Mishra et al. 2001) into five hierarchal groups. Strong positive correlations were observed between leaf area and leaf yield, leaf alkaloid; root alkaloid content with leaf vinblastine yield, and leaf vinblastine content with leaf vincristine and vinblastine yields, suggesting that simultaneous improvement for these pairs of traits should be possible through selection (Mishra et al. 2001; Singh et al. 2002; Singh et al. 2010; Sharma et al. 2012; Chaudhary et al. 2013). Leaf yield and root yield were not correlated with leaf alkaloid concentration and root alkaloid concentration, respectively (Mishra et al. 2001). In another path analysis study, internodal distance traits were analyzed to significantly increase the catharanthine content (Sun et al. 2009). Therefore, it should be possible to combine the high yield of these two plant parts with high concentrations of alkaloids in them. The studies also indicated a narrow genetic base of the populations of *C.roseus* in the Indian subcontinent. No relationship was found between flower color and contents of vindoline and catharanthine in 50 horticultural cultivars which had been bred for flower color (Magnotta et al. 2006).

Genetic diversity using random markers using random amplifies polymorphic DNA (RAPD) and inter simple sequence repeat (ISSR) markers, amplifies fragment length polymorphism (AFLP), proton nuclear magnetic resonography (^{1}H-NMR), phytochemical, morpho-chemical, enzymatic, isoenzymes, microsatellite markers have been used in order to identify *C.roseus* diversity at phenomic, molecular, biochemical and metabolic level (Shaw et al. 2009). Metabolomics and fluxomics have aided in identifying biosynthetic pathways of terpenoid indole alkaloids. Genetic diversity analysis in Catharanthus species has been analyzed in the past by many workers by being an economic marker system (Sharma et al. 2011; Lal et al. 2013; Prasad

2014). Many workers prefer to use simultaneously two arbitrary markers system techniques like RAPD and ISSR to have more accountable clustering of genotypes which can be utilized for accurate genetic diversity assessment and inclusion in a breeding program (Ibrahim et al. 2013). Real-time sequence characterized amplified region (RT-SCAR) marker was developed to overcome the inconsistency of RAPD and non-quantifiable SCAR marker by using SYBR green in SCAR marker amplification developed from tagged RAPD cloned bands (Chaudhary et al. 2012a). Loop-mediated isothermal amplification (LAMP, has been demonstrated in *C. roseus* for the detection of single herb species from a mixture through visible signals (Chaudhary et al. 2012b). Diversity among 32 *C. roseus* accessions obtained from different geographical regions of India, Sri Lanka, Mozambique, and private seed companies from India and Sweden, and one accession each of *C. trichophyllus, C. pusillus, Vinca minor, Thevetia peruviana,* and *Nerium indicum* were assessed using STMS (sequence tagged microsatellite site) markers (Shokeen et al. 2007). *C. trichophyllus, C. pusillus* separated from other *C. roseus* accessions, and within the Catharanthus group, the accessions were generally grouped according to their geographical origin (Shokeen et al. 2007). Identification of molecular markers (RAPD, AFLP, and ISSR) linked with flower characteristics such as different corolla color and eye color have also been screened but could not find any linked DNA marker (El-Domyati et al. 2012). Genetic discrimination of Catharanthus cultivars leaves by using ^{1}NMR spectra of aromatic compounds was made possible by ^{1}NMR. The dendrogram based on the secologanin and polyphenol aromatic region data were in general agreement with the genetic relationships detected by DNA fingerprinting methods (Kim et al. 2007). Pyrolysis mass spectrometry (PyMS) based on thermal degradation of complex materials in a vacuum was demonstrated in eight Catharanthus cultivars for discriminant analysis suggesting that chemical differences in pyrolzate compound among cultivar grouped the cultivars from the same series in one cluster which was earlier not possible by Fourier's IR spectroscopy (Kim et al. 2009).

3.3.2 Interspecific Hybridization and Exploiting Wild Species

The genus Catharanthus includes nine species, two subspecies, and six varieties indigenous to Madagascar, i.e., *C. coriaceus, C. lanceus, C. longifolius, C. makayensis, C. ovalis, C. ovalis* subsp. *grandiflorus, C. ovalis* subsp. *ovalis, C. ovalis* var. *ovalis, C. ovalis* var. *tomentellus, C. pusillus, C. roseus,C. roseus* var. *albus, C. roseus* var. *angustus, C. roseus* var. *nanus, C. roseus* var. *roseus, C. scitulus* and *C. trichophyllus.*Landraces and wild species of Catharanthus possess an underused genetic resource of novel alleles that have great potential for improvement in *C. roseus* for MTIA yield increment and tolerance to biotic and abiotic stresses. Natural interspecific hybridization between *C.roseus* species has been observed in Madagascar. Gene shuffling might have been taken place due to cross-hybridization, in these cohabitating wild species, unopposed by any geographic barrier, interspecific hybrids were successfully generated in nature (Veyret 1974; Markgraf 1976; Plaizier 1981). Major MTIA, i.e., vindoline, catharanthine, ajmalicine, vincristine, and vinblastine could only be detected in four species of *Catharanthus,* i.e., *C. roseus* (L). G Don, *C. longifolius*(Pichon) Pichon, *C. ovalis,* and *C. trichophyllus* (Baker) Pichon.

Later on, artificial hybridization was attempted successfully between *C.trichophyllus, C.-longifolius, C. ovalis* subsp, *C. coriaceus,* and *C. scitulus* with *C. roseus* (Veyret 1974; Markgraf 1976; Plaizier 1981; Sevestre-Rigouzzo et al. 1993; Chaudhary et al. 2013). *C.roseus* has been described geitonogamous, and both self-compatible and self-incompatible mechanisms have been reported by Schnell (1943). Veyret conducted six sets of ever first artificial interspecific hybridization between five species at Botanical Institute, Orsay, France.

C. roseus accession from Fort Dauphin, Madagascar(purple, pink corral with purple eye, pubescent, self-incompatible)was crossed with *C. longifolius* (self-compatible, pink light lilac corolla with white eye from Savannah grasslands. The interspecific hybrid had a lilac corolla and purple eye with intermediate leaf shape, leaf petiole like *C. roseus*, glabrous stem-like *C. longifolius*, quatriangular stem-like *C. roseus*; the hybrid was self-fertile with 97% pollen fertility (Veyret 1974).

C. trichophyllus, self-compatible, chromosome (2n = 16), red corolla, small purple eye and spirally twisted fruits accession from Manasamody massif rockeries, the forest of Ankarafantsika, Saint Marie Islands was crossed with *C.roseus* accession rom south of Madagascar. The hybrid had chromosome number ($2n = 16$), light purplish red corolla and purple eye with the lower side of corolla colored at the distal end, petiolate obovate leaf, quadrangular stem, straight fruits, self-fertile with 95% pollen fertility. All other species of Catharanthus are otherwise white on the lower side. *C. trichophyllus* posses' novel epiallo-yohimbine derivatives (Davidoud et al. 1989).

C. ovalis subsp. *grandifloras* was endemic to Isalo Massif region, trichomeless, red corolla with white eye, self-compatible, $2n = 16$ was crossed with *C.roseus*. The hybrid had an erect habit, but branches drooped after reaching a certain height, obtuse elliptical leaf, intermediate leaf anatomy, trichomeless plants floriferous with purple corolla and purple eye, self-compatible, with 54% pollen fertility.

C. ovalis subsp ovalis was crossed with self-compatible, $2n = 16$ endemic to Isalo Massif, Savannah with a prostrate habit, with the lilac pink corolla, white eye, trichomeless, quadrangular stem was crossed with *C. trichophllus*. Hybrid had the rigid stem-like donor but was quatriangular like the female parent, and lower side of the corolla was distally colored pinkish-purple, self-fertile with 66% pollen fertility.

C. longifolius x *C. ovalis* subsp. *grandifolius*. The hybrid was erect in growth habit, quadrangular stem, intermediate leaf shape, large lilac corolla with white eye, self-compatible, with 93%pollen fertility.

C. trichophyllus x *C. coriaceous,* where the donor was endemic to Intremo. The hybrid obtained after cross-hybridization was self-fertile with 91%, pollen fertility, shrub habit, quadrangular stem, small pubescent leaf with short trichomes similar to recipient parent, small bright pink flower.

All these self-fertile hybrid illustrated their close genetic relationship. As the chromosome number for all species was 16, bivalent were primarily observed with few quadrivalents in a cross with *C.ovalis ssp grandiflorous.* (Anderson 1931). Authors could not find any report of intergeneric hybridization between *C. roseus* and *C. pusillus* to date, which validates their distinct separate origin with genetic incompatibilities.

3.3.3 Inbred Development: Cleistogamy

The lack of pure lines has led to variations in the alkaloid content and a fall in the price of raw plant material produced by the farmers. Self-pollination, brought about by an increase in the length of the gynoecium (ovary or style) in self-pollinating strains, was found to occur after 1–2 days after anthesis. Maintenance of genetic purity is not automatically ensured in such genotypes. For exploitation of heterosis, the need is to develop self-pollinating high alkaloid content strains, preferably with an early-expressing phenotypic marker, to facilitate their seed multiplication, identification, and maintenance of genetic purity (Kulkarni 1999). Self-pollinating *C. roseus* strains developed through cleistogamy facilitate maintenance of genetic purity without manual selfing and seed production without dependence on pollinators. This helps in the production of inbreds. These inbreds can further be used in hybrid seed production. It would also ensure pollen containment which is an important requirement in the development of transgenics. An ethyl methane sulphonate (EMS) induced mutant in which corolla abscised before the opening of the corolla, i.e., at the flower bud stage, was identified. By crossing this mutant with self-pollinating strains, cleistogamous strains were developed. These strains constitute pyramiding of recessive alleles for closed corolla, normal plant height (or a tightly linked gene inhibiting closed corolla), and self-pollination (Kulkarni and Baskaran 2013a).

3.3.4 Hybrid Breeding: Male Sterility

Though Schnell (1943) has described both self-compatible and incompatible strains, Vyvret observed all self-fertile interspecific hybrids supported later by studies of Levy et al. (1983). Lack of self-incompatibility, dioecy, or male sterility in any Catharanthus species makes hybrid seed production commercially not feasible. Male sterility is a condition in which functional anthers, pollens, or male gametes are not produced using either natural mechanism or artificially induction.

Levy et al. (1983) studied heterosis for leaf yield, root yield, and ajmalicine content in roots, in crosses involving three pure lines differing in flower color, pink corolla, white corolla, and red-eye (white corolla and red-eye in the center). The F_1 hybrids (pink x red-eye) performed better for leaf and root weight with an increase of 29% and 24%, respectively, over the better parent (red-eye). However, no heterosis was observed for ajmalicine content in roots. They suggested breeding of pure line cultivars with high herbage and root yield and high content of alkaloids till the availability of sterile male lines for exploiting the observed heterosis for the leaf and root yield. They also opined that the absence of heterosis for ajmalicine content in roots did not favor the development of hybrid cultivars of *C. roseus* for use by the pharmaceutical industry. *C. trichophyllus,* which contains high vindoline, catharanthine, serpentine in leaves; serpentine and ajmalicine in roots as compared to *C.roseus,* was utilized for interspecific introgression of MTIA to develop new breeding material. F_1 hybrids, as predicted, had remarkable heterosis for catharanthine and vindoline from leaves and ajmalicine and serpentine from roots. Interspecific reproductive gametophytic, unilateral incompatibility was observed between *C. roseus* x *C.tricophyllus,* which was otherwise compatible when *C. trichophyllus* was the female parent. F_1 hybrids showed a threefold increase in catharanthine leaves and a two-fold increase in roots from the better parent (Sevdstre-Rigouzz et al. 1993).

Though male-sterile lines were developed in due course of time, being protected by patents, it was put to the application by few breeders. Bowman (2000) from Goldsmith Incorporation, utilized streptomycin to develop pollen less lines ('13,861–1') governed by nuclear recessive allele *msGS* inheritance. It does not set seeds due to abnormal tapet al. cell development and does not carry any undesired pleiotropic effect (PatentUS006166306A). To exploit the vigor and to avoid self-pollination in normally self-pollinating strains, Flaschenreim and Blowers (2017) developed an EMS mutagenized male-sterile line H7487D, male sterility being governed by single recessive all ele*msms*. The male-sterility gene was introgressed by crossing H7487D x 'Pacifica Really Red' and by selfing the F_1 for five generations to obtain the male-sterile inbredP7998D for further use in the hybrid production. They further generated hybrid PAS1157216, by crossing male-sterile inbred P7998D as female with 'Pacifica Really Red' as donor governed by nuclear *Msms* homozygous allele introgressed in the desired female inbred line by backcrossing the F_1 four times to female (recurrent parent) followed by selection for recovery of other morphological and physiological traits in addition to the introgressed male-sterility locus and finally selfed (BallHorticultural company, Patent No US2017/0118933A1). Transfer of male sterility by backcrossing for a dominant allele is simplified, but the method can also be used for a recessive allele or additive alleles further confirmed by progeny test. Modified backcrossing using different recurrent parent/multiple recurrent parents during the backcrossing can also be used to obtain desirable traits from each parent.

A functional 'male-sterile' mutant in which anthers failed to dehisce and demonstrated its utility in the production of hybrid seeds. The mutant (EMS1-10) was derived from parent Nirmal, had a wavy leaf margin, and was phenotypically normal for other morphological traits except for indehiscent anthers. The mutant has small anthers, lesser pollen grains, but was fertile with good fruit and seed set on artificial selfing; thereby no restorer was required. EMS1-10 mutant can be used for hybrid seed production and was tested in a seed production block with a row of the functionally male-sterile line surrounded by its parental variety. Considerable fruit set, a good number of seeds per fruit was observed in the functionally male-sterile line used as the female parent. This male- sterility also was characterized by the recessive gene for three early-expressing traits, which can mark its early identification before transplantation in the field. The mutant can be multiplied by artificial selfing, stem cutting, or micropropagation (Sreevalli et al. 2003; Kulkarni and Baskaran 2008). A partially male-sterile mutant due to in situ pollen germination has also been reported (Mishra and Kumar 2001). These male-sterile lines have been immensely useful for Catharanthus hybrid breeding.

3.3.5 Exploiting Vigor by Polyploidy

Polyploids, natural or induced, generally have larger biomass with large leaves, stems, roots, flowers, and fruits. Induced polyploids in the past have resulted in enhancement of secondary metabolites and are preferred where natural variation is limited, vegetative parts are more economical, diploid with small haploid content, and could be vegetatively propagated (Lavania 2005; Lin et al. 2011; Kulkarni et al. 2016). *C. roseus* has a low chromosome number and its vegetative parts, leaves and roots, and not seeds, are of economic importance. Therefore, many attempts have been made to induce and evaluate autotetraploid.

Autotetraploidy has been successfully induced in *C.roseus* (Janaki Ammal and Bezbaruah 1963; Dnyansagar and Sudhakaran 1977; Gogitidze and Laptev 1981; Mohan Kumar 1980; Kulkarni et al. 1984a, 1987; Krishnan et al. 1985; Handique and Sharma 1990; Chauhan and Raghuvanshi 1980). To induce tetraploidy, either seed (Janaki Ammal and Bezbaruah 1963; Dnyasagar and Sudhakaran 1977; Handique and Sharma 1990; Gogitidze 1981) or apical buds of young seedlings (Janaki Ammal and Bezbaruah 1963; Gogitidze 1981; Kulkarni et al. 1984b) were treated with different concentrations of colchicine solutions ranging from 0.01 to 1.0%. Gogitidze (1981) found that colchicine treatment of apical buds was more effective in inducing tetraploidy than seed treatment. He also found that stamen length and pollen grain diameter were the two most reliable characters to identify tetraploids.

Autotetraploids were found to be more vigorous in growth with broader leaves, larger stomata, flowers, pollen grains, and embryos but had low pollen fertility, poor fruit set, and low seed production as compared with diploids (Janaki Ammal and Bezbaruah 1963; Gogitidze and Laptev 1981; Mohan Kumar 1980).In tetraploids, pollen fertility is low (32–43%), and 80–90% flower generally falls off after pollination leading to poor seed set (17.5–22.5%). Prefertilization abnormalities such as undeveloped ovules, lack of normal organization or delayed organization of the embryosac, and delayed fertilization played an important role in the breakdown of seed (haplontic sterility) in the autotetraploid (Dnyansagar and Sudhakaran 1977). The embryo grows faster during the early period, while in later stages, the growth seems to slow down, whereas endosperm development is slow initially and active later. The tetraploid seeds took a longer time (28–32 days) to mature than the diploid seeds (24 days). Seed development in autotetraploid was similar to the diploid plants; however, polyploidization altered the growth of embryo and endosperm therein several pre-fertilization abnormalities such as reduced pollen fertility and germination, improper megagametophyte development was observed (haplontic sterility). The embryo is more prominent in size in the tetraploid seed than diploid (Dnyansagar and Sudhakaran 1970).

In some studies (Dnyansagar and Sudhakaran 1970; Gogitidze and Laptev 1981; Mohan Kumar 1980; Chauhan and Raghuvanshi 1980), tetraploids were found to have significantly higher leaf yield, root yield, and total alkaloids content while in other studies (Kulkarni et al. 1984a; Krishnan et al. 1985) they were found to be at par with diploids for leaf yield and root yields and had low ajmalicine content and harvest index. Tetraploids performed better than diploids at closer plant spacings (30 cm × 30 cm), especially in the absence of nitrogen application, suggesting that tetraploids could be grown on soils of low fertility as a rain-fed crop for obtaining higher yields than diploids (Kulkarni et al. 1987). Further studies showed that tetraploids had higher nitrogen utilization efficiency than diploids, particularly under low fertility (Kulkarni et al. 1995).

Artificially induced tetraploid using 0.2% colchicine for two days for 6 h. in *C. pusillus* yielded increased in pollen size, stomatal aperture, length, and breadth of guard cell, in autotetraploid and quadrivalent were observed in higher frequency but did not interfere in the normal separation at anaphase (Chauhan and Raghuvanshi 1978; Goswami et al. 1996; Xing et al. 2011).

MTIA contents of vindoline, catharanthine, and vinblastine were found to be higher in tetraploid lines than in diploids and corresponded with higher levels of expression of the genes of *tdc*, *g10h*, *sls*, *str*, *dat*, and *prx1* in the tetraploid lines as compared with the diploids (Xing et al. 2011). HPLC analysis showed tetraploidization could increase the contents of terpenoid indole alkaloids in *C. roseus*. Thus, tetraploidization could be used to produce higher alkaloids lines for commercial use. QRT-PCR results showed that the expression of enzymes involved in terpenoid indole alkaloids biosynthesis pathway had increased in the tetraploid plants (Xing et al. 2011). Further studies showed that tetraploids had higher nitrogen utilization efficiency than diploids (Kulkarni et al. 1995). Induced autotetraploid were also found to be highly resistant to die back, and collar and root rot (devastating diseases in rainy season) and yielded four and five times more leaf and root total alkaloids, respectively, than diploids.

3.4 Breeding for Specific Targets and Targeted Trait Genetics

3.4.1 Breeding for Alkaloid Content

The alkaloid content in the Catharanthus plant varies with genotype, environment, soil fertility, and phenological conditions. For a given genotype, it varies spatially and temporally among the different plant organs, tissues, and phenological stages. Alkaloid content among different plant parts exhibits large variation depending upon the site of biosynthesis. Total alkaloid content is maximum in root bark (4.5–9.0%), fibrous roots up to 9%, root cortex (0.14–1.34%), leaves (0.32–1.16%), stems (0.26–1.16%), flowers (0.005–0.84%), fruit (0.40%), seeds (0.18%) and pericarp (1.14%) (Joy 1998). *C. roseus* MTIA biosynthesis pathway explorations define the site of synthesis, site of storage, and 30 enzymatic steps for the biosynthesis of major Catharanthus alkaloids. Vindoline starting from 2-*C*-methyl-D-erythritol 4-phosphate (MEP) pathways in internal phloem associated parenchyma cells leading to loganic acid, which is then transferred to the upper epidermis and converts into secologanin. Secologanin and tryptamine combine in the cytosol to generate strictosidine, a precursor to both tabersonine and catharanthine in the cytosol. Tabersonine converts into deacetyl vindoline in laticifers cells, exported to idioblast where vindoline is generated. Catharanthine is transported to vacuoles via CrTPT2 transporters, where it conjugates with vindoline to produce vinblastine. Vinblastine is converted to vincristine by an unknown mechanism (Courdavault et al. 2014; Qu et al. 2015; Pan et al. 2016). The subcellular compartmentation separates the enzymes from their precursors and substrates. Alkaloid storage cells have a much lower vacuolar pH than other cells in the plant. These observations indicate that alkaloid synthesis and storage might take place in different cells.

Ajmalicine is exclusively found in Catharanthus root, whereas serpentine is in aerial parts also (Arens et al. 1978). Alkaloids are chiefly concentrated in the collar bases region and primarily localized in root bark tissue, and are not in stelar cylinder. Alkaloids concentration enhances till 45–60 days of germination and at flowering nearly 70–80 days of sowing it reaches a constant level become stabilized. In one generation of selfing of superior plants, there was an increase of 0.5–0.9% of ajmalicine and 1.2–1.7% of serpentine (Arens et al. 1978). Thamm (2014) mutagenized *C. roseus* by EMS and screened 3600 plants, identifying one plant with high ajmalicine and low catharanthine and vindoline content. RNA sequencing and comparative bioinformatics of mutant and wild-type plants showed upregulation of strictosidine beta-glucosidase (SGD) and the transcriptional repressor, Zinc finger Catharanthus transcription factor (*ZCT1*), in the mutant line. The increased SGD activity in mutants seemed to yield a larger pool of uncharacterized SGD reaction products that are channeled away from catharanthine and vindoline towards biosynthesis of ajmalicine when compared to the wild type.

Catharanthine content to date explored in the Catharanthus varieties has been highest up to 3.79 mg/gm dry leaf weight in Pacifica Polka Dot and minimum up to 0.9 mg in Cooler Pink (Sun et al. 2009). Through mutant analysis, it is established catharanthine is secreted on the leaf surface. The concentration of catharanthine and vindoline was found to be maximum in young leaves. Catharanthine is exclusively stored at upper leaf epidermis waxy tissue, in root tips protoderm and cortical cells, whereas vindoline is present in laticifer cells. The 3, 4 anhydrovinblastine is accumulated in fully expanded leaves and reduced in older leaves (Roepke et al. 2010). In hairy root cultures, alkaloid accumulation generally is in a linear relationship with biomass accumulation, and ajmalicine and catharanthine amount obtained ranges from 4.3 to 3.8 mg/g dry weight, respectively (Benyammi et al. 2016).

Vindoline and bisindole alkaloids accumulate only in green tissues and are not found in the root or cell suspension cultures (Endo et al. 1987). In seedlings of *C. roseus* the expression of MTIA enzymes was under strong developmental control modulated by tissue-specific and/or light-dependent factors. The light-regulation of vindoline biosynthesis was shown to be mediated by phytochrome. In *C. roseus* leaves, the concentration of vindoline, catharanthine, and 3',4'-anhydrovinblastine (AVLB) are age-dependent. As the leaf matures, the concentrations of catharanthine and vindoline decrease while the concentration of anhydroblastine increases (Roepke et al. 2010). Genotypes which lack tabersonine16-hydroxylase activity are very low in vindoline like Mediterranean deep orchid. Hence tabersonine16-hydroxylasegene-specific expression in the seedling stage can help in the selection of vindoline rich genotypes (Schröder et al. 1999; Magnotta et al. 2006).

Kulkarni et al. (1999), while evaluating M_3 and M_4 progenies of EMS and NEU mutagenized seeds of variety 'Nirmal' analyzed alkaloid yields of three macromutants, i.e., dwarf, semidwarf, and wavy leaf margin mutants. Both dwarf and semidwarf mutants had lower root yield but significantly higher root alkaloid contents than the parental variety in both the M_3 and M_4 trials. The semidwarf mutant gave about 23% higher root alkaloid yield per hectare than the parental variety. The wavy leaf mutant showed a significantly higher (20%) content of total leaf alkaloids than the parental variety. All these three mutants were controlled by monogenic recessive genes. The F_1 plants of dwarf x semidwarf were semidwarf, and the segregation data of F_2 and backcrosses generations of this cross showed the allele for semidwarf was dominant over the dwarf. The segregation data of crosses involving semidwarf and dwarf mutants with a wavy leaf margin indicated that the gene for wavy leaf margin was inherited independently of the genes for the dwarf and semidwarf characters, and so it is possible to recover progenies by cross high leaf alkaloid with high root alkaloid progenies. Alkaloid content in leaves is found directly correlated with herbage yield, which significantly increases on reducing the internodal growth.

Sixty-four cultivars were screened for the variation of vindoline, catharanthine, and serpentine content, and the highest concentration of serpentine was found in cultivars Cooler Rose Hot (461 ± 46 μg/g DW). Concentrations of vindoline (2082 ± 113 μg/g DW) and catharanthine (2903 ± 384 μg/g DW) were highest in Pacifica Peach (Chung et al. 2011). Catharanthine content was found to be 3.79 mg/g dry leaf weight in Pacifica Polka Dot, and through path analysis, it was found to be positively correlated with internodal distance (Sun et al. 2009) MTIA alkaloid profiles of 50 different cultivars of flowering *C. roseus* concluded that breeding for changes in flower color or growth habit does not change, the alkaloid patterns or alkaloid contents in young plant shoots (Xing et al. 2011). Using path coefficient and direct and indirect effect analysis the relative role of agronomics traits in decided alkaloid yield was analyzed. The accretion of the number of leaves resulted in higher dry matter yield with higher alkaloid yield. While vindoline, catharanthine, and vincristine has direct effect, vinblastine was reported to have a negative indirect effect on alkaloid yield (Roy et al. 2013).

While there have been significant efforts of screening for alkaloid content in the accessions and cultivars to select for high MTIA content with little success, efforts through mutation breeding have unraveled the accomplishments by three independent groups. Curiously, all these three cases emerged while breeding for disease resistance was found to be very high in total alkaloid content. Bowmen in 1996 (Goldsmith Incorporation), in the process of screening several hundred species x species, cultivar x species, and cultivar x cultivar hybrids for Phytophthora resistance in 1992 got three experimental hybrids lines, 13,516, 13,517, and 13,518, which expressed good resistance, from further '15,667' line was developed, which had resistance to Phytophthora, and accumulated high 6.26% leaf alkaloid content. Today it is well established that catharanthine is exclusively stored in the upper leaf waxy surface, which imparts resistance to both diseases and pests (Roepke et al. 2010). Since the resistance gene directly influences the total alkaloid produced, one can intentionally select for increased alkaloid content by screening lines that carry the resistance gene. Kulkarni et al. 2003 through NEU-induced mutagenesis, developed Dhawal, which is dieback resistant and has 2.22% root alkaloid and 1.40% leaf alkaloids, besides a good herbage yield of 2,557 kg/ha. CIM Sushil, which was also developed through EMS-induced mutagenesis, has ~5% total alkaloids and 0.2% vindoline content with an herbage yield potential of 2,418 kg/ha (Gupta et al. 2020; Mall et al. 2021). Dhawal and CIM Sushil varieties from CSIR-CIMAP, India (Table 3.1) both are dwarf types, with much reduced internodal distance, more laticifers, high chlorophyll content wavy leaves, which are glabrous in nature. The glabrous, shiny and waxy leaf surface is an indirect selection trait for the high alkaloid content of catharanthine and can be screened in early development stages. Breeding for semidwarf type and upright growth habits should be practiced for targeted alkaloid breeding.

3.4.2 Breeding for Corolla Color: Ornamental Market Perspectives

Early genetic studies were restricted mainly to the inheritance of flower color and pubescence. Three basic corolla color patterns are most commonly observed in *C. roseus*, namely pink, white, and red-eye (white corolla with red center). Flory (1944) attributed epistatic interaction between two genes *R* and *W*, to explain flower color differences in these three phenotypes. In a cross between a pink flower and a white flower, he observed a ratio of 9 pink: 3 red-eyes: 4 white plants, with genotypes *R-W-* being pink, *R-ww* red-eye, and *rrW-* and *rrww* being white-flowered. He also observed that red pigmentation in stems of pink-flowered plants was

Table 3.1 Released *Catharanthus roseus* varieties and hybrids protected by plant variety patents

Variety	Characteristics	Phenotype
'Nirmal' 1989, CSIR-CIMAP, India	Pure line selection, Erect habit, Plant height 75-85 cm, Pubescent leaves, white corolla, white orifice Total leaf alkaloid content 0.8–1.0% & total root alkaloid content 1.6–1.8% Resistance to die back & collar rot diseases	
'Dhawal' 1997, CSIR-CIMAP, India Raghvendra Narayana Rao Kulkarni et al. 15/4/2003 US6548746B1	Developed by chemical mutagen treatment in C. Roseus variety 'Nirmal' Plant height 65–75 cm; green stem, pubescent, wavy, light green to greyish green leaves & white corolla, white orifice Total leaf alkaloid content high 0.89–1.4% & total root alkaloid content 1.60–2.22% High level of resistance to die back disease	
'Prabal' 2001 CSIR-CIMAP, India	Pure line selection White corolla with small red orifice Total leaf alkaloid content 1.65–1.95% & total root alkaloid content 1.5–1.85% Winter hardiness	
'Ili' (USPP18,315P3) 18/12/2007 Sushil Kumar, etal CSIR-CIMAP, India	Developed Catharanthus roseus 'Nirmal' through chemical mutagenesis. Recessive monogenic Leaf less inflorescence. Hyper branching. Higher flower producing frequency Color at upper corolla surface white & on lower corolla surface albino white Total leaf alkaloid content 1.9%	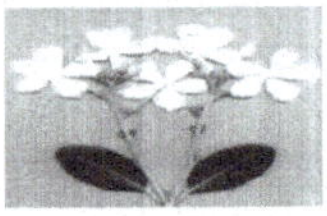
'CIM-Sushil' 2019 CSIR-CIMAP, India	Dwarf, spreading, or bushy growth habit & plant height is 35.5 cm Shorter internodes & petiole, Small, dark green leaves, white corolla & white orifice Higher Vindoline content (0.2%) than others & ~ 5% total alkaloid content in leaves	
'SAKURAI 15' (USPP18154P2) 30/10/2007 Osamu Kitajima, Japan	Hybrid of *C. pacifica* (unnamed)♀ X *C. roseus* (unnamed)♂ Plant is perennial evergreen, sterile and upright; height 35 cm & width also 35 cm; pale green stem, dark-green leaves; red–purple corolla with irregular lobules Disease & pest resistance and heat tolerance	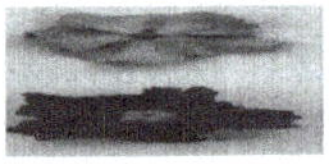
'SAKURAI 01' (USPP18195P2) 13/11/2007 Osamu Kitajima, Japan	Hybrid of *C. pacifica* ♀ *X C. roseus* ♂ Sterile, evergreen & upright; height 30 cm & width 40 cm; pale green stem, dark green leaves; mid pink color corolla in combination of irregular lobules and weekly sinuate margin on each tepal	
'SUNNICHIHOHO' (USPP20541P2) 8/12/2009 Naoto Takamura, Kiyoshi Miyazaki, Takuro Ishihara, Japan	Originated by self-pollination on C. roseus (03-CAT-20) Trailing & vigorous; height ~16.6 cm & width ~34.2 cm. Freely branching and flowering Large, pinkish-white corolla with red–purple orifice. Flowering for long time	

(continued)

Table 3.1 (continued)

Variety	Characteristics	Phenotype
'SUNNICHIRORO' (USPP20559P2) 15/12/2009 Naoto Takamura, Kiyoshi Miyazaki, Takuro Ishihara, Japan	Hybrid of C. roseus (03CAT-7) ♀ X C. roseus (03CAT YM) ♂ Trailing habit with vigorous growth. Height ~17.6 cm & width ~38.7 cm. Freely branching & flowering. Large salverform flowers have red–purple color corolla	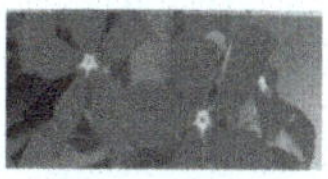
'SUNNICHIROIN' (USPP22594P2) 20 /03/2012 Naoto Takamura, Kiyoshi Miyazaki, Japan	Hybrid of C. roseus (Casp7-1) ♀ X C. roseus (03–22-4) ♂ Trailing with vigorous growth. Height ~12 cm & width ~42.5 cm. Basal & free branches, freely arranged flowers, large red–purple corolla Tolerant to wind, rain, and temperature ranges about 5 ° C to 35–40 °C	
'SUNNICHI TAREPA' (USPP25110P2) 25/11/2014 Tetsuya Kako, Shimane, Japan	Hybrid of C. roseus (*Casp5-1)* ♀ X C. roseus (*Cacv10-1)* ♂ Upright to outwardly spreading & vigorous. Height ~28.3 cm & width ~39.4 cm Branches are basal and free with freely arranged flowers Plant flowers for long period and flowers are large, red–purple color	
'SUNNICHILAVE' (USPP25111P2) 25/11/2014 Tetsuya Kako, Shimane, Japan	Hybrid of C. roseus (*Casp5-1)* ♀ X C. roseus (*03–20-1)* ♂ Trailing, vigorous. Height ~12 cm & width 36.8 cm. Multiple free branches from base Large, round, salverform lavender color flowers Tolerant to wind, rain, and temperature ranges	
'SUNNICHITARO' (USPP25112P2) 25/11/2014 Tetsuya Kako, Shimane, Japan	Hybrid of C. roseus (*Casp2-1)* ♀ X C. roseus (*Cacv10-1)* ♂ Upright, vigorous. Height ~23.5 cm & width ~36.2 cm. Basal and free branches. Single salverform flowers with light red–purple corolla Tolerant to wind, rain & temperature (from 5 °C to about 35–40 °**C**)	
'SUNCATHAMIHO' (US2015/0089699P1) 26/03/2015 Masahiro Yamada, Japan	Hybrid of C. roseus (MC21-15) ♀ X C. roseus (MC21-P2) ♂ Upright & vigorous. Height ~ 25 cm & width also ~ 25 cm. Basal branches. Single flower, pale-pink, deeply bitten, incurvate corolla. Blooms for a very long period	
'SUNCATHA 2243' (US2015/0089697P1) 26/03/2015 Masahiro Yamada, Japan	Hybrid of C. roseus (MC21-52) ♀ X C. roseus (MC21-5) ♂ Upright, vigorous. Height ~ 23 cm & spread also ~ 23 cm. Basal branching red-pink (RHS 60D), deeply bitten, hastate & asymmetric corolla	
'SUNNICHI TAHOME' (USPP25250P2) 27/01/2015 Tetsuya Kako, Shimane (JAPAN)	Hybrid of C. roseus (*Casp2-1)* ♀ X C. roseus (*Cacv10-1)* ♂ Upright to outwardly spreading & vigorous. Height ~22.8 & width ~38.5 cm. Basal & free branches, freely arranged flowers with white color corolla and red purple orifice	
'PAS1052940' (US2016/0198658A1) 14/07/2016a Denis R. Flaschenriem, CA (US)	Hybrid of M6788D♀ X M0239D♂ Semi-dwarf, semi-compact, upright habit. Height ~35 cm & width ~56 cm. Early flowering. single, flat flowers, white(607C) corolla Drought tolerant, disease resistant	

(continued)

Table 3.1 (continued)

Variety	Characteristics	Phenotype
'PAS1053041' (US2016/US0198660A1) 14/07/2016b Denis R. Flaschenriem, CA (US)	Hybrid of M6466♀ X *M1767D♂* Semi-determinate, semi-dwarf, upright, early. Height 37 cm & width 54 cm. Straight stem, many branches, dark green leaves; overlapping baby pink corolla & dark pink orifice Herbicide tolerance, disease & pest resistance	
'PAS1053042' (US2016/US0198662A1) 14/07/2016e Denis R. Flaschenriem, CA (US)	Hybrid of M6819D♀ X M3281D Upright, semi-determinate, semi-dwarf, semi-compact, early. Height 40 cm & width 55 cm. Stem straight, multiple branches, dark green leaves Flowers are single, flat, round shape with red (215C) & overlapping corolla	
Var. 'PAS1053043' (US2016/US0198659A1) 14/07/2016c Denis R. Flaschenriem, CA (US)	Hybrid of M5602D♀ X *M6948D♂* Upright, semidwarf, early. height 38 cm & width 53 cm. Straight stem, branches with anthocyanin coloration. Flat, round flowers, burgundy color (220C) corolla with overlap Disease & pest resistance, herbicide tolerance	
'PAS1053044' (US2016/US0198661A1) 14/07/2016d Denis R. Flaschenriem, CA (US)	Hybrid of M3861D♀ X M4861D♂ Semi-determinate, upright, compact, dwarf, early. Height 29 cm & width 52 cm Straight stem with multiple branches and leaves are medium green Single, flat, intermediate shape, lavender color (2060C) corolla, white orifice	
'PAS1053045' (US2016/US0198663A1) 14/07/2016f Denis R. Flaschenriem, CA (US)	Hybrid of M2246D♀ X M0224D♂ Upright, semi-determinate, semi-dwarf, semi-compact & early. Height 41 cm & width 59 cm. Straight stem with multiple branches and dark green leaves Single, flat, round flowers, violet (2060C), overlapping corolla, white orifice	
'SUNCATHA 2460' (US2016/0295769P1) 6/10/2016 Masahiro Yamada, Japan	Hybrid of C. roseus (FS22-9) ♀ X C. roseus (CL)♂ Upright to outwardly spreading and uniformly mounding; vigorous; height ~35.3 cm & width ~56 cm. Freely basal branching & with plenteous flowers Free, single, small flowers; light red–purple corolla with apparent white orifice	
'SUNCATHA 2316' (US2016/0295767P1) 6/10/2016 Masahiro Yamada, Japan	Hybrid of C. roseus (S22-P-27) ♀ X C. roseus (SIR22-16) ♂ Upright to outwardly spreading and mounding; vigorous growth; height ~39.3 cm & width 58.6 cm. Branches are free and basal, with freely arranged, numerous attractive flowers Double-type flower & pale pink in corolla	
'SUNCHITA 2439' (US2016/0295766P1) 6/10/2016 Masahiro Yamada, Japan	Hybrid of C. roseus (FS22-9) ♀ X C. roseus (CL) ♂ Upright to outwardly spreading, mounding & vigorous. Height ~33 cm & width ~59.7 cm. Free, basal branches; free, single, salverform, small flowers with dark reddish-purple corolla & dark red orifice Tolerant to wind, rain, and temperature ranges from 5 °C to about 35–40 °C	

(continued)

Table 3.1 (continued)

Variety	Characteristics	Phenotype
'SUNCATHA 224' (USPP27478P3) 20/12/2016 Masahiro Yamada, Japan	Hybrid of C. roseus (FS-S) ♀ X C. roseus (NCPS) ♂ Upright to outwardly spreading and mounding; vigorous growth height ~24.3 cm & width ~52 cm; basal branches; free, single, small flowers; red–purple corolla, red orifice Tolerant to wind, rain, and temperature ranges from 5 °C to about 35–40 °C	
'SUNCATFE 185' (US2017/0295690P1) 12/10/2017 Masahiro Yamada, Japan	Hybrid of C. roseus (FS22-3) ♀ X C. roseus (CL) ♂ Compact, upright to outwardly spreading, uniformly mounded, vigorous; branches are compact & free; small, star shaped flowers; pink corolla with red purple orifice	
'SUNCATFE 43' (US2017/0295691) 12/10/2017 Masahiro Yamada, Japan	Hybrid of C. roseus (FS22-9) ♀ X C. roseus (CL) ♂ Compact, upright to outwardly spreading, uniformly mounding, vigorous; height ~26 cm & width ~50 cm; freely, basal branches; free, small, star shaped flowers; white corolla	
'SUNCATSO 23' (US2017/0295689P1) 12/10/2017 Masahiro Yamada, Japan	Hybrid of C. roseus (S22-P-30) ♀ X C. roseus (SIR22-10) ♂ Upright to outwardly spreading, mounding; vigorous growth; height 32 cm & width ~50 cm; double-type, star-shaped, salverform flowers; pink corolla, marked red–purple orifice	
'SUNCATFE 23' (US2019/0098818P1) 28/01/2019 Masahiro Yamada, Japan	Hybrid of C. roseus (FS22-5) ♀ X C. roseus (CL) ♂ Upright, outwardly spreading, uniformly mounding; height ~22 cm & width ~45 cm; basal & free branches. small, star-shaped flowers; white corolla with red–purple orifice Tolerant to wind, rain & temperature (from 5 °C to about 35–40 °C)	
'SUNCATFE 291' (US2019/0223342P1) 18/07/2019 Masahiro Yamada, Japan	Hybrid of C. roseus (FS8239) ♀ X C. roseus (VP8507) ♂ Compact, semi-upright, uniformly mounding; vigorous; height ~18 cm & width ~32 cm; small, star-shaped flowers; light red–purple corolla flowers with dark red–purple orifice Tolerant to wind, rain, and temperature ranges from 5 °C to about 35–40 °C	
'ELD STJARNA' (US2019/0223344P1) 18/07/2019 Gyosuke Sasaoka, Japan	Developed through naturally occurring branch mutation in C. roseus (MP-mtl) Compact, semi-upright, uniformly mounding, vigorous; height ~19 cm & width ~38 cm; basal, free branches; small, salverform, star shaped flowers; red–purple corolla Tolerant to wind, rain & temperature (from 5 °C to about 35–40 °C)	

associated with flower color. Thus stem color indicated flower color prior to flowering. According to Simmonds (1960), two more genes (*A* and *B*) are involved in the determination of flower color. Gene *A* is a basic color gene that is complementary to gene *R* of Flory (1944), without both of which the flower color is white. Gene *B* is a co-pigmentation gene that blues the pigment in pink and red-eye flowers resulting in violet and purple eye flowers, respectively. In addition to the above three colors, Milo et al. (1985) identified flower color pale pink center and assigned gene *I* to this corolla color which is epistatic to gene *R*. The gene *I* produce pigment in the center of the corolla. Still, in smaller amounts than gene, *W*. Kulkarni et al. (1999a) studied the inheritance of pubescence on leaves and fruits in a cross involving the variety Nirmal with pubescent leaves and follicles and white corolla and a germplasm line PS-1 with glabrous leaves and follicles and pink corolla. Pubescence was found to be governed by a single recessive gene. Further, they found that the difference in corolla color of Nirmal and PS-1 was due to a single gene *R* which was inherited independently of the gene governing pubescence. Anthocyanin production in *C. roseus* flowers from both field-grown and in vitro culture. 3-o-glucosides and the 3-o-(6-o-p-coumaroyl) glycosides of hirsutidin, malvidin, and petunidin were the prime anthocyanins observed in the *C. roseus* plant grown either in vivo or in vitro (Piovan and Filippini 2007).

Novel corolla colors orange, red, and white eye OR and magenta MJ genetics and their inheritance pattern was studied by crossing an accession posing these corolla colors with a white flower variety Nirmal and MJ x OR. The phenotype of F_1 and segregation data of F_2 and backcrosses suggested the involvement of three new independently inherited genes (O^m, *J*, and *E*) and their epistatic interaction between previously reported *A, R, W, B, I,* and *O* explains the occurrence of all novel *C. roseus* corolla colors. Gene *E* determines the presence or absence of red-eye, i.e., *R* allele produces pigment in the red-eye region only in the presence of gene *E*; in the absence of *E* flowers will have white eye. Another gene *O*, responsible for the production of orange-red colored corollas, did not express in the presence of *W* allele (Sreevalli et al. 2002; Kulkarni et al. 2005a; Kulkarni et al. 2005b; Kulkarni et al. 2016). Reciprocal crosses between violet corral parent x white corolla parent showed involvement of single pair of the gene with violet color was incompletely dominant over white and cytoplasmic genes was not involved in the inheritance of these colors (Abdelmageed and Abdelrahman 2017).

Prediction of flower color was possible to predict by the flower, stem, and leaf ^{1}H-NMR profiling and multivariate data analysis of chemomarkers. Flower color is characterized by a specific metabolite pattern of anthocyanins, flavonoids, organic acids, and sugars. The correlation was observed between a plant with orange flowers with a high level of kaempferol analogs, red-flowered plants with high levels of malic acid, fumaric acid, and asparagine, pink and purple flowers with sucrose, glucose,2,3-dihydroxybenzoic acid, and in purple-flowered plants with a high level of quercetin, analogs were detected in flowers, leaves. This aids the ornamental sector Catharananthus breeders for early selection of targeted line in the seedling stage itself, saving time, resource, and space (Pan et al. 2014). Today more than 324 cultivars have been developed with a vast range of corolla and eye color and corolla lamina shape and size (Table 3.2). Suncatso 23 variety is a double flower variety (Table 3.1).

3.4.3 Breeding for Plant Height

Plant height is the most important decisive factor for the biomass yield component. Extensive literature is available on the identification and regulation of genes and quantitative trait loci controlling plant height and signaling pathways under regulation by gibberellins and brassinosteroids. The reduction in plant height in semidwarf varieties led to an improved harvest index (grain: straw ratio) and to enhanced biomass production as their short stature confers lodging resistance under high nitrogen fertilization, leading to increase biomass production. The plant height suppressive genes with significant revolutionary effect have been reported previously from wheat *Reduced height1* (*Rht1*), maize *Dwarf8* (*D8*), barley *Slender1* (*Sln1*), rice *semidwarf1* (*sd1*) which ushered the green revolution. Semidwarf cultivars are not only tolerant to lodging but also have higher herbage yield and harvest index in response to fertilizer application. During the induced mutagenesis program on Catharanthus plant height, mutants like dwarf and semidwarf (developed through N-nitroso N-Ethyl Urea,0.06%v/v) were isolated from the M_2 generation, which was 60 and 40% shorter with a higher percentage of total alkaloid in their roots than their parental variety, Nirmal. Plant height mutant 'bushy' (*by*) through EMS 0.8% v/v mutagenesis was isolated and was found to be 30% shorter with 20% lesser width than the parental variety Nirmal. The double-mutant (*by-dw1*) recombinant, which was recovered from the F_2 generation of the cross (NEU 6–15) x (EMS 24–5) with very distinct morphotypes from its parental mutants and variety, Nirmal was used for determining the individual and combined effects of plant height reducing genes *dw1* and *by*. The double-mutant recombinant was found to be 63–68% and 25–29% shorter than its parental mutants (EMS 24–5) and (NEU 6–15) and reduced plant height by 73–77% shorter than parental variety Nirmal, respectively. The double-mutant recombinant was also found to have a higher content of total alkaloids in the roots than parental mutants as well as the variety, Nirmal. The dwarf and semidwarf mutants were found to be conditioned by allelic monogenic recessive genes (Kulkarni et al. 1999b). The genetics of plant height reducing genes using the cross of induced bushy mutant (EMS 24–5) with its parental variety, Nirmal and the dwarf mutant was found to be under the control of a monogenic recessive gene *by*, which was inherited independently of the recessive gene *dw1* involved in the dwarf mutant. The effects of the recessive gene *by* on plant height were discernable only after the age of 4 months, while those of *dw1* and combined effects of *dw1* and *by* became apparent one to 2 weeks after germination (Kulkarni et al. 2009). All the mutants were Gibberellins sensitive, depending on their responsiveness to exogenously applied GAs having defects in the synthesis of GAs (Milach and Federizzi 2001; Kulkarni et al. 2009; Kulkarni et al. 2013b). Genetic analyses indicated that these genes encode negative regulators of gibberellin (GA) responses. The mutations within these genes fall into two categories. The first comprises semi dominant gain-of-function mutations, such as *GAI*, *D8*, *Sln1*, and *Rht1*, which lead to GA-insensitive dwarf phenotypes. The second category consists of recessive loss-of-function mutations, such as *sln1* and *slr1*, which cause elongated phenotypes with constitutive GA responses. The homologous proteins encoded by these genes belong to a plant-specific GRAS family of transcription factors. 'Sunnichilave' is a trailing cultivar, while the majority of the released cultivars are with an upright habit. CSIR-CIMAP released 'Dhawal' and CIM Sushil; both are dwarf varieties (Table 3.1). Later majority of the cultivars in the ornamental series were either semidwarf or dwarf, as that makes them ideally suitable for bedded plants. Catharanthus cultivars of the series 'Carpet' are very short, 3–4 inches in height (Table 3.2).

3.4.4 Breeding for Salt, Heat, Drought, and Light

Plants being sessile in nature, synthesize secondary metabolites for their defense from multi-stresses. Under stress conditions, plants send stress signal both way from leaves to root and root to shoot via phytohormone signaling and secondary metabolite biosynthesis which in turn trigger onset of various physiological and metabolic regulatory pathways.

Though Catharanthus is reported to be salt tolerant up to 2000 ppm (Plaizier 1981) and was actually cultivated in Hainan in China in salt marshes for production of medicinal alkaloids (Zhonghua et al. 2011). High salinity limits its optimum growth and yield of alkaloids. To

Table 3.2 *Catharanthus roseus* cultivars released under Group (series) or as independent nonseries cultivars and in commercial landscape market

Group	Characteristics	Cultivars
Blockbuster	Upright, mounded, floriferous, well branched. Height 10–12″ & width 12–16″. Extra-large, round shape, overlapping petals, show great color intensity even when exposed to full sun	Blockbuster Apricot, Blockbuster Burgundy, Blockbuster Crimson, Blockbuster Dark Red, Blockbuster Icy Water Melon, Blockbuster Peppermint, Blockbuster Punch, Blockbuster Purple, Blockbuster Red W/Eye, Blockbuster White
Carpet	Dwarf, Height 3–4″, Width ~ 24″.Long blooming period	Dawn Carpet, Rose Carpet, Pink Carpet, Magic Carpet
Cobra	Compact; height 6–12″ & width 6–9″; Flower 1.9 -2.3" Drought tolerant. Suitable to grow in xerophytic conditions	Cobra Apricot, Cobra Orchid W/Eye, Cobra Passion Fruit, Cobra Peppermint, Cobra Purple, Cobra Purple W/Eye, Cobra Red, Cobra Red W/Eye, Cobra Rose, Cobra Strawberry Red, Cobra White
Cooler	Compact, low growing. Height 10–14″ & width 6–8" Vivid color flowers. Highly tolerant to cool & wet conditions	Cooler Apricot, Cooler Blush, Cooler Coconut, Cooler Deep Orchid, Cooler Grape, Cooler Hot Rose, Cooler Icy Pink, Cooler Lavender Halo, Cooler Mixture, Cooler Orchid, Cooler Peppermint, Cooler Peppermint Improved, Cooler Pink, Cooler Raspberry Red, Cooler Red, Cooler Rose, Cooler Strawberry
Cora	Plant height 14–16″ & width 22–25″. Glossy green leaves & huge, bold, bright colors flowers in which some have eyes and some don't. Flowering early & uniform First series that is resistant to 'sudden death' & also resistant to aerial phytophthora (aerial blight). Heat tolerant	Cora Apricot, Cora Burgundy, Cora Deep Lavender, Cora Lavender, Cora Mix, Cora Pink, Cora Punch, Cora Violet, Cora White, Cora Cascade Apricot, Cora Cascade Bright Rose, Cora Cascade Cherry, Cora Cascade Lavender Eye, Cora Cascade Lilac, Cora Cascade Magenta, Cora Cascade Peach Blush, Cora Cascade Polka Dot, Cora Cascade Shell Pink, Cora Cascade Strawberry, Cora Cascade Violate, Cora Cascade White
Heat wave	Plant height 14–20″. Early flowering Extraordinary tolerance to heat and drought Good fit for hanging baskets and pots	Heat Wave Blue W/Eye, Heat Wave Burgundy, Heat Wave Cherry, Heat Wave Deep Rose, HeatWave Formula Mixture, Heat Wave Grape, Heat Wave Mid Night Mix, Heat Wave Orchid, Heat Wave Peach, Heat Wave Peppermint, Heat Wave Pink, Heat Wave Raspberry, Heat Wave Red, Heat Wave Rose, Heat Wave Santa Fe, Heat Wave Southwest Mix, Heat Wave White
First kiss	Compact plant Plant height 10–14″. Large flowers (~2″). Very heavily blooms	First Kiss Apricot, First Kiss Blueberry, First Kiss Blush, First Kiss Cherry Red, First Kiss Coral, First Kiss Icy Pink, First Kiss Orchid, First Kiss Peach, First Kiss Polka Dot, First Kiss Raspberry, First Kiss Rose, First Kiss Ruby, First Kiss Sunrise, First Kiss Think Pink, First Kiss White
Jaio	Compact, heat & drought tolerant, Disease resistant	Jaio Dark Red, Jaio Scarlet Eyes

(continued)

Table 3.2 (continued)

Group	Characteristics	Cultivars
James `N Jellies	Erect, Height 14–16″, width 10–12″. Heat & drought tolerant	JAMS `N JELLIES American Pie Mix, JAMS `N JELLIES Blackberry
Little	Dwarf, bushy; height 10–12″. Best for decorative purposes	Little Blanche, Little Bright Eye, Little Delicata, Little Linda, Little Pinkie
Mega bloom	Upright, mounded Height 14″ & width 16″. Large, star-shaped flowers Disease resistant Heat & humidity tolerant	Mega Bloom Apricot, Mega Bloom Burgundy, Mega Bloom Dark Red, Mega Bloom Grape, Mega Bloom Icy Pink, Mega Bloom Lavender, Mega Bloom Mix, Mega Bloom Orchid Halo, Mega Bloom Peach Pink, Mega Bloom Pink, Mega Bloom Pink Halo, Mega Bloom Polka Dot, Mega Bloom Purple, Mega Bloom Raspberry, Mega Bloom Red, Mega Bloom White
Mediterranean	Mounded, cascading, spreading, trailing & vigorous height 4–6″ & spread 20–30” Multiple colors large flowers	Mediterranean Apricot, Mediterranean Broad Eye, Mediterranean Cherry Red Halo, Mediterranean Dark Red, Mediterranean Deep Rose, Mediterranean Deep Orchid, Mediterranean Halo Mix, Mediterranean Lilac, Mediterranean Mix, Mediterranean Peach Improved, Mediterranean Peach XP, Mediterranean Pink, Mediterranean Polka Dot, Mediterranean Red, Mediterranean Rose, Mediterranean Rose Red, Mediterranean Rose Hot, Mediterranean White Broad Eye, Mediterranean White, Mediterranean Strawberry
Nirvana	Upright, mounded, and cascading plant habit Plant height 10″ and width 14″. Dark green leaves, Resistant to aerial phytophthora. Tolerant to heat and drought	Nirvana Cascade Lavender W/Eye, Nirvana Cascade Orchid, Nirvana Cascade Pink Splash, Nirvana Cascade Rose, Nirvana Cascade Shell Pink, Nirvana Cascade White, Nirvana Cascade Pink Blush, Nirvana Cascade Red, Nirvana Cascade Dark Red
Pacifica	Upright, mounded Very floriferous, early, well branched 14–20″ & width 6–8″. Open pollinated, largest (2″) flowers with overlapping petals Tolerant to heat, humidity, and other types of stresses Need low maintenance and require less watering	Pacifica Apricot, Pacifica Blush, Pacifica Burgundy, Pacifica Burgundy Halo, Pacifica Cherry Halo, Pacifica Cherry Red, Pacifica Coral, Pacifica Dark Red, Pacifica Deep Orchid, Pacifica Icy Pink, Pacifica Lilac, Pacifica Lipstick Mix, Pacifica Magenta Halo, Pacifica Mix, Pacifica Orchid Halo, Pacifica Peach, Pacifica Pink, Pacifica Polka Dot, Pacifica Punch, Pacifica Punch Halo, Pacifica Raspberry, Pacifica Really Red, Pacifica Red, Pacifica Rose Halo, Pacifica Pure White
Pretty	Compact plant. Height 12″. Multiflowered	Pretty Aureo-Variegata, Pretty Bowles, Pretty pink
Soiree	Perennial, compact, evergreen, trailing Height 10″ & width 18″. Star-shaped flowers High level of tolerance to urban pollution	Soiree Crown Pink, Soiree Crown Rose, Soiree Double Orchid, Soiree Double Pink, Soiree Double White, Soiree Flamenco Cheesy Pink, Soiree Flamenco Pink Twist, Soiree Flamenco Salmon Glow, Soiree Flamenco Senorita Pink, Soiree Kawaii Blue Berry Kiss, Soiree Kawaii Coral, Soiree Kawaii Coral

(continued)

Table 3.2 (continued)

Group	Characteristics	Cultivars
		Reef, Soiree Kawaii Lavender, Soiree Kawaii Light Purple, Soiree Kawaii Pink, Soiree Kawaii Red, Soiree Kawaii White Peppermint
Solar	Early flowering Better tolerant to disease, heat & humidity Perfect for bedding & potting	Solar Apple Blossom, Solar Apricot, Solar Blueberry, Solar Blush Pink, Solar Cherry W/Eye, Solar Formula Mixture, Solar Fresh Red, Solar Lilac, Solar Orange W/Eye, Solar Orchid W/Eye, Solar Pink, Solar Raspberry W/Eye, Solar Red, Solar Red W/Eye, Solar White
Stardust	Compact; height 10–12″; star-shaped, with white eye	Stardust Orchid, Stardust Pink, Stardust Mix, Stardust Rose
Sunstrom	Upright, mounding; height 8–10″, width 10–12″, Glossy green leaves, and large flowers. Resistant to aerial phytophthora Tolerant to both hot and cool conditions	Sunstorm Apricot, Sunstorm Bright Red, Sunstorm Blush, Sunstorm Lilac, Sunstorm Orchid, Sunstorm Pink, Sunstorm Red W/Eye, Sunstorm Violet W/Eye, Sunstorm White W/Eye
Tattoo	Mounded, upright; height 10–14″, width 6–8″. Under warm & moderate light conditions display best color contrast, but with increased temperature and light, colors are lightened	Tattoo Pie Mix, Tattoo Black Cherry, Tattoo Blue Berry, Tattoo Papaya, Tattoo Raspberry, Tattoo Tangerine
Titan	Upright, mounded, bushy, vigorous; height 14–16″, width 10–12″. Early, large, profuse, high seed quality flowers, two weeks early than OP Vinca species. Excellent performance in sunny, hot, dry conditions, low watering intake	Titan Apricot, Titan Blush, Titan Burgundy, Titan Cotton Candy Mix, Titan Dark Red, Titan Icy Pink, Titan Lavender Blue Halo, Titan Lilac, Titan Mix, Titan Polka Dot, Titan Punch, Titan Pure White, Titan Rose, Titan White
Tropicana	Fast growing & early flowering; height 14–20″; large, round shape flowers. Tolerant to heat and humidity	Tropicana Apricot, Tropicana Blush, Tropicana Bright Eye, Tropicana Pink, Tropicana Rose
Tutti fruit	Plant has large flowers. Tolerant to drought and heat	Tutti Frutti Agastache, Tutti Frutti Rose
Valiant	Upright, height14-20″,width 10–12″. Aerial Phytophthora resistance; tolerance to heat, drought, salt, cool/wet conditions	Valiant Apricot, Valiant Burgundy, Valiant Lilac, Valiant Magenta, Valiant Mix, Valiant Orchid, Valiant Punch, Valiant Pure White
Victory	Compact habit, branches are low, early & basal; height 10–14″ & width 6–8″; large, round-shape flowers with overlapping petals Disease tolerant	Victory Apricot, Victory Blue, Victory Bright Eye, Victory Carmine, Victory Carmine Rose, Victory Cranberry, Victory Deep Apricot, Victory Deep Pink, Victory Grape, Victory Lavender, Victory Light Pink, Victory Pure White, Victory Purple, Victory Red
Viper	'King of Vinca-Vigorous'. Large, mounded; height is 8–20″; floriferous flowers, leaf rot, and bud drop tolerant	Viper Apricot, Viper Grape, Viper Orchid Halo, Viper Pink, Viper Purple, Viper Purple Halo, Viper Red, Viper Red W/Eye, Viper Rose, Viper Watermelon

(continued)

Table 3.2 (continued)

Group	Characteristics	Cultivars
Virtuosa	Sturdy, upright, compact, and uniform with basal branching; height 8.3″ & width 7.8″; large, round flowers; petals are strongly overlapped	Virtuosa Apricot, Virtuosa Cranberry, Virtuosa Deep Purple, Virtuosa Deep Red, Virtuosa Lavender, Virtuosa Mix, Virtuosa Orchid, Virtuosa Peach, Virtuosa Polka Dot, Virtuosa Punch, Virtuosa Rose, Virtuosa White
Vitalia	Mounded; height 10–12″, width 6–8″; uniform OP bright colored flowers. Exceptionally good in hot & dry conditions	Vitalia Blush, Vitalia Cranberry, Vitalia Cherry, Vitalia Icy Pink, Vitalia Mix, Vitalia Pink, Vitalia Polka Dot, Vitalia Scarlet, Vitalia Sweet Heart Mix, Vitalia White
Vitesse	Compact, upright, mounded; basal branching. Height 12–14″ & width 16”	Vitesse Apricot, Vitesse Blush, Vitesse Cranberry, Vitesse Fuchsia, Vitesse Grape, Vitesse Hot Pink, Vitesse Lavender, Vitesse Orange, Vitesse Orchid, Vitesse Peach, Vitesse Peppermint, Vitesse Pink, Vitesse Purple, Vitesse Raspberry, Vitesse Red W/Eye, Vitesse Rose, Vitesse Strawberry Red, Vitesse Strawberry Twist, Vitesse Tropical Orange, Vitesse White
Non-series		Apricot Delight, Blue Pearl, Dwarf Little Salmon, Fanfare Orchid, Little Linda, Merry-Go-Round, Morning Mist, Parasol, Pink Panther, Santa Fe, Snow Flake, Kurari White, Kermesiana, Papion Silver Blue, Flappe Lilac, Flappe Coconut

utilize coastal, high salinity barren lands breeding for salt tolerance has its own economic benefits. *C. roseus* is adapted to high temperatures and light radiation and is able to grow in mildly saline conditions (Mishra and Gupta 2006).

Eight mutants (*gsr1* to *gsr8*), tolerant to salinity (250 mM NaCl) and high temperature (45 °C), stress has been successfully isolated (Rai et al. 2001a, 2003; Kumari et al. 2013a, b). These mutants (*gsr1* to *gsr 6*) accumulated more proline and glycine betaine constitutively as well as underwater stress and transpired lower amounts of water underwater stress than their parental variety. The F1 of wild mutants (gssr1 to gsr6 failed to grow on 250 mM salt stress, and complementation analysis revealed different genes were mutated in these six mutants. Although mutants' most conspicuous morphological traits were inherited as monogenic recessive traits, the mutants exhibited pleiotropic effects for several other traits (Rai et al. 2003). Three of these mutations were thought to be in loci/genes that have very large and wide regulatory roles in the regulatory gene network of *C. roseus* for metabolism, development, and adaptation to the environment. The pleiotropies displayed by the mutants were considered to result from changes in gene expression affecting various kinds of functions responsible for the achievement of plant morphology. The mutants were found to be hypomethylated at repeat sequences, up-regulated and down-regulated for many genes resulting in pleiotropic alterations for several traits (Kumari et al. 2013a, b). Similarly, vindoline and catharanthine-rich mutants may be obtained if serpentine route gene functions can be knocked out (Singh et al. 2008).

The contents of catharanthine, VLB, VCR, and serpentine in two of these *gsr* mutants (*gsr3* and *gsr6*) were found to be higher than those in their parental variety and correlated well with the expression profiles of TIA biosynthetic pathway genes, strictosidine synthase, deacetoxyvindoline 4-hydroxylase, and deacetyl vindoline 4-O-acetyl transferase (Dutta et al. 2005). CSIR-CIMAP released 'Illi variety was developed through EMS mutagenesis of variety Nirmal. The variety is unique in leafless inflorescence governed by a monogenic recessive gene and could tolerate 250 mM of NaCl (Kumar et al. 2007, US20050251887P1).

Secondary metabolites protect from oxidative damage and lipid peroxidation by scavenging reactive oxygen species in heat and drought stress. 'Metallothionein type II gene from *C. roseus* can enhance drought resistance and improved quality (Zu et al. 2006). Similarly, Glutathione S-transferase (GST) gene promises to enhance drought resistance in *C. roseus* (Zu et al. 2005). The cultivar series 'Cobra' has drought tolerance (Table 3.2). PAS1052940 is also drought tolerant tolerance to herbicide glyphosate was explored in sequential step wise screening for 1 mM to 35 mMglyphosate and 5-enolpyruvylshikimic acid 3-phosphate synthase (EPSP) activity. Glyphosate tolerant cell cultures in the presence of 10 mM glyphosate inhibited the 5-enolpyruvyl shikimic acid-3-phosphate (EPSP)synthase, blocking the synthesis of aromatic amino acids and contained undetectable serpentine or ajmalicine (Cresswell et al. 1988). PAS1053041 cultivar is uniquely herbicide-tolerant (Flaschenriem DR2016, Table 3.1). Cultivar series 'Cora,' 'Nirvana,' 'Valiant,' 'Jaio,' 'Heat Wave,' 'Tuttifruti,' 'Pacifica' has been developed for tolerance to heat and drought hence are very popular in the landscape industry (Table 3.2).

Light plays an important regulatory factor. Catharanthine biosynthesis is completely repressed in the absence of light. When light is provided for just 15 h ajmalicine production is increased, and in the dark, it is 80% secreted in the extracellular medium but in the presence of light amount secreted is only 18% (Moreno et al. 1995). 'Titan' series cultivars have been developed for excellent temperate environments along with drought tolerance and are uniquely-tall type (Table 3.2).

3.4.5 Breeding for Disease and Pest Resistance

Major pests and diseases causing yield losses in Catharanthus are foliar blight *(Phytophthora nicotinanae),* die back (*Pythium aphanidermatum*), leaf spot (*Alternaira alternata*), footrot and wilt (*Sclerotium rolfsii* and *Rhizoctonia solani*), Mycoplasma, Phytoplasma yellow disease, Spiroplasma virus, Aster yellow, mycoplasma like an organism, chlorotic stunt virus and Little leaf, which can be controlled by breeding disease-resistant varieties, the timely spray of broad spectrum and systemic fungicides and following intercultural cultivation practices with other crops (Zaidi et al. 1984). *Pythium aphanidermatum* Edson Fitzp causing dieback disease (Kulkarni 1984, 1987, 2003) and little leaf caused by Phytoplasma strains have been reviewed as major pathogens besides *Colletotrichum dermatum, Macrophomina phaseolus,* and *Phyllosticta vicaeae majoris* (Tiwari et al. 1978), causing severe twig blight. Twig blight started with the appearance of small pin-head size, brown necrotic spots expanding in a circular pattern covering both sides of leaves caused by *Colletotrichum dermatium* and *Macrophomina phaseolina* (secondary invader) affecting the twig while *Phyllosticta vincae majoris* on the leaves (Tiwari et al. 1978). *Onconetopia nigrans,* a sharpshooter insect, damages by transmitting mycoplasma with chlorotic wilting and vein clearing symptoms, yellowing of leaves caused by leaf mosaic virus, and phytoplasma infestation inducing extreme shortening of nodes and internodes reducing flower and seed set. The prophylaxis care includes regular spray of oxytetracycline, gibberellic acid, tetracycline, etc. A recent report of folivory of *C.roseus* using Manduca sexta caterpillars demonstrated that both local and systemic induction of MTIA biosynthesis results from induction of jasmonic acid and ethylene signaling pathways which lead to abundant accumulation of reactive strictosidine aglycone as the first line of defense to aggressors and further downstream synthesis of MTIA in newly developing leaves from further attacks (D`uge de Bernonville 2017).

Strictosidine β-glucosidase involved in terpenoid indole alkaloids (TIAs) biosynthesis pathway showed significant upregulation in experimentally and naturally infected *C. roseus* lethal yellows, caused by *Spiroplasma citri*, and this gene can be used as a host biomarker to increase the fidelity of *S.citri* detection and in breeding programs to develop stable disease-resistance varieties (Nejatet al. 2012).

Catharanthus is highly susceptible to two diseases that is leaf blight (*Phytophthora nicotianiae*) and die back and collar/root rot (*Pythium aphanidermatum*), due to which 80–100% mortality occurs on the spread of fungal mycelia, especially in the rainy season. Disease-resistance breeding in Catharanthus has achieved success in these two diseases (Pareek et al. 1981; Kulkarni, 1984). Dieback resistant variety 'Nirmal' was developed as a pure line selection from a single plant that survived dieback epidemic. EMS-induced mutagenesis of Nirmal further was subjected to develop 'Dhawal' which had much higher resistance to dieback (Kulkarni 1999; Kulkarni et al. 2016). Induced tetraploids were evaluated for resistance to die back under field conditions and were found to be eight times more resistant than the diploids. The resistance of the tetraploids to dieback disease appeared to increase with the age of plants and horizontal in nature (Kulkarni 1984). Pathogen stress in tetraploids yielded four and five times more leaf and root total alkaloids, respectively than diploids in dieback epidemics. The tetraploid lines (13–155,18–218 and 17–366) were more resistant than diploid lines when resistance was measured in terms of plant mortality, infection rate, and area under disease curve, respectively (Kulkarni and Ravindra 1988). The resistance of autotetraploid was equivalent to protection provided by 5.76 kg per hectare of the fungicide Captafol or soil solarization (Kulkarni and Ravindra 1997; Kulkarni et al. 1992). These studies suggested that tetraploids can be an effective solution to die back, especially in the absence of any chemical control. Inheritance of resistance to dieback in *C. roseus* was studied using a dieback-resistant dwarf mutant, *Dw*, with a green stem and a susceptible accession, 'OR,' with a purple stem. From both quantitative and qualitative analyses of data, resistance to dieback appeared to be monogenic, with a broad-sense heritability of 0.85, inherited independently of genes governing dwarfness and stem pigmentation, and resistance was non-specific but stable (Kulkarni and Baskaran 2003).

C. roseus is highly susceptible to *Phytophthora spps.*, especially *Phytophthora nicotianae* Breda de Haan (aerial Phytophthora), which cause stem and foliage blight. Chlamydospores survive in the infested field soils for many years; hence resistant genotypes are the only solution to save from huge economic losses as well as from the use of pesticides. Dastur in 1916 first reported *P. nicotianae* causing foliage blight in *C. roseus* from India, and later it became widespread where *C. roseus* was popular as a mono-crop bedding plant. The first Phytophthora resistant Catharanthus variety was developed by pedigree analysis using epiphytotic selections for resistance to *Phytophthora parasitica* governed by monogenic dominant allele by Bowman (1996). The resistance source plant with light blue color corolla was initially infertile and had to be additionally crossed with a fertile genetic background. The source plant had high alkaloid content with 2.0–9.5% total alkaloids and was also resistant to aphids(*Myzus persicae*) and mites (*Tetranychus urticae*) (Goldsmith seed incorporation, Patent US5491285).

Another *P. nicotiana* resistant line developed was broad-spectrum FD0922, from the Cora Lavender, which could survive 70 days after infection. Virulent isolates(40) from pathogen infested recovered field plants with broken resistance Cora and Nirvana cultivars along with certain avirulent (control) strains were screened on Cora Lavender, Cora Burgandy, and Titan Blush as control. The evaluation disease parameter was symptom severity 50 days after infection, survival time in a number of days until symptom severity score of 5isreached, and the area under disease progress curve (prime criteria) accounts for both virulence and aggressiveness of the isolates, and thus breeding line FD0922 was found to be only18.2% susceptible with symptoms. Isolate by cultivar interaction was significant with differential virulence. Highly virulent isolates were mefenoxam sensitive and mating type A2. Sunstorm deep pink was observed to be most susceptible (Hershberger et al. 2016, Syngenta Seeds BV patent WO2016/120191 A1). In search of tolerance the 'Cora', 'Nirvana', 'Valiant', 'Jaio' cultivar series have been launched. 'PAS1052940' is disease tolerant while 'Viper' is tolerant to leaf rot and bud drop (Table 3.2). The catharanthine stored on the upper leaf surface protects the plants with its antifungal, anti-insect properties. In a direct selection of glabrous waxy surface rich genotypes may be a vital lead to biotic stress-tolerant genotypes.

3.5 Development of Genetic Screens: Catharanthus Mutants

Mutagenesis is adopted by the plant breeders when available genetic resources have limited variability or a narrow genetic base for the targeted trait. There have been commendable achievements in the history of plant breeding when bottlenecks were overcome with random and targeted mutation breeding genetic tools. The first registered plant variety through mutation breeding was developed in 1929 by generating Chlorina F1 in *Nicotinana tabacum*in Indonesia. Since then, effective genetic screens have been developed in various crops which assist in high throughput screening of desired mutations. Using physical, chemical, combined mutagens, and somaclonal variations, 3,365 varieties belonging to 239 plant species have been registered across 70 countries over the globe in Mutation Variety Database maintained jointly by FAO and IAEA (access dated 18 July,2021 https://mvd.iaea.org). Irradiations have been used to induce plant mutations to explore improved quality by altering metabolite profile, improved yield, and resistance/tolerance to biotic and abiotic stresses, which can impart greater resilience to climate change. Induced mutations for many millennia have been extensively exercised in food crops; 70% of released mutant varieties are developed through ionizing radiation gamma rays. Recently ion beams radiations such as fast neutrinos or UV rays have emerged as effective tools(Kumawat et al. 2019). Since physical mutagens generated chromosomal aberrations or large DNA deletions, point mutation generated by chemical mutagens is the preferred choice in plant breeding. Mutation breeding produces macromutants in low frequency and is often an expression of recessive mutations; hence remains hidden and is to be advanced through selfing. It is successful in self-pollinated crops, but for cross-pollinated crops, it has to be artificially selfed until advanced mutant generations. Backcrossing with the parental unmutated genotype is essential to transfer the desirable genetic background. Since Catharanthus is herkogamous and cross-pollinated, its sib mating or manual selfing is essential(Kulkarni et al. 2016).

Venkateswarlu (1988, 1989) studied the effect of gamma rays (30- 60 kR) and EMS (0.1% for 6, 12, 18, and 24 h) on M_1 parameters and the frequency and spectrum of chlorophyll mutations in M_2 generation. In general, low dose/duration of mutagen treatment enhanced the growth of plants and percentage of alkaloids in leaves and roots as compared with untreated control. The frequency of chlorophyll mutations in M_2 increased with an increase in the dose of mutagen. The spectrum of chlorophyll mutations generated were *xantha*, *albin*a, *chlorina*, *viridis*, *maculata,* and *tigrina* types. Mogili et al. (1984) utilized EMS-induced mutations in the wild population of Catharanthus. Pink flowers were mutagentically more sensitive than white flowers. Combination treatment of chemical and physical mutagen enhanced the frequency of chlorophyll mutation.

Reduced plant height mutants, 'dwarf,' 'semidwarf, and 'bushy,' about 60, 40, and 30% shorter than their parental variety, Nirmal, have been reported in *C.roseus* (Kulkarni et al. 1999; 2009). The 'dwarf' and 'semidwarf mutants were due to monogenic recessive genes ($dw_1$and dw_2, respectively), which were allelic to each other and had a significantly higher content of root alkaloids than parental variety. The 'bushy' mutant which was governed by an independently inherited non-allelic recessive gene (*by*), however, had similar contents of leaf as well as root alkaloids as the parental variety, Nirmal. The double-mutant recombinant ($bydw_1$) was 30% shorter than the shortest parental mutants and exhibited a 20% higher content of root alkaloids than the better parent. All the three mutants and the double-mutant recombinant ($bydw_1$) had similar contents of leaf alkaloids. Higher root alkaloids content has been found to be related to thin root morphology in hairy root cultures of *C. roseus* (Palazon et al.1998). An extremely tall mutant, about 90% taller than the parental variety, Nirmal and controlled by epistatic inhibitory interaction between two independently inherited dominant genes, has also been reported. The mutant, however, had similar leaf and root yields as well as contents of leaf and root alkaloids as the parental variety (Kulkarni and Baskaran 2013a).

As economically important alkaloids are present in the leaves, altered leaf morphology may suggest altered alkaloid contents. Three leaf mutants way leaf margin, 'necrotic leaf' (a lesion mimic mutant), and 'nerium leaf' (resembling leaf lamina of another Apocynaceous plant, *Nerium oleander*) exhibited higher contents of leaf alkaloids than their respective parents (Kulkarni et al. 1999; Baskaran et al. 2013). Further, enhanced contents of leaf alkaloids of 'necrotic leaf' and 'nerium leaf' mutants over their parental variety were found to be due to recessive alleles at different loci, and 13 out of 14 double-mutant recombinants for parental mutant traits 'necrotic leaf' and 'nerium leaf' developed by crossing the two mutants had a significantly higher content of leaf alkaloids than parental mutants (Kulkarni and Baskaran 2014). Kulkarni et al. (2016) suggest 'necrotic leaf' trait could be used as a seedling marker trait for enhanced content of leaf alkaloids. Extract of *Pythium* (a soil-borne pathogen of *C. roseus*) is well known to be an elicitor of alkaloid production in cell and tissue cultures of *C. roseus* (Nef et al. 1991). Therefore, constitutive expression of self-defense reactions in the 'necrotic leaf' mutant may have induced enhanced production of alkaloids similar to that elicited by *Pythium*.

Mutations affecting the gynoecium have also been reported. In *C. roseus* flower, the stigma is about 0.5 mm below the anthers, typical of reverse herkogamy. Mutants with short style (about 1/3 length of normal style) and long style with stigma 2.5 mm above the tip of the cone of anthers, with partial and high pollen sterility, respectively, have been reported (Mishra and Kumar 2003; Kulkarni and Baskaran 2008). The short style mutant trait was inherited as a recessive trait, while the long-styled mutant constituting 'pin' flower in contrast to 'thrum' flower in normal plants appeared to be under the control of inhibitory, epistatic interaction between two independently inherited genes *P* and *T*, with gene *T* being inhibitory to gene *P*. Accordingly, genotypes *P–T-, T-pp, or pptt* produce normal 'thrum' flowers whereas *P-tt* produces mutant 'pin' flowers.

Six glycolytic insensitive salinity responses were isolated on treating variety Nirmal with EMS and NEU (0.00–0.2%), which could tolerate salt up to 250 mM. One such mutant had monocotyledonae which upon supplementation with the natural cytokinin Kinetin develop two cotyledons. This mutant $Cdd^{-}$was governed by a recessive monogenic (Rai and Kumar 2000).

A recessive mutant producing heterocarpous flowers, with one (3%), two (82%), and three (15%) carpels and high fertility has been reported with a possibility of genetic engineering for fruit size, carpel number, and seed number per plant (Rai and Kumar 2001a; 2001c). As *C. roseus* is also valued as an important garden plant because of its variously colored flowers, mutants affecting flower color, flower density, floral persistence, etc., would also be of interest. A mutant described as 'leafless inflorescence,' in which flowers are borne on nodes without leaves, has been reported, and the locus '*lli*' has been mapped (Chaudhary et al. 2011). The mutant produced more flowers per plant than its parent and further enhancing its horticultural value. New, improved horticultural genotypes were developed by crossing this mutant with other genotypes with different flower colors and plant habits (Kumar et al. 2012).

EMS (0.2%) induced mutagenesis in Nirmal led to the isolation of salt tolerant monocotyledonous caducous closed corolla (corolla abscising before anthesis) *Cdd* mutant, which could germinate on 200 mM NaCl medium. Monocotyledonae of the seedling and germinability of seeds on 200 mM NaCl coinherited recessively. Both the trait were suggested to be linked due to pleiotropy (Rai and Kumar 2000). Six *gsr* mutants (*gsr-1 to gsr-6*) that tolerated salt stress due to loss of the normal glycophytic salinity response were reported by Rai et al. (2003). They represented six complementation groups in all of which different genes have mutated. They followed the Mendelian monogenic recessive transmission pattern. These mutants accumulated more proline, and glycine betaine transpired lower amounts of water. *Gsr-5 and Gsr-7* were also found to be heat resistant and as well as salt tolerant (Rai et al. 2001b).

A gamma-ray (10–45 krad) induced mutant *Abo*$^{-}$ of variety Nirmal, which was heat, salt, and water stress-tolerant and abscisic acid (ABA) accumulation were 500 fold were reported by Rai et al. (2001b). The mutant seeds germinated at 45 ℃, embryo tricotyledonous, short stature, lanceolate leaves, poorly developed vascular bundles, trichotomous leaf surface with a large majority of stomata closed. The genetic lesion *abo* responsible for the above pleiotropy was recessive and inherited in Mendelian fashion. A mutant *Rpt*$^{-}$(for the absence of regulated pollen tube growth) was isolated through EMS-induced mutagenesis, whose pollen grains germinated to produce pollen tube while still inside the anthers by Mishra and Kumar (2001). A six petal flower mutant (Singh et al. 2001) was studied for chromosomal behavior in the meiotic system of *C. roseus*. Distorted metaphase I orientation, anaphase I with laggard, precocious anaphases separation, and meiotic apolarity with 70% pollen sterility were observed. Kulkarni and Baskaran (2008) also reported six corolla lobe mutant EMS13-2 from the parental variety Dhawal, which also had thick abnormal leaves and indehiscent anthers. A heterocarpous mutant *Hcf*$^{-}$with tricarpellary gynoecium and a higher number of floral meristems by Rai and Kumar (2001a). The *Hcf*$^{-}$ mutant had a high density of flowers on the shoots, larger sepals, smaller petals, carpels had common style and stigma with three nectarines at the ovary base. The mutation was recessive and inherited in Mendelian fashion.

Mutation Breeding is a two-step art. First the induction, and the second is detection. Generally, most economic detection way is through phenotype identification which comes on its own cost of time and land resource. Recently techniques like targeting induced local lesions in genomes (TILLING) traditional chemical mutagenesis is followed by high throughput screening for point mutations permit directed identification of mutations in specified genes of mutagenized population genome by using next generation sequencing. Thus allelic variation in gene expressions profile is correlated with chemotype and phenotype profiling, an allelic series at genetic loci for the trait of interest in germplasm or induced mutants is established. Precision breeding techniques such as zinc finger nuclease (ZFN) technology, oligonucleotide-directed mutagenesis (ODM), CIS genesis and intragenesis, RNA dependent DNA methylation, and reverse breeding, have evolved for precisely targeted mutagenesis in the genome. These techniques, as per the Court of Justice of the European Union (CJEU) states that organisms modified by the new targeted mutagenesis techniques will fall under EU GMO legislation (Holme et al. 2019). The zinc finger nuclease technique is site-directed nuclease mutagenesis that induces a mutation at the precisely defined locus of the target genome. TALEN and CRISPER-cas9 are the emerging tools of this category that have already been accomplished in maize (Holme et al. 2019). However, TILLING created a limited density of mutational events at a defined target locus. CRISPER-cas9(clustered regularly interspaced short palindromic repeats and CRISPER-associated protein 9 fused to a cytidine deaminase could serve as a localized mutagenic agent to produce high-density mutant populations. These techniques though not yet tested in Catharanthus, open a vast potential to target for MTIA breeding using allelic variation in natural or mutagenized populations using large number of isolated mutants and TIA pathway genes already cloned in *C. roseus*.

In recent years, mutations have been induced using high-energy (220 meV) and low-energy (30 keV) ion beams, respectively, and many ornamental crop cultivars with unique color characteristics have been developed in Japan and China. The spectrum of mutations induced by ion beam radiation was found to be different from that of gamma rays, and novel mutants have been identified in ornamental plants (Pathirana 2011). Reverse genetics techniques based on genetic screens of mutant resources can be used for discovering allelic variation in natural or mutagenized populations and elucidation of MTIA pathways and targeted trait genetics using available gene and genome sequence as reference.

3.6 DUS Guidelines for Variety Release: Registered and Patented Varieties

Guidelines have been established for testing *C. roseus* new advanced breeding line's approval as a new variety by testing for distinctness, uniformity, and stability characteristic for one growing cycle and one location trial (UPOV 2004). The test is conducted on at least 40 plants for seed propagated varieties and at least 20 plants for vegetatively propagated varieties. Expression of the characteristics may be qualitative (QL), quantitative (QN), or pseudo-qualitative (PQ) with an acceptable probability of at least 95% (Mishra et al 2001b). These grouping characteristics may be mandatory (always examined for DUS for the international harmonization of variety description*) or standard test characteristics (breeders can select those characteristics suitable for their particular circumstances). Grouping characteristics assess varieties of common knowledge along with candidate varieties for the defined state of expression of a specified trait or in combination to exclude varieties of common knowledge from its distinctness and to organize the similar varieties in the trial in one group. Example varieties are usually taken as references. Catharanthus grouping characteristics internationally practiced are as follows:

1. Plant growth habit (QN): Upright (Kermesiana), semi upright, horizontal (Dawn Carpet)
2. Plant Height(QN*): Short (Dawn Carpet), medium (Little Bright Eye), tall (Kermesiana)
3. Plant width (QN*):Narrow (Kermesiana), medium Papion Silver Blue),broad (Papion Silver Blue)
4. Stem anthocyanin coloration (QN*): Absent, weak, medium (Little Bright Eye), strong (Pink Carpet), very strong (Kermesiana)
5. Number of primary branches (QN): Few (Pretty in Pink), medium (Little Bright Eye), many
6. Number of secondary branches (QN): Few (Kermesiana), medium (Little Bright Eye), many (Pretty in Pink)
7. Leaf shape (PQ): Liner, oblong (Little Bright Eye), elliptical (Peppermint Cooler)
8. Leaf length (QN*): Short, medium (Little Bright Eye), long
9. Leaf width (QN*): Narrow, medium (Little Bright Eye), broad (Parasol)
10. Leaf variegation (QL*): Absent/present
11. When non-variegated leaf: Green color intensity (QN*): Light(Papion Silver Blue), medium(Little Bright Eye), dark
12. Petiole length (QN): Short (Pretty in Pink), medium (Little Bright Eye), long
13. Flower diameter (QN*): Small, medium ((Little Bright Eye), large (Parasol)
14. Flower arrangement of petals (PQ*): Free (Kururi White), touching (Flappe Coconut), slightly overlapping (Flappe Lilac), strongly overlapping (Peppermint Cooler)
15. Main color of the upper side of corolla (PQ*): White, pink, red, or purple, or as per RHS chart
16. Eye zone (QL*): Absent Papion Silver Blue), present (Papion Silver Blue)
17. Size of eye zone relative to flower size (QN*): Small (Peppermint Cooler), medium (Pretty in Pink), large (Dawn Carpet),

18. Flower: number of colors of eye zone (QL*): One, two, or more
19. Varieties with one color of eye zone (QL): Sharp/diffuse
20. Varieties with more than one eye zone: Color of inner eye zone(PQ*): As per RHS color chart
21. Varieties with more than one eye zone: Color of outer eye zone(PQ*): As per RHS color chart
22. Color of flower receptacle (PQ*): White, yellow, pink, red, purple
23. Pet al. width (QN): Narrow, medium, broad
24. Pet al. lobbing (QL*): Absent/present.

Catharanthus roseus is widely cultivated in Spain, the United States, and China for its pharmaceutical compounds. The first international register of Catharanthus roseus was released on 50 cultivars (Snoeijer W 2001). A variety was developed from 'leafless inflorescence,' mutant in which flowers are borne on nodes without leaves named '*lli*' (Kumar et al. 2005). A double flowered *C. roseus* variety "Taoyuan No 1 Rose Girl was developed where the additional petals originate from the apex of the corolla and did not develop at the expense of stamen or carpels by Chen and Yeh (2012). Prabal variety was registered for its unique intense red-eye and white corolla (Dwivedi et al. 2001). Flaschenriem (2016) developed a series of ornamental Catharanthus cultivars which were very successful in the landscaped industry such as PAS1052940, PAS1053041-PAS1053045.

So far, there have been 33 varieties that have been globally released and developed for enhanced alkaloid content (Table 3.1). For the landscape industry, there have been rigorous efforts by the private sector breeders, especially by the Netherland, The United States, Japan, Belgium, Australia, and France, registering 324 cultivars denominated under 27 Groups (series) and 11 nonseries cultivars (Table 3.2) (Ball Horticultural, United States). They were named as per the International Code of Nomenclature for Cultivated Plants (IPCN), initially outlined by William Stearn, and first published from Wageningen University in 1953. This IPCN underwent subsequent nine editions, i.e., 1958 (Utrecht 1961updating, Utrecht 1969, Edinburgh 1980 Seattle 1995, Edinburgh 2004, Toronto and Wageningen 2009 and recently in Beijing 2016 editions (IPCN 2016).

3.7 Next Generation Breeding

Till 2006, there were no genomic resources or expressed sequence tags (ESTs) reported from this plant. Murata et al. (2006) sequenced two cDNA libraries from leaf base and the root tips to generate 5,023 ESTs, of which 3,553 could be annotated. This work served as the first platform to mine microsatellite markers as the first prerequisite for molecular breeding in *Catharanthus roseus*. Using the rapidly growing ESTs databases, comparative mapping, and genomics, efforts to develop molecular markers for this plant were initiated. Currently, 22,867 ESTs, 88,543 genomic sequences, and 118 GSS are available in the public domain. Medicinal Plant Genomic Resource consortium indicates 19,899 EST clusterings to 20,460 contigs, 115GSS, and 749 nucleotides sequences, as of 14 May 2010 (http://medicinalplantgenomics.msu.edu). Currently, 138 gene sets, and 404 microsatellite sequences, of which 311 are available in the public domain. Sequence read archive (341) of the transcriptome of different cultivars under different treatment conditions and of different tissues (shoot, flower, root, leaf, seedling), 25,298 proteins, 95 pop sets, 215GEO datasets,49 Bioprojects with 36 biosamples (NCBI databases,18 July 2021) and croFGD *C. roseus* functional genomics database (She et al. 2019) are available in the public domain for functional genomics pathway gene discovery. Plant breeding assisted with omics technology such as single-cell metabolomics, Smart cell consortium, Phytometasyn, Medicinal Plant Genomic Resource, and CathaCyc (Rai et al. 2017) have triggered the pace of genetic selection of 'right' genotypes for generation advancement and inclusion in the breeding program.

Plastome markers of potential utility for *Catharanthus* breeding and phylogenetic analyses were identified from the plastome of Pacifica Punch Halo, yielding 41 *C. roseus*-specific simple sequence repeats (SSRs). Fifty-six SSR markers localized in intergenic regions than in genic regions were tested. Furthermore, five intergenic regions with the high divergence between *C. roseus* and three other euasterids I taxa were identified as candidate markers (Ku et al. 2013).

The whole genome of cultivar Sunstorm Apricot using 50xIlluminaHiSeqsequencingby Michigan State University has generated 523 Mb consisting of 79,302 scaffolds with scaffold N5026,249 bp, but this assembly still has 68,690 gaps (Kellner et al. 2015). Many transcriptomes (341) of Catharanthus are available in the public domain and croFGD *C.roseus* Functional Genomics Database (She et al. 2019) for functional genomics of pathway genes discovery.

A genome-wide strategy based on the identification of low copy evolutionarily conserved sequences within transcribed sequences of *C. roseus* can be used as primer annealing sites for PCR amplification of intercalated introns that are subsequently sequenced in order to ascertain polymorphism between mapping parents (comparative anchor tagged sequence markers, Fredslund et al. 2006).

Microsatellite and STMS marker resource (423) have been developed from enriched genomic libraries (Shokeen et al. 2005, 2007, 2011; Joshi et al. 2011). Bhatia and Shokeen (2009) have described a method for the isolation of microsatellites in *C. roseus* through hybridization with biotin-labeled oligoprobe followed by their capture with streptavidin-coated magnetic beads. Jhang et al. (2012) developed and validated a set of 350 unigene derived microsatellite primers pairs markers, of which 80 were found to be co-transferable to other Apocynaceses species like *Rauwolffia serpentina*, *R.tetraphylla,* and *R. vomitoria* besides *Nerium* and *Tabernomontana* used in the study. Development of EST-based SSR markers has been reported by many workers (Pathak 2009; Fartayal 2010; Jhang et al. 2012; Mishra et al. 2011). These markers have also been checked for their co-transferability across genus in the family. The high reproducibility and their co-transferability make them the ideal choice of genetic diversity assay Kumar et al. (2014) in silico developed 2,520 SSR markers from the transcriptome of 26 days old seedlings and have validated a subset of 48 markers. These validated SSR marker resources can be used for enriching the saturation in the available linkage map.

A genetic linkage map is the second prerequisite to map the gene/QTLs (quantitative trait loci) controlling qualitative and quantitative traits, integrating molecular information that is essential for marker-assisted selection and map-based cloning. The first framework genetic linkage map of *C.roseus* was constructed by using 240 DNA and 6 morphological markers on F_2 population of Pink Delhi X *gsr8*.The 125 DNA markers (79 RAPD, 7 ISSR, 2 EST-SSR, 37 other PCR based DNA markers, and six morphological markers mapped in the study resulted in 14 linkage groups generating a total map length of 1131.9 cM with an average map length and distance between two markers equal to 80.9 cM and 8.6 cM respectively (Gupta et al. 2007). A linkage map was also developed using F_2 mapping population by crossing high alkaloid content mutant EMS18X PS-1 with a population size of 120 individuals with 92 markers resulting in nine linkage groups spanning 284.4 cM (Pathak 2009). The third genetic linkage map was developed using 423 codominant markers on *C.roseus* CrN1(Nirmal) × CrN82(Kew) derived 111 F_2 individuals. A total of 114 markers were mapped on eight linkage groups that spanned a 632.7 cM region of the genome with an average marker distance of 5.55 cM (Shokeen et al. 2011). These linkage maps of *C. roseus* will help localize the loci concerned with TIA yield, disease resistance, and ornamental features on the map and will serve as a platform for MAS and molecular breeding programs. The TIA high yielding mutant *gsr8* (glycophtic salinity tolerant) was hybridized with Pink Delhi, and *C. trichophyllus,* and recombinant inbred lines (RILs) were stabilized till F_5 generation. Using bulk segregant analysis with 695 and 200 polymorphic markers, respectively, 12 quantitative traits, i.e., ajmalicine, serpentine, catharanthine, and total alkaloids high/low in roots and vindoline, vincristine, vinblastine high/low in leaves were found to be linked with 150 markers in two different RILS populations. A genetic map of *gsr8(lli)* x Pink Delhi of 179 markers on eight linkage groups covering 1786.5 cM mapped 20 QTLs on 4 linkage groups for catharanthine for leaf and root, vindoline from leaf, vinblastine from leaf, and ajmalicine and serpentine from roots (Chaudhary et al. 2012, 2013).

The accurate evaluation of genes in breeders' germplasm is of great importance for the selection of parental lines and development of new breeding populations. Having gene information for specific target loci (deduced from markers) can be extremely useful for breeders to efficiently use germplasm. EcoTILLING examines natural genetic variation in populations and has been successfully utilized in Vigna, wheat, barley Populas, melon, etc. EcoTILLING will aid in the discovery of natural variants to rapidly screen through many samples with a gene of interest to identify naturally occurring SNPs and /or small INDELS. EcoTILLIING reduces the time and effort for SNP discovery generally required by weeding out identical haplotypes. Therefore, this method does not require one to sequence all individuals in a population to identify polymorphisms, which can be burdensome expensive, and time-consuming. Many workers have advocated the use of functional genomics and gene sequence information for gene discovery in Catharanthus (Goossens and Rischer 2007; Roepke et al. 2010; Ozber et al. 2020). Knowledge about the identity and map location of MTIA important genes and QTL provides the basis for parental selection and marker-assisted selection (MAS) in Catharanthus breeding. Several mutants have been developed and characterized to discover the allelic variation in natural populations. This large set of mutants can be typed by EcoTILLING for polymorphism in a specific gene and to discover the allelic variation in natural populations.

Space breeding initiated by People's Republic of China in 1987, explored the space particle radiation, microgravity induced regulatory gene expression, and mutagenic effects of space radiation ionization and vaccum resulted in more than 300 distinct genotypes in 70 species like disease-resistant, high yielding hybrid japonica rice a green pepper which single fruit weight of half a Kg, high-quality disease-resistant tomato, large pod in brassica rape, large fruit size in white lotus, large bulb in lily (Li et al. 1999; Ou et al. 2010; Jiang et al. 2020). Space breeding is in use by the US, Russia, and PR China. Though the cost of space breeding is very high and its regulatory guidelines are still obscure, it opens a fast new avenue to explore novel variations beyond the prospects of ground breeding.

MTIA linkage with SNP markers based on sequenced genes and whole-genome sequence can support MTIA targeted breeding in Catharanthus. Phylogeny and systematics studies of the Apocynaceae revealed four pop sets based on molecular and morphological evidence (Simoes et al. 2007) and one pop set based on trnL intron and trnL-F spacer sequences (Potgieter and Albert. 2001). Progress in Catharanthus molecular breeding is limited by the ability to predict plant phenotype based on its previous data on the phenotypic performance over the years and over the locations, especially for complex adaptive traits. Suitably constructed crop growth and development models, statistical models, and simulatory tools, combined with increasingly abundant genomic information, will enable the deployment of prediction-based breeding methods like Genomic selections to bridge this predictability gap. Marker-assisted selection guided by genomic selections using high-density markers across the whole genome along with pedigree, multi-environment, and multitrait data can well predict the genomic breeding values of untested populations. Marker trait association estimation, especially for genes with small effects, which are often escaped through association analysis but by genomic selection coupled with marker-assisted selection, estimates all marker effects in all loci and capture small QTL effects (Desta and Oritz 2014).

3.8 Conclusions and Future Prospects

Literature mining validates that *C. roseus* is still the only viable resource of valuable anti-cancerous MTIA alkaloids and besides huge efforts, *in planta* cultivation is the only sustainable alternative for MTIA alkaloids production. The base raw material still comes from elite high MTIA rich Catharanthus plant varieties. *C. roseus* is a diploid, 79% cross-pollinated, self-compatible, herkogamous, short genome medicinal, and ornamental plant which has fetched great breeding efforts with many bottlenecks as well as accomplishments. Today breeder has developed pure lines, inbreds, male-sterile lines, more than 324 cultivars released and competing in the open market. Pre-breeding for targeted trait diversity to hybrid development had already been established. Linkage maps have been generated, and QTLs have been mapped for MTIA component traits. Excellent genetic screens and mutant resources set a platform for exciting functional genomics open for characterization. Growing knowledge from all prospective disciplines of Catharanthus study along with genome sequencing has enabled a better understanding of biosynthetic pathway genes regulating MTIA accumulation in a spatially and temporally coded manner at target sites. This helps to strategically make a breeding plan to break the bottlenecks MTIA genetic improvement in *C. roseus*. Breeding for a higher content of vindoline and catharanthine in *C. roseus* is one of the major challenges to meet its growing demands at the domestic and international levels. Accumulation of high amounts of total alkaloid and specific alkaloid either in leaf or root or both in certain accessions reflects reproductive fitness achieved under the biotic and abiotic stress imposed on their parent population during evolution. Extensive hybridization to explore the heterotic combining ability and its fixation by polyploidization is a prospect for a breeder to venture. Pyramiding disease resistance, Phytophthora and Phytoplasma, and dieback resistance by utilizing well adapted local cultivars and landraces along with tolerance to heat and drought are much sought for successful adoption of the Catharanthus varieties under diverse agroclimatic zones. The chapter has created a compendium of all cultivars released so far in Catharanthus, both for ornamental and as well for medicinal prospects. The chapter briefs targets for Catharanthus breeders for profitable and agronomically viable high productivity. It reviews the desirable genes and patents information to set the pace for targeted Catharanthus breeding. MTIA metabolite breeding can be realized by utilizing recombinational approaches and perhaps mutation breeding and an extensive repertoire of genetic screens along with homologous and heterologous expression by genetic transformation. Wide hybridization with *C. pusillus* should be undertaken for molecular characterization to understand the failure of accumulation of MTIA.

Acknowledgements Authors are grateful to the Director, CSIR-CIMAP, for encouragement, support, and facilities. They are also grateful to the former directors Dr. Sushil Kumar and Dr. SPS Khanuja for their valuable guidance, motivation, and critical suggestions. This is CSIR-CIMAP Communication No: CIMAP/SUB/2021/120. Authors declare no conflicts of interest.This work was funded by CSIR-CIMAP MLP14 project. This chapter is respectfully dedicated to Dr. R N Kulkarni and Prof. Sushil Kumar, the eminent front runners in Catharanthus breeding.

References

Abdelmageed AH, Abdelrahman ME (2017) Genetics of flower color in Periwinkle *Catharanthus roseus* (L) G. Don. Int J Genet Genom 5:36–41

Allorge L, Phillipson PB, Razakamalala R. *Catharanthus makayensis* L. Allorge, Phillipson & Razakamal (2015) (Apocynaceae), a New Species from Madagascar. Candollea 70:61–66

Almagro L, Fernández-Pérez F, Pedreño MA (2015) Indole alkaloids from *Catharanthus roseus*: bioproduction and their effect on human health. Molecules 20:2973–3000

Ambrin G, Ali HM, Ahmad A (2020) Metabolic regulation analysis of ajmalicine biosynthesis pathway in *Catharanthus roseus* (L.) G. Don suspension culture using nanosensor. Processes 8:589

Anderson A (1931) Studien uberdie embroyology der familen Celastraceae, Oleaceae and Apocynaceae. Act Univ Lund Adv 2(27):1–112

Arens H, Stockigt J, Weiler EW, Zenk MH (1978) Radioimmunoassays for the determination of the indole alkaloids ajmalicine and serpentine in plants. Planta Med 34:37–46

Baskaran K, Srinivas KVNS, Kulkarni RN (2013) Two induced macro- mutants of periwinkle with enhanced contents of leaf and root alkaloids and their inheritance. Ind Crops Prod 43:701–703

Benyammi R, Paris C, Khelifi-Slaoui M, Zaoui D, Belabbassi O et al (2016) Screening and kinetic studies of catharanthine and ajmalicine accumulation and their correlation with growth biomass in *Catharanthus roseus* hairy roots. Pharmaceut Biol 54:2033–2043

Bhatia S, Shokeen B (2009) Isolation of microsatellites from *Catharanthus roseus* (L.) G. Don using enriched libraries. Methods Mol Biol 547:289–302

Boke NH (1948) Development of perianth in *Vinca rosea* L. Am J Bot 35:413–423

Boke NH (1949) Development of the stamens and carpels in *Vinca rosea* L. Amer J Bot 36:535–547

Bowman RM (1996) Phytophthora resistance gene of Catharanthus and its use in plant breeding for improved disease resistanceand increased alkaloid levels. PCT Int. Appl. WO 9519698 A1 19950727, US Patent 5,491,285

Bowman RN (2000a) Method of producing hybrid *Catharanthus* using male sterility. US Patent 6, 166, 306 A

Bowman RN (2000b) Catharanthus seed bearing male-sterile allele, plant derived there from, production of F1 hybrid seed from the plant, and transformation with the sterile gene. Jpn Kokai Tokkyo Koho JP 2000125683 A 20000509

Bruneton J (1995) Pharmacognosy, Phytochemistry, Medicinal Plants. Lavoisier, Paris

Carew DP (1966) Growth of callus tissue of *Catharanthus roseus* in suspension culture. J Pharmaceut Sci 55:1153–1154

Chang CJ (1998) Pathogenicity of Aster Yellows Phytoplasma and Spiroplasma citri on Periwinkle. Phytopathology 88:1347–1350

Chaudhary S, Sharma V, Prasad M, Bhatia S, Tripathi BN et al (2011) Characterization and genetic linkage mapping of the horticulturally important mutation leafless inflorescence (*lli*) in periwinkle *Catharanthus roseus*. Sci Hortic 129:142–153

Chaudhary S, Pandey R, Tripathi BN, Goel R, Kumar S (2012) Detection and mapping of quantitative trait loci for the contents of the terpenoid indole alkaloids, vindoline and catharanthine, in the leaves of *Catharanthus roseus* using bulk segregant analysis. J Hort Sci Biotechnol 87:179–185

Chaudhary AA, Yadav D, Hemant JSS, Asif M (2012b) Real time sequence characterized amplified region (RT-SCAR) marker: development and its application for authentication and quantification of *Catharanthus roseus* L. Don. J Med Plants Res 6:4748–4754

Chaudhary S, Pandey R, Sharma V, Tripathi BN, Kumar S (2013) Detection and mapping of QTLS affecting contents of pharmaceutical alkaloids in leaf and root of *Catharanthus roseus*. Agric Res 2:9–23

Chaudhary AA, Hemant, Mohsin M, Ahmad A (2012a) Application of loop-mediated isothermal amplification (LAMP)-based technology for authentication of *Catharanthus roseus* (L.) G. Don. Protoplasma 249:417–422

Chauhan AKS, Raghuvanshi SS (1978) Cytomorphological studies of artificially induced tetraploids of *Catharanthus pusillus*. Phyton 18:217–219

Chauhan AKS, Raghuvanshi SS (1980) Cyto taxonomy and induced poly ploidy in Apocynaceae. Biol Memoirs 5:163–209

Chen CM, Yeh DM (2012) Taoyuan No. 1 Rose Girl': A Double-flowered Periwinkle, *Catharanthus roseus*. HortScience 47:1175–1176

ChemFaces (2021) https://www.chemfaces.com

Choi YH, Tapias EC, Kim HK, Lefeber AW, Erkelens C et al (2004) Metabolic discrimination of Catharanthus roseus leaves infected by phytoplasma using 1H-NMR spectroscopy and multivariate data analysis. Plant Physiol 135(4):2398–2410

Chung IM, Kim EH, Li M, Peebles CA, Jung WS et al (2011) Screening 64 cultivars *Catharanthus roseus* for the production of vindoline, catharanthine, and serpentine. Biotechnol Prog 27:937–943

Collu G, Unver N, Peltenburg-Looman AMG, van der Heijden R, Verpoorte R et al (2001) Geraniol 10-hydroxylase, a cytochrome P450 enzyme involved in terpenoid indole alkaloid biosynthesis. FEBS Lett 508:215–220

Consiglio T, Schatz GE, McPherson G, Lowry PP, Rabenantoandro J et al (2006) Deforestation and plant diversity of Madagascar's littoral forests. Conserv Biol 20:1799–1803

Courdavault V, Papon N, Clastre M, Guivarc'hNG, St-Pierre B et al (2014) A look inside an alkaloid multisite plant: the *Catharanthus* logistics. Curr Opin Plant Biol Phytochem 19:43–50

Cresswell RC, Fowler MW, Scragg AH (1988) Glyphosphate-tolerance in *Catharanthus roseus*. Plant Sci 54:55–63

Curry HA (2012) Naturalising the exotic and exoticising the naturalized: horticulture, natural history and the rosy periwinkle. Environ History 18:343–365

Dastur JF (1916) Phytophthora on *Vinca rosea*. Mem Dep Agric India Bot Ser 8:233–242

Davidoud E, Kan C, Quiron JC, Das BC, Husson HP (1989) Epiallo-yohimbine derivatives isolated from in vitro hairy-root cultures of *Catharanthus trichophyllus*. Phytochemistry 28:1383–1387

De Luca V, St-Pierre B (2000) The cell and developmental biology of alkaloid biosynthesis. Trends Plant Sci 5:168–173

Desta ZA, Ortiz R (2014) Genomic selection: genome-wide prediction in plant improvement. Trends Plant Sci 19(9):592–601

Dnyansagar VR, Sudhakaran IV (1966) Karyotype and behaviour of nucleolus in mitosis in *Vinca rosea* Linn. syn. *Catharanthus roseus* (L) G. Don. J Biol Sci 9:16–24

Dnyansagar VR, Sudhakaran IV (1977) Seed development in diploid and tetraploid of *Vinca rosea* syn. *Catharanthus roseus* (*Lochnera rosea*). Proc Indian Nat Sci Acad 43 (Part B):133–141

Dovrat A, Goldschmidt J (1978) Cultivation aspects of *Catharanthus roseus* for roots. Acta Hort 73:263–268

Drapeau D, Blanch HW, Wilke CR (1987) Economic assessment of plant cell culture for the production of ajmalicine. Biotechnol Bioeng 30(8):946–953

De Padua DS, Barrion AA, Casal CMV, De La Cruz Ma PR (1992) Karyomorphology of chichirica *Catharanthus roseus* (L.) Don. Philip J Sci 121:299–303

Duge de Bernonville T, Clastre M, Besseau S, Oudin A, Burlat V et al (2015) Phyto-chemical genomics of the Madagascar periwinkle: unravelling the last twists of the alkaloid engine. Phytochemistry 113:9–23

Duge de Bernonville T, Carqueijeiro I, Lanoue A, Lafontaine F, Sânchez Bel P et al (2017) Folivory elicites a strong defence reaction in metabolomic and transciptomic analyses reveal distinct local and systemic responses. Sci Rep 7(1):1–14

Dutkiewicz EP, Su CH, Lee HJ, Hsu CC, Yang YL (2021) Visualizing vinca alkaloids in the petal of *Catharanthus roseus* using functionalized titanium oxide nanowire substrate for surface-assisted laser desorption/ionization imaging mass spectrometry. Plant J 105:1123–1133

Dutta A, Batra J, Pandey-Rai S, Singh D, Kumar S et al (2005) Expression of terpenoid indole alkaloid biosynthetic pathway genes corresponds to accumulation of related alkaloids in *Catharanthus roseus* (L.) Don. Planta 220:376–383

Dwivedi S, Singh M, Singh AP, Sharma S, Uniyal GC et al (1999) Genetic variability, heritability and genetic advance foralkaloid yield attributing traits in 26 genotypes of periwinkle *Catharanthusroseus*. J Med Arom Plant Sci 21:320–324

Dwivedi S, Singh M, Singh AP, Sharma S, Uniyal GC et al (2000) Assessment of genetic divergence for its purposeful exploitation in periwinkle *Catharanthusroseus* (Apocynaceae). J Genet Breed 54:95–99

Dwivedi S, Singh M, Singh AP, Uniyal GC, Khanuja SPS et al (2001) Registration of a new variety Prabal of *Catharanthus roseus*. J Genet Breed 23:104–106

Ehrenworth AM, Peralta-Yahya P (2017) Accelerating the semisynthesis of alkaloid-based drugs through metabolic engineering. Nat Chem Biol 13:249–258

El-Domyati FME, Ramadan AM, Gadalla NO, Edris S, Shokry AM et al (2012) Identification of molecular markers for flower characteristics in *Catharanthus roseus* producing anticancer compounds. Life Sci J 9 (4):5949–5960

El-Sayed M, Verpoorte R (2002) Effect of phytohormones on growth and alkaloid accumulation by *Catharanthus roseus* cell suspension cultures fed with alkaloid precursors tryptamine and loganin. Plant Cell Tiss Org Cult 68:265–270

Endo T, Goodbody AE, Misawa M (1987) Alkaloid production in root and shoot cultures of *Catharanthus roseus*. Planta Med 53:479–482

Fartyal D (2010) Functional Characterization of novel EST-derived microsatellite markers in *Catharanthus roseus* and their cross transferability. MTech Thesis. G.B Pant Engineering College, Pauri Garhwal Uttarakhand.

Favali MA, Musetti R, Benvenuti S, Bianchi A, Pressacco L (2004) *Catharanthus roseus* L. plants and explants infected with phytoplasmas: alkaloid production and structural observations. Protoplasma 223 (1):45–51

Fehr WR (1987) Principles of cultivar development: crop species. *Agronomy Books* 2, New York https://lib.dr.iastate.edu/agronbook/2

Ferreres F, Pereira P, Valentao P, Andrade PB, Seabra RM, Sottomayor M (2008) New phenolic compounds and anti-oxidant potential of *Catharanthus roseus*. J Agri Food Chem 56:9967–9974

Flaschenriem DR (2016a) *Catharanthus* variety PAS1052940, U.S. Patent 9,451,747

Flaschenriem DR (2016b) *Catharanthus* variety AS1053041, U.S. Patent 9,451,749

Flaschenriem DR (2016c) *Catharanthus* variety PAS1053043, U.S. Patent 9,451,748

Flaschenriem DR (2016d) *Catharanthus* variety PAS1053044, U.S. Patent 9,451,750

Flaschenriem DR (2016e) *Catharanthus* variety PAS1053042,U.S. Patent 9,451,751

Flaschenriem DR (2016f) *Catharanthus* variety PAS1053045, U.S. Patent 9,451,752

Flaschenriem DR, Blowers AD (2017) Male sterility in *Catharanthus*. U.S. Patent 10212908

Flory WS Jr (1944) Inheritance studies of flower colour in periwinkle. Proc Am Soc Hort Sci 44:525–526

Fredslund J, Madsen LH, Hougaard BK, Nielsen AM, Bertioli D et al (2006) A general pipeline for the development of anchor markers for comparative genomics in plants. BMC Genomics 7:1–10

Fulzele DP, Namdeo AG (2018) Cost effective pilot-scale ajmalicine production by Catharanthus roseus cell suspension cultures in a 100 L bioreactor. Biotechnol J 21:1–15

Gogitidze TSR, Laptev YuP (1981) Comparative evaluation of experimental polyploids of *Catharanthus roseus* and their parent forms. Genetika 17:563–564

Gogitidze TSR, Maisuradze NI (1980) Approaches to Selection of *Catharanthus roseus*. Khimiko Farmatsevticheskii Zhurnal 14:56–59

Gogitidze TSP (1981) Induced polyploidy in *Catharanthus roseus*. Subtropicheskie-Kul'-tury 5:136–139

Goossens A, Rischer H (2007) Implementation of functional genomics for gene discovery in alkaloid producing plants. Phytochem Rev 6:35–49

Goswami R, Tyagi BR, Rani A, Uniyal GC, Kumar S (1996) Colchicine induced autotetraploids in periwinkle *Catharanthus roseus*. J Med Arom Plant Sci 18:38–45

Guimara˜es G, Cardoso L, Oliveira H, Santos C, Duarte P et al (2012) Cytogenetic characterization and genome size of the medicinal plant *Catharanthus roseus* (L.) G. Don. *AoB* Plants 2012: pls002. https://doi.org/10.1093/aobpla/pls002.

Gupta S, Pandey-Rai S, Srivastava S, Naithani SC, Prasad M et al (2007) Construction of genetic linkage map of the medicinal and ornamental plant *Catharanthus roseus*. J Genet 86:259–268

Gupta AK, Shukla AK, Singh P, Mall M, Yadav S et al (2020) CIM-Sushil: a high vindoline yielding variety of *Catharanthus roseus* (L.) G. Don. J Med Arom Pant Sci 42(1–2):51–63

Handique AK, Sharma CM (1990) Colchicine induced hypertrophy in *Catharanthus roseus* (L.) G. Don and *Nicotiana rustica* L. Crop Res Hisar 3:287–290

Hershberger A, Gawenda I,Thomas D, Trolinger J, Van Bentem SDLF et al (2016) Noval Phytophthora resistant *Catharanthus roseus* plant. Application PCT WO2016120191A1, Syngenta International Ag

Holme IB, Gregersen PL, Brinch-Pedersen H (2019) Induced genetic variation in crop plants by random or targeted mutagenesis: convergence and differences. Front Plant Sci 10:1–468

Hong JC, Zheng DF, Xu BY, Qiang JZ, Zhang L et al (2008) Analysis of karyotype of *Catharanthus roseus* (Apocynaceae). J Trop Subtrop Bot 16:169–172

Husain A (1993) Medicinal plants and their cultivation. Central Institute of Medicinal and Aromatic Plants, Lucknow, India

Ibrahim MM, Abou El-Nasr THS Abdel-Samea NS, Aboud KA (2013) Efficiency of RAPD and ISSR markers in assessment of genetic diversity in some *Catharanthus roseus* L. cultivars grown in Egypt. World Appl Sci 26:1407–1415

International Atomic Enertgy Agency (IAEA) Mutation Variety Database (MVD v2.0.5575.27348(27348) https://mvd.iaea.org)

Janaki Ammal EK, Bezbaruah HP (1963) Induced teteraploidy in *Catharanthus roseus* (L.) Don. Proc Indian Acad Sci Sect B 57:339–342

Jhang T, Gautam TP, Shukla S, Fatayal D, Annula et al (2012) Development of microsatellite marker resource for genome analysis of *Catharanthus roseus*. In: Proceedings of international plant conference on "Molecular Mapping and Marker Assisted Selection", University of Veterinary Medicine, Veterinärplatz 1, 1210 Vienna, Austria, Vienna, Int Plant Conf Assoc, p 39

Jiang L, Zhang B, Chen W, Qiu T, Wang L et al (2020) Space breeding in modern agriculture. Am J Agric Res 1:5

Johnson IS, Armstrong JG, Gorman M, Burnett JP Jr (1963) The Vinca alkaloids: a new class of oncolytic agents. Cancer Res 23:1390–1427

Joshi RK, Kar B, Nayak S (2011) Exploiting EST databases for the mining and characterization of short sequence repeat (SSR) markers in *Catharanthus roseus* L. Bioinformation 5:378–381

Joy PP, Thomas J, Mathew S, Skaria BP (1998) Medicinal plants. Kerala Agricultural University

Jussieu AL de (1789) Genera Plantarum. Parisiis [Peris]: Apud viduan Herissant. https://doi.org/10.5962/bhl.title.284

Kellner F, Kim J, Clavijo BJ, Hamilton JP, Childs KL et al (2015) Genome-guided investigation of plant natural product biosynthesis. Plant J 82:680–692

Kim SW, Kim JH, Liu JR (2009) Genetic discrimination of *Catharanthus roseus* cultivars by pyrolysis mass spectrometry. J Plant Biol (NY, US) 52:462–465

Kim SW, Ban SH, Jeong S, Chung Hw, Ko SM et al (2007) Genetic discrimination between Catharanthus roseus cultivars by metabolic fingerprinting using 1H NMR spectra of aromatic compounds. Biotechnol Bioprocess Eng 2:646–652

Knuth P (1909) Handbook of flower pollination, vol III. Clarendon Press, Oxford

Krishnan R, Naragund VR, Vasantha Kumar T (1979) Evidence for outbreeding in *Catharanthus roseus*. Curr Sci 48:80–82

Krishnan R, Chandravadana MV, Mohan Kumar GN, Ramachander PR (1985) Effect of induced autotetrapoidy on alkaloid content and root weight in *Catharanthus roseus* (L.) G. Don. Herba Hung 24:43–51

Ku C, Chung WC, Chen LL, Kuo CH (2013) The Complete plastid genome sequence of Madagascar periwinkle *Catharanthus roseus* (L.) G. Don: plastid genome evolution, molecular marker identification, and phylogenetic implications in Asterids. PLoS One 18:8:e68518

Kulkarni RN (1984) Relative resistance of diploid tetraploid plants of *Catharanthus roseus* plants to die back disease. Curr Sci 53:577

Kulkarni RN (1999) Evidence for phenotypic assortative mating for flower colour in periwinkle. Plant Breed 118:561–564

Kulkarni RN, Baskaran K (2003) Inheritance of resistance to *Pythium* dieback in the medicinal plant, periwinkle. Plant Breed 122:184–187

Kulkarni RN, Baskaran K (2008) Inheritance of pollen-less anthers and 'thrum' and 'pin' flowers in periwinkle. J Hered 99:426–431

Kulkarni RN, Baskaran K (2013a) From Herkogamy to Cleistogamy – development of cleistogamy in periwinkle. J Hered 104:140–148

Kulkarni RN, Baskaran K (2013b) Individual and combined effects of genes producing opposite effects on plant height in periwinkle (*Catharanthus roseus*). J Crop Sci Biotechnol 16:123–129

Kulkarni RN, Baskaran K (2014) Increasing total leaf alkaloid concentrations in periwinkle (*Catharanthus roseus*) by combining the macro-mutant traits of two induced leaf mutants ('necrotic leaf' and 'nerium leaf'). J Hort Sci Biotechnol 89:513–518

Kulkarni RN, Ravindra NS (1988) Resistance to *Pythium aphanidrmatum* in diploids and induced autotetraploids of *Catharanthus roseus*. Planta Med 54:356–359

Kulkarni RN, Ravindra NS (1997) Integrheration of host resistance with fungicide in the control of dieback of periwinkle. Trop Agri 74:321–323

Kulkarni RN, Dimri BP, Rajagopal K, Suresh N, Chandrashekar RS (1984) Variability for quantitative characters in Periwinkle *Catharanthus rose*us. Indian Drugs 22:61–64

Kulkarni RN, Chandrashekar RS, Dimri BP (1984) Induced autotetraploidy in *Catharanthus roseus*: a prliminary report. Curr Sci 53:484–486

Kulkarni RN, Kalra A, Ravindra NS (1992) Integration of soil solarization with host resistance in the control of die back and collar rot of periwinkle. Trop Agri 69:217–222

Kulkarni RN, Chandrashekara RS, Chadrashekara G (1995) Nitrogen utilization efficiency of diploid and induced autotetraploids strains of periwinkle. Trop Agri 72:249–251

Kulkarni RN, Baskaran K, Suresh N (1999) Inheritance in periwinkle: leaf pubescence and corolla color. J Herbs Spices Med Plants 6:85–88

Kulkarni RN, Baskaran K, Chandrashekara RS, Kumar S (1999) Inheritance of morphological traits of periwinkle mutants with modified contents and yields of leaf and root alkaloids. Plant Breed 118:71–74

Kulkarni RN, Sreevalli Y, Baskaran KS (2001) The mechanism and inheritance of intraflower self-pollination in self-pollinating strains of periwinkle. Plant Breed 120:247–250

Kulkarni RN, Sreevalli Y, Baskaran K (2005) Allelic genes at two loci govern different mechanisms of intra-flower self- pollinating strains of periwinkle. J Hered 95:71–77

Kulkarni RN, Baskaran K, Sreevalli Y (2005b) Genetics of novel corolla colours in periwinkle. Euphytica 144:101–107

Kulkarni RN, Baskaran K, Sreevalli Y (2008) Genetics of corolla colour in periwinkle: relationship between genes determining violet, orange-red and magenta corolla. J App Hort 10:20–23

Kulkarni RN, Baskaran K, Shyamaprasad DV, Kulkarni SS (2009) Individual and combined effects of plant height reducing genes in periwinkle. Euphytica 170:309–316

Kulkarni RN, Baskaran K, Jhang T (2016) Breeding medicinal plant, periwinkle [*Catharanthus roseus* (L) G. Don]: a review. Plant Genet Res 14:283–302

Kulkarni RN, Rajagopal K, Chandrashekar RS, Dimri BP, Suresh N et al (1987) Performance of diploids and induced autotetraploids of *Catharanthusroseus* under different levels of nitrogen and plant spacing. Plant Breed 98:136–140

Kulkarni RN, Baskaran K, Chandrashekhra RS, Khanuka SPS, Darokar MP et al (2003) 'Dhawal', A high alkaloid producing periwinkle plant. US006548746B1, United States Patent Office

Kumar S, Chaudhary S, Kumari R, Sharma V, Kumar AA (2012) Development of improved horticultural genotypes characterized by novel over-flowering inflorescence trait in periwinkle *Catharanthus roseus*. Proc Natl Acad Sci India Sect B Biol Sci 82:399–404

Kumar S, Shah S, Garg V, Bhatia S (2014) Large scale in-silico identification and characterization of simple sequence repeats (SSRs) from *de novo* assembled transcriptome of *Catharanthus roseus* (L.) G. Don. Plant Cell Rep 33:905–918

Kumar S, Rai SP, Kumar SR, Singh D, Srivastava S et al (2005) Plant variety of Catharanthus roseus named LLI. U.S. Plant Pat Appl Publ US 20050251887 P1 20051110

Kumar S, Rai SP, Rai SK, Singh DV, Srivastava R et al (2007) Plant variety of *Catharanthus roseus* named 'Ili' US Patent PP18315

Kumari R, Yadav G, Sharma V, Sharma V, Kumar S (2013) Cytosine hypomethylation at CHG and CHH sites in the pleiotropic mutants of Mendelian inheritance in *Catharanthus roseus*. J Genet 92:499–511

Kumari R, Sharma V, Sharma V, Kumar S (2013) Pleiotropic phenotypes of the salt-tolerant and cytosine hypomethylated *leafless inflorescence*, *evergreen dwarf* and *irregular leaf lamina* mutants of *Catharanthus roseus* possessing Mendelian inheritance. J Genet 92:369–394

Kumawat S, Rana N, Bansal R, Vishwakarma G, Mehetre ST et al (2019) Expanding avenue of fast neutron mediated mutagenesis for crop improvement. Plants 8:164

Lal S, Kn M, Chaturvedi SP (2013) Random amplified polymorphic DNA (RAPD) fingerprinting for evaluation of intraspecific *Catharanthus roseus* populations collected from Gujarat, India. Intl J Biol Pharm Allied Sci 2:373–385

Lavania UC (2005) Genomic and ploidy manipulation for enhanced production of phyto-pharmaceuticals. Plant Genet Resour Characteriz Utiliz 3:170–177

Levy A, Milo J, Ashri A, Palevitch D (1983) Heterosis and correlation analysis of the vegetative components and ajmalicine content in the roots of the medicinal plant—*Catharanthus roseus* (L.) G. Don. Euphytica 32:557–564

Li JG, Wang PS, Zhang J, Jiang XC (1999) Development and prospect of plant mutation breeding induced by aviation and spaceflight in China. Hang tian yi xue yu yi xue gong cheng= Space Medicine and Medical Engineering 12:464–468

Lin X, Zhou Y, Zhang J, Lu X, Zhang F et al (2011) Enhancement of artemisinin content in tetraploid Artemisia annua plants bymodulating the expression of genes in artemisinin biosynthetic pathway. Biotechnol App Biochem 58:50–57

Magnotta M, Murata J, Chen J, De Luca V (2006) Identification of a low vindoline accumulating cultivar of *Catharanthus roseus* (L.) G. Don by alkaloid and enzymatic profiling. Phytochemistry 67:1758–1764

Mahroug S, Courdavault V, Thiersault M, St-Pierre B, Burlat V (2006) Epidermis is a pivotal sit of at least four secondary metabolic pathways in *Catharanthus roseus* arial organs. Planta 223:1191–1200

Mahroug S, Burlat V, St-Pierre B (2007) Cellular and sub-cellular organisation of the monoterpenoid indole alkaloid pathway in *Catharanthus roseus*. Phytochemistry Rev 6:363–381

Mall M, Singh P, Kumar R, Shanker K, Gupta AK et al (2021) Phenotypic, genetic and expression profiling of a vindoline-rich genotype of *Catharanthus roseus*. S Afr J Bot 139:50–57

Markgraf F (1976) Flora de Madagascar et des comores, 169e famille- Apocynacées:135:139–156

McKnight TD, Roessner CA, Devagupta R, Scott AI, Nessler CL (1990) Nucleotide sequence of a cDNA encoding the vacuolar protein strictosidine synthase from *Catharanthus roseus*. Nucleic Acids Res 18:4939

Milach SCK, Federizzi LC (2001) Dwarfing genes in Plant improvement. Adv Agro 73:35–63

Milo J, Levy A, Akavia N, Ashri A, Palevitch D (1985) Inheritance of corolla colour and anthocyanin pigments in periwinkle *Catharanthus roseus* (L) G. Don. Zeitschrift Für Pflanzenzüchtung 95:352–360

Mishra P, Kumar S (2000) Emergence of periwinkle *Catharanthus roseus* as a model system for molecular biology of alkaloids: Phytochemistry, pharmacology, plant biology and in vivo and in vitro cultivation. J Med Arom Plant Sci 22:306–337

Mishra P, Kumar S (2001a) A monogenic recessive mutant with precocious *in situ* pollen germination in periwinkle *Catharanthusroseus*. J Med Arom Plant Sci 22/4A &23/1A:277–279

Mishra P, Kumar S (2003a) Manifestation of heterostyle character by induction of recessive *hsf* mutation responsible for thrum type herkogamous flowers in *Catharanthusroseus*. J Med Arom Plant Sci 25:2–7

Mishra P, Kumar S (2003b) Flower organ size (FOS) gene in *Catharanthus roseus* with pleiotropic roles in flower growth. J Med Arom Plant Sci 25:705–711

Mishra P, Singh M, Dwivedi S, Kumar S (2001b) Descriptors of periwinkle *Catharanthus roseus*. J Med Arom Plant Sci 22:268–272

Mishra P, Uniyal GC, Sharma S, Kumar S (2001c) Pattern of diversity for morphological and alkaloid yield related traits among the periwinkle *Catharanthus roseus* accessions collected from in and around Indian subcontinent. Genet Resour Crop Evol 48:273–286

Mishra RK, Gangadhar BH, Yu JW, Kim DH, Park SP (2011) Development and characterization of EST based SSR markers in Madagascar periwinkle (*Catharanthus roseus*) and their transferability in other medicinal plants. Plant Omics J 4:154–162

Misra N, Gupta AK (2006) Effect of salinity and different nitrogen sources on the activity of antioxidant enzymes and indole alkaloid content in *Catharanthus roseus* seedlings. J Plant Physiol 163:11–18

Mogili T, Sathaiah G, Kumar P, Vidyavati (1984) Studies of Ethyl Methanesulfonate induced mutations in Periwinkle. Life Sci Adv 3:71–73

Mohan Kumar GN (1980) *Comparative studies on growth and alkaloid of autotetraploids and their diploid progenitors* in Catharanthus roseus (L.) G. Don. M. Sc. (Hort) Thesis, University of Agricultural Sciences, Bangalore, India

Moreno PR, Van der Heijden R, Verpoorte R (1993) Effect of terpenoid precursor feeding and elicitation on formation of indole alkaloids in cell suspension cultures of *Catharanthus roseus*. Plant Cell Rep 12:702–705

Moreno PRH, van der Heijden R, Verpoorte R (1995) Cell and tissue cultures of *Catharanthus roseus* (L.) G. Don: a literature survey II. Updating from 1988 to 1993. Plant Cell Tiss Org Cult 42:1–25

Morton JF (1977) Periwinkle. In: Major medicinal plants. botany, culture and uses. Charles C, Thomas Publ, Springer-field, 11, USA, pp 237–241

Murata J, Bienzle D, Brandle JE, Sensen CW, De Luca V (2006) Expressed sequence tags from Madagascar periwinkle (*Catharanthus roseus*). FEBS Lett 580:4501–4507

Nef C, Rio B, Chrestin H (1991) Induction of catharanthine synthesis and stimulation of major indole alkaloids production by *Catharanthusroseus* cells under non-growth altering treatment with *Pythium vexans* extracts. Plant Cell Rep 10:26–29

Nejat N, Vadamalai G, Dickinson M (2012) Expression patterns of genes involved in the defense and stress response of spiroplasma citri infected madagascar Periwinkle *Catharanthus roseus*. Int J Mol Sci 13:2301–2313

Noble RL, Beer CT, Cutts JH (1958) Role of chance observations in chemotherapy: V*inca rosea*. Ann NY Acad Sci 76:882–894

O'Connor SE, Maresh JJ (2006) Chemistry and biology of monoterpene indole alkaloid biosynthesis. Nat Prod Rep 23:532–547

Ou X, Long L, Wu Y, Yu Y, Lin X et al (2010) Spaceflight-induced genetic and epigenetic changes in the rice (*Oryza sativa* L.) genome are independent of each other. Genome 53:524–532

Ozber N, Watkins JL, Facchini PJ (2020) Back to the plant: overcoming roadblocks to the microbial production of pharmaceutically important plant natural products. J Ind Microbiol Biotechnol: Official J Soc Ind Microbiol Biotechnol 47:815–828

Palazon J, Cusido RM, Gonzalo J, Bonfill M, Morales C (1998) Relation between the amount of *rol*C gene product and indole alkaloid accumulation in *Catharanthus roseus* transformed root cultures. J Plant Physiol 153:712–718

Palmer E (2019) Pfizer scrambles to fill void after Teva stops making chemo drug often given to children. FiercePharma. Published October 15, 2019. https://www.fiercepharma.com/manufacturing/pfizer-scrambles-to-fill-void-after-teva-stops-making-chemo-drug-often-children. Accessed 16 Oct 2019

Pan Q, Mustafa NR, Tang K, Choi YH, Verpoorte R (2016) Monoterpenoid indole alkaloids biosynthesis and its regulation in *Catharanthus roseus*: a literature review from genes to metabolites. Phytochem Rev 15:221–250

Pan Q, Dai Y, Nuringtyas TR, Mustafa NR, Schulte AE et al (2014) Investigation of the chemomarkers correlated with flower colour in different organs of Catharanthus roseususing NMR-based metabolomics. Phytochem Anal 25:66–74

Pandeya SC, Jose L (1970) Population Differences in *Catharanthus roseus* D Part 3 Cytological distinction. Indian Sci Congr Assoc Proc 57

Pareek SK, Singh S, Srivastava VK, Mandal S, Maheshwari ML et al (1981) Advances in periwinkle cultivation. Indian Farming 31:18–21

Pathak R (2009) Development of framework genetic linkage map and Phylogenetic Analysis of *Catharanthus roseus* by RAPD, URP, ISSR and SSR markers. Masters Thesis. Sam Higginbottom University of Agriculture , Technology and Sciences, Allahabad UP

Pathirana R (2011) Plant mutation breeding in agriculture. CAB Reviews:Persp in Agriculture, Veterinary Science, Nutrition and Natural Res No. 032

Pezzuto JM (1997) Plant-derived anticancer agents. Biochem Pharmacol 53:121–133

Piovan A, Filippini ER (2007) Anthocyanins in *Catharanthus roseus in vivo* and *in vitro*: a review. Phytochem Rev 6:235–242

Plaizier AC (1981) A revision of *Catharanthus roseus* (L.) G. Don (Apocynaceae). Mededelingen Landbouwhogeschool 81:1–12

Potgieter K, Albert V (2001) Phylogenetc relationship within Apocynaceae sl based on trnL intron and trnL-F spacer sequences and propagule characters. Ann MO Bot Gard 88(4):523–549

Potier P (1989) The synthesis of Navelbine prototype of a new series of vinblastine derivatives. Semin Oncol 16:2–4

Prasad MP (2014) Detection of genetic diversity of Catharanthus species using RAPD PCR. Indian J Adv Plant Res 1:24–28

Qu Y, Easson MLAE, Froese J, Simionescu R, Hudlicky T et al (2015) Completion of the seven-step pathway from tabersonine tothe anticancer drug precursor vindoline and its assembly in yeast. Proc Natl Acad Sci USA 112:6224–6229

Rai SP, Kumar S (2000) Induced mutation to monocotyledony in periwinkle, *Catharanthus roseus* and suppression of mutant phenotype by kinetin. J Genet 79:97–104

Rai SP, Kumar S (2001a) Heterocarpus flowers resulting from a recessive mutation in periwinkle *Catharanthusroseus*. Curr Sci 80:1581–1584

Rai SP, Kumar S (2001b) A tricotyledonous seedling mutant with Mendelian inheritance in periwinkle *Catharanthus roseus*. J Med Arom Plant Sci22/23 (4A/1A):267–268

Rai SP, Kumar S (2001c) Loss of negative control on rooting in the hyper-rooting Rrg- mutant of periwinkle *Catharanthus roseus*. J Med Arom Plant Sci 22/23 (4A/1A): 300–302

Rai SP, Luthra R, Kumar S (2001e) Differential expression of proteins in monogenic salt resistant mutants with and without thermal stress in periwinkle *Catharanthusroseus*. J Med Arom Plant Sci 22/4A &23/1A:287–290

Rai SP, Luthra R, Gupta MM, Kumar S (2001b) Pleiotropic morphological and abiotic stress resistance phenotypes of the hyper-abscisic acid producing Abo-mutant in the periwinkle *Catharanthus roseus*. J Biosci 26:57–70

Rai SP, Luthra R, Kumar S (2003) Salt-tolerant mutants in glycophytic salinity response (GSR) genes in *Catharanthusroseus*. Theor App Genet 106:221–230

Rai A, Saito K, Yamazaki M (2017) Integrated omics analysis of specialized metabolism in medicinal plants. Plant J 90:764–787

Rajasekaran LR, Nambisan KMP, Shanmugavelu Kg, Selvarajan M (1981) Studies on floral biology of *Catharanthus roseus* G Don. In: IV All India Workshop on Medicinal and Aromatic Plants. Ann Rep. ICAR Scheme on the Improvement of Yield of *Catharanthus roseus*. Department of Horticulture, Agricultural College and Research Institute, Madurai, India

Rendle AB (1971) The classification of floweringplants, II. Cambridge University Press, Cambridge

Roepke J, Salim V, Wu M, Thamm AM, Murata J et al (2010) Vinca drug components accumulate exclusively in leaf exudates of Madagascar periwinkle. Proc Natl Acad Sci USA 107:15287–15292

Roller U (1978) Selection of plants and plant tissue cultures of *Catharanthus roseus* with high content of serpentine qand ajmalicine. In: Alfermann AW, Reinhard E (eds) Production of natural compounds by cell culture methods. Proceedings of International Symposium Plant Cell Culture, München, KRG, pp 95–104

Le Roux M, Guéritte F (2017) *Catharanthus roseus* alkaloids to the discovery of vinorelbine (Navelbine®). Navelbine® and Taxotère®, Elsevier, pp 87–149

Rowinsky E (2003) The Vinca alkaloids. In: Kufe DW, Pollock RE, Weichselbaum RR, Bast RC, Gansler TS, Holland JF, Frei E (eds) Cancer medicine, 6th edn., vol 2. BC Decker Publisher, Hamilton, ON, Canada

Roy AM, Banerjee N, Mandal S (2013) Studies on alkaloid yield in mutants of *Catharanthus roseus* (L.) G. Don. through character correlation and path analysis. Elixir Int J 19193–19198

Schnell L (1943) Self-sterility in *Vinca rosea*. Proc Oklahoma Acad Sci 23:21

Schröder G, Unterbusch E, Kaltenbach M, Schmidt J, Strack D et al (1999) Light-induced cytochrome P450 dependent enzyme in indole alkaloid biosynthesis: tabersonine 16-hydroxylase. FEBS Lett 458:97–102

Schutz FA, Bellmunt J, Rosenberg JE, Choueiri TK (2011) Vinflunine: drug safety evaluation of this novel synthetic vinca alkaloid. Exp Opinion Drug Safe 10:645–653

Seca AM, Pinto DC (2018) Plant secondary metabolites as anticancer agents: successes in clinical trials and therapeutic application. Int J Mol Sci 19:263

Sevestre-Rigouzzo M, Nef-Campa C, Ghesquiere A, Chrestin H (1993) Genetic diversity and alkaloid production in *Catharanthus roseus*, C. *trichophyllus* and their hybrids. Euphyitca 66:151–159

Sharma A, Girdhar A, Sharma P, Joshi N, Sharma S et al (2011) Genetic diversity among different medicinal plants of semi-arid region revealed by RAPD. Natl J Life Sci 8(1):7–10

Sharma V, Chaudhary S, Srivastava S, Pandey R, Kumar S (2012) Characterization of variation and quantitative trait loci related to terpenoid indole alkaloid yield in a recombinant inbred line mapping population of *Catharanthus roseus*. J Genet 91:49–69

Shaw RK, Acharya L, Mukherjee AK (2009) Assessment of genetic diversity in a highly valuable medicinal plant *Catharanthus roseus* using molecular marker. Crop Breed App Biotechnol 9:52–59

She J, Yan H, Yang J, Xu W, Su Z (2019) croFGD: *Catharanthus roseus* functional genomics database. Fron Genet 10:238

Shokeen B, Sethy NK, Choudhary S, Bhatia S (2005) Development of STMS markers from the medicinal plant Madagascar periwinkle [*Catharanthus roseus* (L.) G. Don.]. Mol Ecol Notes 5:818–820

Shokeen B, Sethy NK, Kumar S, Bhatia S (2007) Isolation and characterization of microsatellite markers for analysis of molecular variation in the medicinal plant Madagascar periwinkle (*Catharanthus roseus* (L.) G. Don). Plant Sci 172:441–451

Shokeen B, Choudhary S, Sethy NK, Bhatia S (2011) Development of SSR and gene-targeted markers for construction of a framework linkage map of *Catharanthus roseus*. Ann Bot 108:321–336

Simmonds NW (1960) Flower colour in *Lochnera rosea*. Heredity 14:253–261

Simoes AO, Livshultz T, Conti E, Endress ME (2007) Phylogeny and systematics of the Rauvolfioideae (Apocynaceae) based on molecular and morphological evidence. Ann MO Bot Gard 94:268–297

Singh SN, Vats P, Suri S, Shyam R, Kumria MM et al (2001) Effects of an antidiabetic extract of *Catharanthus roseus* on enzymatic activities in streptozotocin induced diabetic rats. J Ethnopharmacol 76(3):269–277

Singh M, Singh AP, Singh V, Srivastava A, Dwivedi S (2002) Identification of morphologically distinct genotypes of *Catharanthus roseus* by SDS-PAGE and peroxidase of leaf proteins. J Med Arom Plant Sci 24:68–73

Singh D, Mehta SS, Neoliya NL, Shukla YN, Mishra M (2003) New possible insect growth regulators from *Catharanthus roseus*. Curr Sci 84(9):1184–1186

Singh D, Rai SK, Pandey-Rai S, Srivastava S, Mishra K et al (2008) Predominance of the serpentine route in monoterpenoid indole alkaloid pathway of *Catharanthus roseus*. Proc Indian Natl Sci Acad 74:97–109

Singh M, Srivastava A, Singh AP, Singh V, Srivastava Aet al(2010) Genetic analysis of important morpho-ecomonic traits in periwinkle *Catharanthus roseus* (L.) G. Don. Indian J Genet 67:62–65

Singh J (1996) Ajmalicine (Raubasine): A medicinally important alkaloid from *Catharanthus roseus* (*Vinca rosea*). In: Handa SS, Kaul MK (eds) Supplement to Cultivation and Utilization of Medicinal Plants, Jammu-Tawi. Regional Research Laboratory, India, pp 199–206

Snoeijer W (2001) International register of *Catharanthus roseus*. Leiden/Amsterdam Centre for Drug Research, Division of Pharmacognosy, Leiden

Sreevalli Y, Baskaran K, Chandrashekara RS, Kulkarni RN (2000) Preliminary observations on the effect of irrigation frequency and genotypes on yield and alkaloid concentration in periwinkle. J Med Arom Plant Sci 22–23:356–358

Sreevalli Y, Baskaran K, Kulkarni RN, Kumar S (2000) Further evidence for the absence of automatic and intra-flower self-pollination in periwinkle. Curr Sci 79:1648–1649

Sreevalli Y, Kulkarni RN, Baskaran K (2002) Inheritance of flower colour in periwinkle: orange-red corolla and white eye. J Hered 93:55–58

Sreevalli Y, Baskaran K, Kulkarni RN (2003) Inheritance of functional male sterility in the medicinal plant periwinkle. Indian J Genet 63:365–366

SreevalliY KRN, Baskaran K, Chandrashekara RS (2004) Increasing the content of leaf and root alkaloids of high-alkaloid-content mutants of periwinkle through nitrogen fertilization. Ind Crops Prod 19:191–195

Stearn WT (1975) A synopsis of the genus *Catharanthus* (Apocynaceae). In: Taylor RW, Farnsworth NR (eds) The Catharanthus alkaloids. botany, chemistry, pharmacology, and clinical use. Marcel Dekkar, New York, pp 9–44

St-Pierre B, Laflamme P, Alarco AM, De Luca V (1998) The terminal *O*-acetyltransferase involved in vindoline biosynthesis defines a new class of proteins responsible for coenzyme A-dependent acyl transfer. Plant J 14:703–713

Sun X, Chen Y, Pan J, Wang Y, Sun K et al (2009) Analysis of catharanthine content and agronomic traits in *Catharanthus roseus*. Zhongguo Zhong Yao Za Zhi 34:128–31

Svoboda GH, Johnson IS, Gorman M, Neuss N (1962) Current status of research on the alkaloids of *Vinca rosea* Linn. J Pharm Sci 51:707–720

Svoboda GH, Blake DA (1975) The phytochemistry and pharmacology of *Catharanthus roseus* (L.) G. Don. In: Taylor RW, Farnsworth NR (eds) The Catharanthus alkaloids. botany, chemistry, pharmacology, and clinical use. Marcel Dekkar, New York, pp 45–83

Talha M, Radwan AS, Negm S (1975) The effect of soil moisture deficit on growth and alkaloidal content of *Catharanthus roseus* G. Don. Curr Sci 5:614–616

Taylor W, Farnsworth N (eds) (1975) The Catharanthus alkaloids: botany, chemistry, pharmacology, and clinical use. Marcel Dekker, New York, p 323

Thamm NK (2014) Induction and characterization of *Catharanthusroseus* mutant altered in monoterpenoid indole alkaloid biosynthesis.http://hdl.handle.net/10464/5682

Union for the protection of new varieties of plants (UPOV) (2004) *Catharanthus roseus* (L).G. Don.

Guidelines for the conduct of test for distinctness, uniformity and stability. TG/214/1Internationa, Geneva, p 21
Tiwari DP, Lodh AR, Jacob AI (1978) Disease of *Catharanthus rosus Vinca rosea)* in Jabalpur. MP Natl Acad Sci Lett 1(5):163–164
Tyler VE (1988) Medicinal plant research: 1953–1987. Planta Med 54:95–100
Van Bergen M, Snoeijer W (1996) *Catharanthus* G. Don: the madagascar periwinkle and related species. Wageningen Agricultural University Papers 96:1–20
Van der Heijden R, Jacobs DI, Snoeijer W, Hallard D, Verpoorte R (2004) The *Catharanthus* Alkaloids: pharmacognosy and biotechnology. Curr Med Chem 11:1241–1253
Vazquez Flota FA, Loyola Vargas VM (1994) A *Catharanthus roseus* salt tolerant line II. Alkaloid production. J Plant Physiol 144:613–616
Vazquez Flota F, De Carolis E, Alarco AM, De Luca V (1997) Molecular cloning and characterization of desacetoxyvindoline-4-hydroxylase, a 2-oxoglutarate dependent-dioxygenase involved in the biosynthesis of vindoline in *Catharanthus roseus* (L.) G. Don. Plant Mol Biol 34:935–948
Venkateswarlu M, Susheelamma BN, Kumar P, Subhash K (1988) Studies of induced mutation frequency in *Catharanthus roseus* (L) G. Don by gamma rays and ems individually and in combination. Indian J Genet Plant Breed 48:313–316
Venkateswarlu M, Suryanarayana N, Subhash K (1989) Effect of gamma rays and EMS alone and in combination on growth and alkaloid production in periwinkle. Adv Plant Sci 2:28–32
Verma P, Mathur AK, Srivastava A, Mathur A (2012) Emerging trends in research on spatial and temporal organization of terpenoid indole alkaloid pathway in *Catharanthus roseus*: a literature update. Protoplasma 249:255–268
Verpoorte R, Van der Heijden R, Schripsema J, Hoge JHC, Ten Hoopen HJG (1993) Plant cell biotechnology for the production of alkaloids: present status and prospects. J Nat Prod 56:186–207
Veyret Y (1974) Quelques données pour la biosystématique de pervenches malgaches (genre Catharanthus G. Don, Apocynaceae). Candollea 29:297–307
Virk SS, Singh OS, Bhullar BS (1988) Assessment of variability for different quantitative characters in periwinkle. Crop Improv 15:138–141
Whitmer S, van der Heijden R, Verpoorte R (2002a) Effect of precursor feeding on alkaloid accumulation by a strictosidine synthase overexpressing transgenic cell line S1 of *Catharanthus roseus*. Plant Cell Tiss Org Cult 69:85–93
Whitmer S van der Heijden R, Verpoorte R (2002b) Effect of precursor feeding on alkaloid accumulation by a tryptophan decarboxylase overexpressing transgenic cell line T22 of *Catharanthus roseus*. J Biotechnol 96:193–203
Xing SH, Guo XB, Wang Q, Pan QF, Tian YS et al (2011) Induction and flow cytometry identification of tetraploids from seed-derived explants through colchicine treatments in *Catharanthus roseus* (L.) G. Don. J Biomed Biotechnol Article ID 793198. https://doi.org/10.1155/2011/793198
Yamamoto K, Takahashi K, Mizuno H, Anegawa A, Ishizaki K et al (2016) Cell specific localization of alkaloids in *Catharanthus roseus* stem tissue measured with imagine MS and single cell MS. Proc Natl Acad Sci USA 113:3891–3896
Yamamoto K, Takahashi K, Caputi L, Mizuno H, Rodriguez-Lopez CE et al (2019) The complexity of intercellular localization of alkaloids revealed by single cell metabolomics. New Phytol 224:848–859
Za′rate R, Verpoorte R (2007) Strategies for the genetic modification of the medicinal plant *Catharanthus roseus* (L.) G. Don. Phytochem Rev 6:475–491
Zaidi AA, Srivastava KM, Singh BP (1984) Effect of periwinkle chlorotic stunt virus (PCSV) on phosphorus fractions of *Catharanthus roseus*. Indian J Plant Pathol 2(1):55–58
Zhao J, Verpoorte R (2007) Manipulating indole alkaloid production by *Catharanthus roseus* cell cultures in bioreactors: from biochemical processing to metabolic engineering. Phytochem Rev 6:435–457
Zhonghua T, Yanju L, Xiaorui G, Yuangang Z (2011) The combined effects of salinity and nitrogen forms on *Catharanthus roseus*: the role of internal ammonium and free amino acids during salt stress. J Plant Nutri Soil Sci 174:135–144
Zu Y, Nie M, Yu J, Tang Z, Wang H et al (2005) Protein and cDNA sequences of lipid transfer protein LTP from *Catharanthus roseus* and use in enhanceddrought-resistance in transgenic plants. Faming Zhuanli Shenqing
Zu Y, Nie M, Yu J, Tang Z, Wang H et al (2006) Metallothionein type II (MT-II) gene from *Catharanthus roseus* and its uses. Faming Zhuanli Shenqing Gongkai Shuomingshu, CN 1824780 A 20060830

Somatic Embryogenesis in *Catharanthus Roseus:* Proteomics of Embryogenic and Non-embryogenic Tissues; and Genome Size Analysis of Regenerated Plant

4

A. Mujib, Moien Qadir Malik, Yashika Bansal, Rukaya Syeed, Bushra Ejaz, and Jyoti Mamgain

Abstract

The asexual, non-zygotic (somatic) embryogenesis was reported over 60 years ago. It is a unique process in which somatic cell transforms into embryo in cultured condition affirming cellular totipotency. Since its first report, somatic embryogenesis has been noted in different plant genera of a dicot, monocot, and lower fern groups. In the present chapter, an attempt has been made in describing somatic embryogenesis and its current status in periwinkle, *Catharanthus* roseus—an immensely important medicinal plant. The plant has anticancerous properties, the yield of alkaloids specific to this disease is however low and needs enrichment. The early embryogenic cells could serve as an important tissue for generating new cell lines with improved yield and related traits. Here, the influence of explants, the involvement of plant growth regulators, and the role of other regulating factors in establishing *Catharanthus* embryogenic tissues are discussed. The somatic embryo development, structure, progress, i.e., maturation, germination of embryo are also discussed with the help of Scanning Electron Microscopic method. The secondary somatic embryogenesis cycle, the embryo irregularities, and the role of protoplast culture in somatic hybridization is briefly deliberated taking *Catharanthus* as an example. Liquid overlaying at varying levels are described of which 0.50 ml overlaying on solid medium improved embryogenic growth and embryo number significantly. Successful artificial seed production, storage at low temperature, and plant regeneration are discussed by optimizing sodium alginate (2.5%) and calcium chloride (100 mM) solution. The proteomic approach has recently been exploited to investigate non-embryogenic, embryogenic tissues, embryo development, and other events at molecular level. In *Catharanthus,* over 1000 new proteins were identified from embryogenic callus, which was absent in non-embryogenic tissues. Some important proteins identified at various stages of embryogenesis are catalase, glutathione S transferase, superoxide dismutase, heat shock protein 70 (HSP70), HSP 90, and cytochrome C. Conservation of embryogenic tissues is important as it produces embryos and plantlets continuously. The cryopreservation protocol of *Catharanthus* is discussed by mentioning the role of osmotica, various cryoprotective

A. Mujib (✉) · M. Q. Malik · Y. Bansal · R. Syeed · B. Ejaz · J. Mamgain
Cellular Differentiation and Molecular Genetics Section, Department of Botany, Jamia Hamdard, New Delhi, India

© The Author(s), under exclusive license to Springer Nature Switzerland AG 2022
C. Kole (ed.), *The Catharanthus Genome*, Compendium of Plant Genomes,
https://doi.org/10.1007/978-3-030-89269-2_4

agents, and the exposure of dehydration. Somatic embryo regenerated plants are true to type when developed directly on explant, but various types of variations are noted in plants developed from callus, suspension, and similar other differentiated tissues. In *Catharanthus,* the genome size or the 2C DNA of somatic embryo regenerated plant was noted to be 1.516 pg, which is similar to field grown plant DNA. No alteration in genome size was noted although the plants were produced through callusing. All this accumulated information on *Catharanthus* will be very valuable in practical and basic applications as it broadens the knowledge of cellular differentiation and dedifferentiation processes; at the same time, the methodology will augur fast multiplication of raw materials for the synthesis and enrichment of alkaloids.

4.1 Introduction

Somatic embryogenesis (SE) is a process where somatic or vegetative cells acquire embryogenic potentiality and form embryos under cultural conditions (Thorpe 2012). It mimics zygotic embryogenesis and a zygotic 'embryo-like' structure forms on committed single or multiple cells based on the principle of cellular totipotency (Osuji et al. 2016; Mendez-Hernandez et al. 2019). During somatic embryogenesis, the sexual gametic fusion is entirely averted; the predetermined somatic cell of explant and callus undergoes molecular, biochemical, and other alterations in acquiring embryogenic ability, very similar to the developmental pathways of zygotic embryogenesis. Various factors trigger the reprogramming of somatic cells into embryos by the involvement of genes, proteins, transcription factors, stress, and other complex networks (Gulzar et al. 2019). The embryogenic cells have unique features, morphologically and biochemically different from non-embryogenic cells/tissues. These cells are small, isodiametric, less vacuolated, darkly stained nuclei with bigger nucleolus, elevated nucleus to cytoplasm ratio with reduced levels of heterochromatin. The embryogenic cells also have dense cytoplasm full of organelles, contain profusely accumulated starch, and all the nucleic and cytoplasmic matters are surrounded by cellulose-made thick cell walls (Campos et al. 2017; Silva de Almeida 2020). Somatic embryos are formed directly on explants (without any preceding callus formation) in which the committed cells of explants transformed into an embryo after receiving a reprogramming signal, commonly called direct embryogenesis. Indirect embryogenesis on the other hand is manifested in two different stages; firstly the callus is initiated from cultivated tissues/explants and followed by embryo formation on callus (Pulianmackal et al. 2014). In dicotyledonous genera, the globular, heart, torpedo, and cotyledonary embryo stages are formed in sequential order. Globular, scutellar, and coleoptilar structures are the common embryo forms of monocot genera; in conifers, juvenile or immature stage, pre-cotyledonary, early and late cotyledonary embryo stages have been reported (Mujib and Samaj 2006; Guan et al. 2016). Mikula et al. (2015) recently noted a few different stages of somatic embryos in lower group of plants like fern, these stages are called (a) linear stage (up to proembryo stage containing many cells), (b) early embryonic leaf stage (until first leaf appearance) and (c) late embryonic leaf stage (along with second leaf). The somatic embryos are formed directly on cultured tissues or in intervening callus termed as primary somatic embryogenesis while second cycle of embryo formation is also noticed on primary embryo structures regarded as secondary somatic embryogenesis (Naing et al. 2013). Like zygotic embryos, the somatic embryos show bipolar morphological attributes with root and shoot ends; but are devoid of any vascular connection with parental entity, unlike zygotic counterparts and other plant parts/organs (von Arnold et al. 2002; Delporte et al. 2014). As soon as the embryos attain cotyledonary stage, the shoot meristem with two leafy primordia starts to appear and embling growth commences (Yang and Zhang 2010). Thus, the process represents and reaffirms the establishment of the totipotency model of plant regeneration. The entire process is unique and is regulated by a variety of complex signaling networks including the expression of specific groups of genes in response to external and internal triggers. Some of the important signaling elements identified are the plant growth regulators especially auxin, and stress induced by osmotic shock, drought, low/high temperature, heavy metal, and similar other in vitro conditions (Nic-Can et al. 2016). Once the cells' fate is decided a distinct initiation, multiplication, maturation, and embryo germination stages are noted in different plants, the last two steps are, however, overlapping in cases (Hu et al. 2017).

This vegetative or somatic embryogenesis process has several practical and fundamental applications. The plant regeneration from single or aggregated somatic cells is the primary objective of this process and is therefore an efficient alternative way of rapid vegetative propagation method (Hofmann 2014; Guan et al. 2016). Besides bulk production of propagules in solid/liquid especially in bioreactor (Mujib et al. 1995, 2014), the process has been exploited in production of artificial seeds, cryopreservation of embryos/axes, protoplast isolation, for developing transgenics, genetic engineering/improvement of traits and in fundamental academic studies (Vikrant and Janardhanan 2018). Although the process has been extensively studied in several model plants like Arabidopsis, carrot, Medicago, coffee, Gossypium, the technique has now been extended to non-model genera of ornamentals, fruits, vegetables, fibers, oil -, timber yielding, medicinals and other diverse groups of plants of economic importance.

4.2 *Catharanthus* Embryogenesis

Catharanthus roseus L.(G.) Don, popularly known as periwinkle, is a medicinal and ornamental plant belonging to the Apocynaceae family. Among various Phyto-compounds isolated, vinblastine and vincristine are valuable compounds as these alkaloids are administered to treat various cancers such as leukemias, solid tumors, Hodgkin's disorder, etc. (Mujib et al. 2012). The alkaloids are extracted from different plant parts especially from aerial shoots of *Catharanthus*, the content is, however, very poor (0.0005%). The plant cell culture technique has been suggested to be an efficient alternative for extraction and enrichment of alkaloids in which embryogenic tissue could serve as a potential source of embryos, plants, and alkaloids (Mehpara and Mujib 2017; Ali et al. 2019). Since the first observation of Steward, the incidence of somatic embryogenesis has been reported in a large number of plants of monocot, dicot, conifer, and in lower group of organisms (Mujib 2016). In *C. roseus,* somatic embryogenesis and plant regeneration was similarly reported (Junaid et al. 2006). The rapid multiplication of *Catharathus* through embryogenesis is important for fast generation of raw materials as the plant demonstrates strong anticancer properties. One basic application of this process is that the embryogenic mother cell can be exploited as a single cell for genetic improvement of cells, in developing transgenics to improve alkaloid yield, and in studies of plant differentiation and development process. The embryogenic tissues grown in suspension or bioreactor may also be used as a good source of viable protoplasts for somatic hybridization (Calovic et al. 2019). Besides bulk production of plants, the same somatic embryos can also be used to make synthetic seeds. The various facts, factors controlling somatic embryogenesis and the uses/applications of embryogenic tissues/embryos have been summarized in *Catharathus* in the following sections.

4.3 Explants, PGRs, Callus and Embryo Differentiation from Embryogenic Callus

The somatic embryogenesis potentiality is primarily dependent on various factors like mother plant's physiology, season of harvesting, type of explants, cultural media composition and endogenous level, and exogenous amendment of plant growth regulators (Boutchouang et al. 2016). In *C. roseus* different plant parts like stem, leaf, root, and hypocotyl of in vitro germinating seeds were used as explants for establishing cultures. The MS medium was added with various plant growth regulators such as *2,4-Dichlorophenoxyacetic acid (*2,4-D), Chlorophenoxyacetic acid (CPA), *2,4,5*-Trichlorophenoxyacetic acid (2,4,5-T), Naphthaleneacetic acid (NAA) either alone or in combinations with cytokinin like 6-Benzylaminopurine (BAP) and N^6-furfuryladenine (Kinetin). The calli were induced from all explants with variable intensities and the most efficient treatment contained 2.0 mg/l 2,4-D which profusely produced callus. The callus derived from stem, leaf, and root was noted to be non-embryogenic (Fig. 4.1a), never produced embryos in different tested conditions while hypocotyl obtained callus transformed into embryogenic callus (Fig. 4.1b), produced embryos (Fig. 4.1c) and plantlets (Fig. 4.1d, e) in various NAA and BAP concentrations. Wu et al. (2020) noted variable and higher rates of embryogenic callus and somatic embryos in *Ormosia henryi* from mature zygotic embryos than the other tested explants like stem, cotyledon, leaf, and hypocotyl. Similar explant, genotype, season dependent somatic embryogenesis was reported in several investigated plants (Saeed et al. 2019; Mostafa et al. 2020). Two–three weeks old hypocotyl callus was placed on media added with various plant growth regulators (PGRs) combinations for differentiation of embryos. Almost all the tested conditions induced embryos at variable numbers. The addition of 1.0 mg/l NAA along with 0.5–2.0 mg/l BAP was more effective in inducing embryos in which over 90 somatic embryos were formed in 40–50 mg embryogenic tissue. A heterogeneous mixture of globular, heart, torpedo, and cotyledonary embryos was differentiated at upper surface of the embryogenic callus. Somatic embryos of different development stages and successful plant regeneration from embryos were observed in different other plant genera (Konar et al. 2018; Al-Sahara et al. 2020).

Fig. 4.1 Callus induction and plant regeneration in *Catharanthus roseus* through somatic embryogenesis. **a** non embryogenic callus, **b** embryogenic callus, **c** developing embryos, **d** embryo derived plant **e** regenerated plant (bar **a–b**: 0.5 cm; **c–d**: 1 cm; **e**: 2 cm)

4.4 Scanning Electron Microscopy (SEM) Study of Embryogenic Callus

The hypocotyl-induced embryogenic callus with differentiated entities was studied through SEM, which reveals morphology, structure, and other features of callus and somatic embryos more clearly. The direct and indirect origin, the tissue, the single cell or group of cells in proembryogenic mass on which the embryos are formed can easily be investigated by the use of SEM (Schlogl et al. 2012; Junaid et al. 2014; Sangra et al. 2019). For SEM study, the embryogenic callus was fixed in 2% glutaraldehyde, post-fixed in 1% osmium tetroxide, dehydrated in ethyl alcohol, later coated with gold palladium and the samples were photographed for study. The SEM study indicated the presence of different, i.e., globular, heart, torpedo, and cotyledonary stages of somatic embryos, typical to dicot groups of plants. In *Catharanthus*, the advanced somatic embryos were cylindrical, elongated structure had apical notch, which later turned into the first pair of rudimentary leaves; the basal part of embryo with root end embedded in callus finally develops into the root. The use of SEM has been used in studying embryogenic callus, embryo development and in describing embryo morphology in many studied materials (Popielarska et al. 2006; Ulisses et al. 2010). The SEM and Transmission electron microscopy (TEM) technologies, along with histological preparations thus, are important in studying the ontogeny and progress of somatic embryo, mimicking zygotic embryogenesis.

4.5 Secondary/Adventive Embryogenesis

Secondary or adventive somatic embryogenesis is a process when the second cycle of embryo formation takes place on primary somatic embryos. The process has several applications as it increases in vitro propagation potential by maintaining repeated continuous cycle of embryogenesis in laboratory for prolonged, extended time (Lelu-Walter et al. 2017; Reeves et al. 2017). In *C. roseus,* the adventive or secondary somatic embryos were frequently formed on primary somatic embryos. The incidence of secondary embryogenesis was more on solid MS compared to liquid culture medium. The same combinations of PGRs induced secondary embryos, on which primary somatic embryos were formed earlier. The various combinations of BAP and NAA were noted to be very efficient in inducing adventive embryos and on average three to four early developing stages (globular, heart) embryos were noted on primary structures. The frequency of advanced embryo forms like cotyledonary was low. The occurrence of adventive embryos, aggregation/clustering of embryos, and repeated callus formation on embryos prevented primary embryo progression and germination into plantlets. Secondary somatic embryogenesis and plant regeneration have been achieved in a number of investigated plant species of dicot plants like black pepper, Brassica, M*edicago*, Siberian ginseng, cassava, and white clover (Kim et al 2012; Yang et al. 2013; Gautier et al. 2018) but unfortunately, the information is not too many in monocotyledonous plants (Mujib et al. 2013).

4.6 Somatic Embryo Irregularities

The somatic embryogenesis has been noted in many plants and the induced embryos are called bipolar entities with shoot and root ends. Besides this typical embryo structure, several abnormal embryos have been observed in culture in many investigated plants (Hashemloian et al. 2008). The abnormality rate of somatic embryo is also very high in cases in contrast to uniform and synchronous embryos present in zygotic embryogenesis. These unwanted abnormalities hinder embryo progression, reduce plant conversion from embryos, and thus the potentiality of embryogenesis in obtaining plants is drastically lowered. In *Catharanthus,* a few distinct classes of embryo abnormality was noted in culture, these are (i) multiple fusing somatic embryos at base, (ii) embryos with shoot ends without any root primordia, (iii) presence of root axis but no shoot axis in somatic embryos and (iv) callusing embryos, i.e., embryos showing callusing. The various triggers causing embryo abnormality need to be identified early and are found to be dependent on genotype and plant growth regulators used in culture. Sangra et al. (2019) recently reported monocotyledonous, polycotyledonous, fasciation, and vestigial cotyledons types of somatic embryo abnormalities in alfalfa in high numbers. These somatic embryo abnormalities are manifested with morphological and physiological disorders caused by epigenetic and other mutations which influence embryo development programs (Xiao 2006; Landey et al. 2015). Various types of auxins are added to culture medium for inducing somatic embryos but these PGRs generate mutations through genetic, epigenetic, and global/ local methylation alteration of DNA in cells (Feher 2015; Fraga et al. 2015). The auxin is are therefore directly linked to the formation of aberrant embryos (Pescador et al. 2008). In *Catharanthus,* auxins like 2,4-D and NAA were used to induce embryos, these perhaps responsible for aberrant somatic embryo formation. The embryos cultured on PGR free medium improved somatic embryo quality; the use of 1.0 mg/l NAA and 0.5 mg/l BA generated less aberrant somatic embryos compared to other higher levels. Similarly, the addition of 1.0 mg/l IBA in medium produced more roots in embryo when placed for rooting, this rooting was even more prominent in activated charcoal added medium. The amendment of GA_3 and ABA had similar promotive influence on embryo maturity and germination. The recovery of embryo irregularities and plantlet germination frequency may partly be restored by the use of other sets of PGRs, and no morphological alterations were observed among regenerated populations.

4.7 Embryo Maturation and Plantlet Development

The maturation and germination of embryos is a crucial step in somatic embryo-based propagation in getting large population of plants. It is primarily dependent on quality embryo formation, synchronous development, and progression of embryos. The typical embryo (without any aberration and secondary embryogenesis incidence) with physiological reserves (similar to seed embryos) regenerated into normal plants. In recent times, various supplements have been amended to media for improving embryo quality and for better germination process; some of the used treatments and compounds are gibberellic acid (GA_3), abscisic acid (ABA), polyethylene glycol (PEG), activated charcoal (AC), sugar, sugar-alcohol, use of freezing at low temperature, etc. (Lipavska and Konradova 2004; Robichaud et al. 2004; Tikkinen et al. 2018; Tvorogova and Lutova 2018). It is believed that PEG, ABA, and similar other compounds reduce osmotic potential by limiting moisture uptake and generates desiccation linked drought stress in improving maturation and germination (Tikkinen et al. 2018). Several genes were expressed due to stress caused by auxins and ethylene homeostatis and other stress related transcription factors in establishing embryogenesis (Cao et al. 2017). In *Catharanthus*, selectively picked advanced cotyledonary somatic embryos (average 5–6 mm length) were transferred to GA_3 added MS medium. The embryos grew in size and turned chlorophyllous within one or two weeks' time—a foremost sign of photosynthetic autotroph. These photosynthetic pigments help make important sugars, proteins, amino acids, lipids, triglycerides like storage products which improve the overall physiology of developing embryos in the germination process. Many of the embryos were coiled, green structures with distinct root end. During these developmental stages, a few common embryo types were found (i) embryos

Table 4.1 Explant/tissue, plant growth regulators (PGRs), and various responses in *C. roseus*

Explant / tissue	PGRs (mg/l)	Responses
Hypocotyl	2,4-D/ CPA (1.0–2.0 mg /l)	Callus and embryogenic callus
Embryogenic callus	BAP (1.5 mg/l) + NAA (1.0 mg/l)	Formation of somatic embryos
Callus with embryos	Liquid overlaying containing BAP (1.5 mg/l) + NAA (1.0 mg/l)	More and more numbers of somatic embryo (proliferation)
Somatic embryos	GA_3 (1.0 mg/l)	Maturation of somatic embryos
Matured somatic embryos	BAP (0.5 mg/l), solid or liquid MS	Germination of embryos into plants

with shoots, without root ends, (ii) embryos with root ends but devoid of shoot primordia, and (iii) the typical bipolar embryos (with distinct shoot and root axis). These matured stages of embryos were germinated into plantlets and the scheme is presented in Table 4.1. In *Allium*, the addition of 0.5 mg/l GA_3 showed positive impact on maturation and germination of embryos (Nasim et al. 2010). Improved embryo quality with better maturation and germination was noted in ABA, maltose added media in many investigated plants (Corredoria et al. 2003; Maruyama et al. 2007; Tikkinen et al. 2018). The various BAP concentrations were noted to be more responsive compared to kinetin in embryo germination process. The rate of plant conversion showing shoot and root growth was more in 0.5 mg/l BAP added liquid MS medium. Several germinated embryos lacked rooting ends or had very little root axes; media added with indole-3-butyric acid (IBA) and indole-3-acetic acid (IAA) promoted root development in cultured somatic embryos. It appears therefore that diverse groups of compounds control development and germination of embryos in in vitro plant development process.

4.8 Liquid Overlaying

In *C. roseus*, the liquid overlaying process improved embryo proliferation, maturation, and germination of embryos remarkably. The liquid overlaying may be described as a technique to use liquid MS over solid medium as a thin layer. Based on the nature of callus and culture, full or half strength liquid medium (composition is same as that of solid medium, except agar) with or without plant growth regulators were used in different investigated plant genera (Nitayadatpat and Te-chato 2005; Pullman and Skryabina 2007). Here, in *C. roseus*, varying levels of liquid overlaying, i.e., 0.25, 0.50, 0.75, and 1.0 ml were poured on a solid medium to improve the growth of embryogenic tissues. Liquid overlaying of 0.50 ml increased embryo number significantly. The other levels (0.75 and 1.0 ml) also had positive influence on embryo maturation and germination of embryos, even liquid overlaying at 0.50 ml reduced embryo to plant recovery time. Improved embryogenic callus growth may be due to better absorption and uptake of nutrient medium by the cultured tissues present on semisolid MS medium. Liquid overlaying helps maintain pH, offers better nutritional and PGRs requirements in in vitro grown tissues (Pullman et al. 2009). Liquid overlaying is considered to be a temporary immersion technique where a thin film of medium is applied which promotes cultural growth in several studied plant species like rubber, Pyrus, Anthurium, etc. (Sushamakumari et al. 2000; Berthouly and Etienne 2005; Nitayadatpat and Te-chato 2005). Liquid overlay technique is a mix of methods, uses the benefits of solid and liquid medium and the limitations of both are discarded.

4.9 Encapsulation of Somatic Embryos in Making Artificial Seeds

Synthetic or artificial seed is produced by encapsulating embryos or other plant parts such as shoot bud, meristem, axillary bud, callus, cell aggregates, in nutrient medium with various gelling agents like sodium alginate, calcium chloride, sodium pectate, carrageenan, potassium alginate (Gantait et al. 2015). This encapsulation technology is important in mass multiplication of plants, which suffer from sexual incompatibilities. Two different types of synthetic seeds have been described in literature; the first type is desiccated and the other one is hydrated synthetic seeds. Both types have technical advantages and limitations. The poor embryo respiratory gas exchange through the protective covers is an important limitation in hydrated synthetic seeds while desiccation technique reduces embryo viability by reducing internal moisture in capsule, In medicinal plant category, the technique has successfully been used in several investigated plants such as in *Tylophora indica* (Gantait et al. 2017), oil palm (Palanyandy et al. 2015), *Rumex vesicarius* (Nandini et al. 2014), *Daucus carota* (Latif et al. 2007), *Allium sativa* (Bekheet 2006), *Adhatoda vesica* (Anand and Bansal 2002), and other medicinal plants. In *C. roseus,* successful somatic embryo encapsulation method was conducted for making synthetic seeds (Mehpara et al. 2012). The advanced-stage somatic embryos (torpedo and cotyledonary) obtained from hypocotyl embryogenic callus were used for encapsulation. Different concentrations of sodium alginate (2.0–4.0%) and calcium chloride (75, 100, 125 mM) were used in the encapsulation process in which the former acts as endosperm, offers nourishment to the developing embryos within beads, while the calcium-made outer cover prevents or minimizes embryos from injuries during handling. All the used combinations produced artificial seeds but the best bead formation with uniform size was noted in 2.5% sodium alginate and 100 mM calcium chloride mixed encapsulation process. The artificial seeds containing embryos germinated and developed into plantlets (Table 4.2) with fair success in BAP and NAA medium, added with 3% sucrose.

The synthetic seeds can also be stored at low freezing temperatures. Different temperature regimes were evaluated for storage ability of synthetic seeds; Storage at 4 °C was noted to be the best temperature for synthetic seeds. Although the longer storage time reduced germination ability, the synthetic seeds showed viability and germinated into plantlets even after 10 weeks of low temperature storage. The 4 °C preserve the encased embryos well and the recovered developing plants were phenotypically identical to the field grown *Catharanthus* plants without any morphological variations.

4.10 Cryopreservation of Embryogenic Tissues

Cryopreservation of cells, tissues, organs, and even DNA is immensely valuable technique for conservation of elite and normal germplasm,

Table 4.2 Synthetic/artificial seed development conditions from somatic embryo and plant regeneration

Tissue	Sodium alginate (A), calcium chloride (B), PGRs (mg/l), and other conditions	Responses
Advanced somatic embryos	A (2.5%) + B (100 mM); 15 min exposure	Synthetic seed or bead formation
Synthetic seeds	4 °C (temperature)	Storage of synthetic seeds up to 10 weeks or more
Synthetic-or stored synthetic seeds	NAA (0.25) + BAP (0.25)	Plant development from synthetic seeds

Table 4.3 Cryopreservation protocol of *Catharanthus roseus*

Tissues	Different conditions with cryogenic agents	Responses
Embryogenic cell suspension	Sucrose and 5% DMSO	Cell viability% high
Survived cells	Sucrose and 5% DMSO	Regrown cell colonies (mean number) were highest
Embryogenic cells suspensions	Sucrose, 5% DMSO, and 5% glycerol	Cell viability% higher
Survived cells	Sucrose, 5% DMSO, and 5% glycerol	Regrown cell colonies were high

very similar to tissue banking (Tahtamouni et al. 2015). Organogenic and embryogenic callus, suspensions, zygotic axes storage at low temperature is even more precious (Kundu et al. 2018). *C. roseus* is a plant with anticancerous properties. In order to extract alkaloids regularly, the embryogenic tissue of *C. roseus* which produced embryos and plantlets was used for preservation. Before being placed in liquid nitrogen (LN), the embryogenic tissues were exposed to ormotica and cryoprotective agents to protect cells from low temperature injury. Sugar and sorbitol were used as osmotica; these compounds provide energy and generate stress by limiting water level during successful preservation (Yun-peng et al. 2012). Various compounds have been used as cryoprotective agents to protect tissues from injuries; these are like polyethylene glycol (PEG), dimethyl sulphonate (DMSO), glycerol, etc. (Reed and Uchendu 2008). In *C. roseus,* PEG, DMSO, and glycerol were used as cryoprotective agents (Table 4.3). Individually, the sugar was more efficient in improving callus biomass compared to sorbitol. The combination of 0.4 M sucrose + 5% PEG showed a maximum of 55.23% viability after cryopreservation. The regrowth % (77.7) was pretty high and showed about nine cell colonies/plate. Varied tissue dehydration period was applied to minimize moisture content in tissues. The highest viability was observed when 2 h dehydration period was used. The survived calli grew in size, transformed into embryos in NAA (0.5 mg/l) and BAP (0.5 mg/l) amended medium. After successful preservation, the survived tissues continuously produced somatic embryos/plants in numbers, very identical to hypocotyl derived embryogenic tissue. Fairly successful cryopreservation was achieved in a number of flowering taxa and lower groups of organisms (Rowntree et al. 2011; Kundu et al. 2018; Coelho et al. 2020). The plant obtained through cryopreservation showed same genome size as that of field grown plants. The process may facilitate in developing transgenic, helps make artificial seeds, and generate mutants for improvement purposes. This established protocol may be exploited to other groups of medicinal plants for extraction of alkaloids.

4.11 Protoplast Culture from Embryogenic Cells

Since first isolation of protoplasts (Cocking 1960) a number of protoplast isolation and fusion information has been reported in a number of plant genera including medicinal plants (Borgato et al. 2007; Thomas 2009). Protoplasts are isolated from diverse plant parts such as leaf, hypocotyl, cotyledon, root, and in vitro grown cultured tissues like callus and suspension (Yang et al. 2007). The leaf mesophyll cells and embryogenic callus/suspension are noted to be the most important source of protoplasts (Guo et al. 2006). Beside somatic hybridization applications, the use of protoplast has advantages in several research areas like differentiation/ dedifferentiation, somatic cell genetics, in integrating foreign DNA and cell organelles. Protoplasts are

Table 4.4 Embryogenic suspension of *Catharanthus roseus* was incubated (12 h) with different enzymes for isolating protoplasts

Tissue used	Different enzymes, PGRs, and other conditions used	Responses (protoplast yield, viability; callus, somatic embryo formation)
Embryogenic suspension	(i) Cellulase, pectinase, and macerozyme	Yield and viability is high
	(ii) Cellulase, pectinase, macerozyme, and driselase	Yield, viability is higher and more
	(iii) Cellulase, pectinase, macerozyme, driselase, and sorbitol (as osmoticum)	Best yield and viability of protoplasts
Isolated protoplasts	0.5 mg/l NAA and 0.5 mg/l 2,4-D	Microcolony and callus formation
Protoplast derived embryogenic callus	1.0 mg/l NAA and 1.5 mg/l 2,4-D	Somatic embryo formation and later plant development

also used for broadening genetic base and improving genetic diversity by facilitating fusion of two or more similar/dissimilar genetic constitutions (Davey et al. 2005).

Cell suspension divided rationally and produced cell aggregates and colonies fast in medium (Shrestha et al. 2007). Thus, the embryogenic cells or tissues especially in suspension were used frequently in obtaining suitable protoplast density (Mahanom et al. 2003). In *C. roseus* hypocotyl-induced embryogenic tissue was used for obtaining protoplasts. The embryogenic callus tissues were kept in liquid MS to obtain a suspension, the suspension was later exposed to various enzymatic treatments and the various steps of protoplast isolation process are presented in Table 4.4. The enzyme combinations of cellulase (2.0%), pectinase (1.0%), macerozyme (0.02%) and driselase (0.50%) generated highest yield of protoplasts (37.2×10^5) with about 65% viability. In 0.6 M sorbitol and 100 mM $CaCl_2$ added medium, the yield of protoplast was more and protoplast viability was even better (70%). The combination of enzymes was noted to be very effective in producing protoplasts in oil palm (Te-Chato et al. 2005). Similarly, the use of enzymes mixture in digestion solution produced viable, good yield of protoplasts in many investigated plants (Pan et al. 2003). The isolated protoplasts divided fast produced micro-colonies (12–13/Petriplate) and callus in 0.50 mg/l NAA and 0.50 mg/l 2, 4-D added medium. The protoplast derived embryogenic callus biomass (fresh weight) was poor compared to hypocotyl obtained embryogenic callus. The protoplast derived embryogenic callus produced somatic embryos, very similar to hypocotyl obtained embryogenic callus. The somatic embryos later germinated and developed into plants. The plant recovery time from protoplast took longer time (about 40 weeks) than the normal hypocotyl based somatic embryogenesis pathway. The protoplast technology using embryogenic culture has thus many advantages including the genetic modifications of cells by uptaking DNA and other foreign genetic materials. The process can overcome the incompatibility of sexual fertilization in making somatic hybrids and cybrids of higher or lower groups of plants/organisms with improved traits.

4.12 Proteomic Study of *Catharanthus* Somatic Embryos

Beside morphological, biochemical and other traditional methods, protein and proteomic approach has recently been exploited to investigate somatic embryogenesis at molecular level in diverse plant groups including medicinals (Tchorbadjieva 2016). Except for a handful of observations, the role of protein/s in acquiring embryogenic ability, formation, and progress of embryo development to platelet information is inadequate and needs updation (Li et al. 2010; Vale et al. 2014). In *Catharanthus,* polyacrylamide gel electrophoresis study of callus and various stages of embryos was conducted and compared in which 38 and 33 kD proteins were noted to be present in torpedo staged embryos, and these proteins were off in other tested samples and embryos (Ilah et al. 2009). The presence or disappearance of specific proteins will reaffirm marker linked embryogenesis in identifying specific stages of tissues and embryos. Currently, a non-gel electrophoresis technique liquid chromatography-mass spectroscopy (LC–MS) has been used to study proteins molecularly (Heringer et al. 2015; Jing et al. 2016; Gulzar et al. 2021a, b). The process is far more accurate, sensitive, and reproducible than the conventional gel technique in identifying proteins even present in low abundance. Gulzar et al. (2019) investigated SE related proteins and a comparative proteomic map was prepared in *C. roseus* using gel free shotgun method. A large number of proteins (over 3500) were found in two contrasting tissues of embryogenic and non-embryogenic callus. In embryogenic callus, over 1000 proteins were identified which were absent in non-embryogenic callus. Similarly, about 980 proteins were observed in non-embryogenic tissues while about 1500 proteins were common and present in both tissues. Some important enzymes/proteins identified at various stages of embryos/embryogenesis are pyruvate dehydrogenase, catalase, glutathione S transferase, superoxide dismutase, heat shock protein 70 (HSP70), HSP 90, succinate dehydrogenase, and cytochrome c. About 17% identified proteins were located in cytosol/cytoplasm, 10% in cytoplasmic membrane, 4% in ribosome, another 4% in nucleus, and over 50% identified proteins' locations are still not known. Most of the identified proteins were housekeeping nature, play different metabolic, molecular roles, and act in stress responses. Some prime and key functions played by these identified proteins are catalytic activity, nucleotide -, metal-, protein and RNA binding activity, and other cellular functions regulating activities. Besides embryogenesis, proteomics has been exploited in other facets of plant science including in investigating proteins during secondary metabolite biosynthesis. Jacobs et al. (2005) studied novel proteins that appeared during alkaloid biosynthesis and identified two isoforms of strictosidine synthase, tryptophan synthase and 12-oxophytodienoate reductase which catalyze various biosynthetic steps of alkaloid synthesis in *C. roseus*.

4.13 Genome Size of Somatic Embryo Regenerated Plant

Somatic embryo regenerated plants are true to type when produced on explant directly, but nucleotide mutations and chromosomal irregularities like aneuploidy, polyploidy have been identified in callus, suspension and other differentiated tissue derived plants (Liu et al. 2013). The aberrations are more prolific in tissues kept for longer time (Krishna et al. 2016). Besides genetic homogeneity, the study of genome is important in investigating taxonomy, the similarity and dissimilarity, and in establishing evolutionary links among plants and plant groups (Doležel et al. 2007). Genome size analyses of inter and intra plant groups can fix taxa, resolve low-level taxonomy issues, draw evolutionary relationships by complementing chromosome numbers and help in speciation (Loureiro et al. 2010; Alizadeh et al. 2015). Because of precision and other advantages, the flow cytometry technique has been efficiently applied in several tissue cultured raised plants for measuring genome size and homogeneity of regenerated populations (Zafar et al. 2019; Malik et al. 2020). In *C. roseus,* the somatic embryo derived plants were achieved from hypocotyl-induced embryogenic callus (not directly on explant), which was maintained in laboratory for months. Many of the produced embryos were also aberrant, thus there is a possibility of induction of somaclonal variants or aberrant population, which may show varied DNA content. The 2C DNA content of the somatic embryo obtained plant was measured and compared with field grown *Catharanthus* plant by using flow cytometry (Fig. 4.2). The genome size or the 2C DNA of regenerated plant was 1.514 pg similar to field grown plant. The 2C DNA content of embryo regenerated plants and field grown plants are about 1482.64 mega

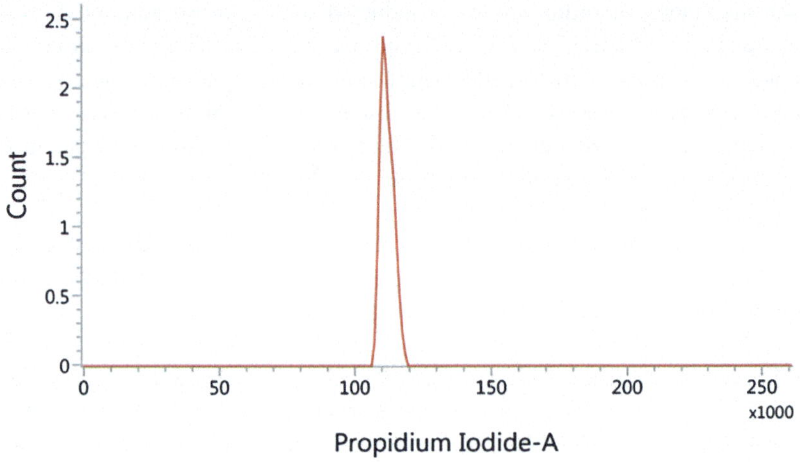

Fig. 4.2 Flowcytometric peak of nuclear 2C DNA content, obtained from in vitro regenerated *Catharanthus roseus* leaf

billion base pairs (Mbp). No alteration in genome size was noted in regenerated populations although the plants were produced indirectly through callusing. Moreover, a large number of signaling elements like PGRs / auxins, ethylene, culture-induced reactive oxygen species often cause injuries and damage cellular DNA. Guimara˜es et al. (2012) studied chromosomes by using DNA/chromatin and fluorescence in situ hybridization (FISH) of rDNA techniques; the genome size was also measured by flow cytometry in *C. roseus* (L.) G. Don cv. Little Bright Eye—a seed derived plant of Australian source. The plant contains $2n = 16$ chromosomes, of which two are metacentric, four are subtelocentric and another two are telocentric chromosomes, a nucleolus organizer region (NOR) was noted to be present in chromosome number 6. The nuclear genome size of *C. roseus* was also estimated and the 1C-value was calculated to be 0.76 pg or equivalent to 738 Mbp. Here, in somatic embryo regenerated *Catharanthus*, the DNA or genome size of embryo regenerated plant was not perturbed by culture-induced stress and the content of DNA was the same as that of field grown plant. The conducted experimental work reconfirms observation of other studies showing genetic homogeneity of embryo regenerated plants (Ali et al. 2016; Raji et al. 2017). The establishment of *C. roseus* genome size will help in cytogenetic mapping, understanding genes and enzymes in alkaloid synthesis, and in drafting total nucleotide sequences (Guimara˜es et al. 2012).

Ethical Approval This article did not involve any experiments on human volunteers or animals.

Acknowledgements The authors are thankful to the Department of Botany, Jamia Hamdard, New Delhi for receiving various facilities. The first author (AM) is grateful to the past and present Ph. D students for their inputs and timely help.

Authors' Contributions All the authors helped in developing the manuscript. AM wrote and edited the manuscript.

Funding Not applicable.

Conflict of Interest There is no conflict of interest in this manuscript, the corresponding author certifies.

References

Ali M, Mujib A, Tonk D, Zafar N (2016) Plant regeneration through somatic embryogenesis and genome size analysis of *Coriandrum sativum* L. Protoplasma 254:343–352

Ali M, Mujib A, Gulzar B, Zafar N (2019) Essential oil yield estimation by Gas chromatography–mass spectrometry(GC–MS) after Methyl jasmonate (MeJA) elicitation in in vitro cultivated tissues of *Coriandrum sativum* L. 3Biotech,9:41, https://doi.org/10.1007/s13205-019-1936-9

Alizadeh M, Krishna H, Eftekhari M, Modareskia M, Modareskia M (2015) Assessment of clonal fidelity in micropropagated horticultural plants. J Chem Pharm Res 7:977–990

Al-Sahara B, Taha RM, Mohamad J, Elias H, Khan A (2020) Somatic embryogenesis and plantlet regeneration in the *Carica papaya* L. cv. Eksotika. Plants 9 (3):360. https://doi.org/10.3390/plants9030360

Anand Y, Bansal YK (2002) Synthetic Seeds: a novel approach of in vitro plantlet formation in Vasaka (Adhatodavasica Nees.). Plant Biotechnol 19(3):159–162

Bekheet SA (2006) A synthetic seed method through encapsulation of in vitro proliferated bulblets of garlic (*Alliumsativum* L.) Arab J Biotech 9: 415–426

Berthouly M, Etienne H (2005) Temporary immersion system: a new concept for use liquid medium in mass propagation. In: Hvoslef-Eide AK, Preil W (eds) Liquid Culture Systems for in vitro Plant Propagation. Springer, Dordrecht, pp 165–195

Borgato L, Pisani F, Furini A (2007) Plant regeneration from leaf protoplasts of *Solanum virginianum* L. (Solanaceae). Plant Cell Tissue Org Cult 88:247–252

Boutchouang RP, Akitio OFZ, Tchouatcheu AGN, Niemenak N (2016) Influence of the position of flowers buds on the tree on somatic embryogenesis of cocoa (*Theobroma cacao* L.). Int J Plant Physiol Biochem 8:7–16. https://doi.org/10.5897/IJPPB2016.0247

Calovic M, Chen C, Yu Q, Orbovic V, Gmitter FR Jr, Grosser JW (2019) New somatic hybrid mandarin tetraploids generated by optimized protoplast fusion and confirmed by molecular marker analysis and flow cytometry. J Am Soc Hortic Sci 144:151–163

Campos NA, Panis B, Carpentier SC (2017) Somatic embryogenesis in coffee: the evolution of biotechnology and the integration of omics technologies offer great opportunities. Front Plant Sci 8:1460

Cao A, Zheng Y, Yu Y, Wang X, Shao D, Sun J, Cui B (2017) Comparative transcriptome analysis of SE initial dedifferentiation in cotton of different SE capability. Sci Rep 7:8583. https://doi.org/10.1038/s41598-017-08763-8

Cocking EC (1960) Enzymatic degradation of cell wall for protoplast formation. Nature 187:927–929

Coelho N, Gonçalves S, Romano A (2020) Endemic plant species conservation: Biotechnological approaches. Plants 9(3):345. https://doi.org/10.3390/plants9030345

Corredoira E, Ballester A, Vieitez AM (2003) Proliferation, maturation and germination of *Castanea sativa* Mill. somatic embryos originated from leaf explants. Ann Bot 92:129–136

Davey MR, Anthony P, Power JB, Lowe KC (2005) Plant protoplast technology: Current status. Acta Physiol Plant 27:117–129

Delporte F, Pretova A, du Jardin P, Watillon B (2014) Morpho-histology and genotype dependence of *in vitro* morphogenesis in mature embryo cultures of wheat. Protoplasma 251:1455–1470

Doležel J, Greilhuber J, Suda J (2007) Estimation of nuclear DNA content in plants using flow cytometry. Nat Protoc 2:22–33

Fehér A (2015) Somatic embryogenesis—stress-induced remodeling of plant cell fate. Biochim Biophys Acta 1849:385–402

Fraga HPF, Vieira LN, Caprestano CA (2015) 5-Azacytidine combined with 2,4-D improves somatic embryogenesis of *Acca sellowiana* (O. Berg) Burret by means of changes in global DNA methylation levels. Plant Cell Rep 31:2165–2176

Gantait S, Kundu S, Ali NM, Sahu NC (2015) Synthetic seed production of medicinal: A review on influence of explants, encapsulation agent and matrix. Acta Physiol Plant 37:98

Gantait S, Vijayan J, Majee A (2017) Artificial seed production of *Tylophora indica* for interim storing and swapping germplasm. Hortic Plant J 3:41–46

Gautier F, Eliášová K, Leplé J-C, Vondráková Z, Lomenech A-M, Metté CL, Label6 P, Costa G, Trontin J-F, Teyssier C, Lelu-Walter M-A (2018) Repetitive somatic embryogenesis induced cytological and proteomic changes in embryogenic lines of *Pseudotsuga menziesii* [Mirb.]. BMC Plant Biol 18:164

Guan Y, Li SG, Fan XF, Su ZH (2016) Application of somatic embryogenesis in woody plants. Front Plant Sci 7:938

Guimara˜es G, Cardoso L, Oliveira H, Santos C, Duarte P, Sottomayor M (2012) Cytogenetic characterization and genome size of the medicinal plant *Catharanthus roseus* (L.) G. Don. AoB Plants pls 002. https://doi.org/10.1093/aobpla/pls002

Gulzar B, Mujib A, Rajam MV, Frukh A, Zafar N (2019) Identification of somatic embryogenesis (SE) related proteins through label-free shotgun proteomic method and cellular role in *Catharanthus roseus* (L.) G. Don. Plant Cell Tiss Org Cult. https://doi.org/10.1007/s11240-019-01563-0

Gulzar B, Mujib A, Rajam MV, Zafar N, Mamgain J, Malik M, Syeed R, Ejaz B (2021a) Shotgun label-free proteomic and biochemical study of somatic embryos (cotyledonary and maturation stage) in *Catharanthus roseus* (L.) G. Don. 3Biotechnology 11:86. https://doi.org/10.1007/s13205-021-02649-3

Gulzar B, Mujib A, Mushtaq Z, Malik MQ (2021b) Old *Catharanthus roseus* culture (14 years) produced somatic embryos and plants and showed normal genome size; demonstrated an increased antioxidant defense mechanism; and synthesized stress proteins as biochemical, proteomics, and flow-cytometry studies reveal. J Appl Genet 62:43–57

Guo JM, Liu QC, Zhai H, Wang YP (2006) Regeneration of plants from *Ipomoea cairica* L. protoplasts and production of somatic hybrids between *I. cairica* L. and sweet potato, *I. batatas* (L.) Lam. Plant Cell Tissue Org Cult 87:321–327

Hashemloian BD, Ataei-Azimi A, Majd A, Ebrahimzadeh H (2008) Abnormal plantlets regeneration through direct somatic embryogenesis on immature seeds of *Vinca herbacea* Waldst. and kit. Afr J Biotechnol 7:1679–1683

Heringer AS, Barroso T, Macedo AF, Santa-Catarina C, Souza GHMF, Floh EIS, Souza-Filho GA, Silveira V (2015) Label-free quantitative proteomics of embryogenic and non-embryogenic callus during sugarcane somatic embryogenesis. PLoS ONE. https://doi.org/10.1371/journal.pone.0127803

Hofmann N (2014) Getting to the root of regeneration: adventitious rooting and callus formation. Plant Cell 26:845

Hu R, Sun Y, Wu B, Duan H, Zheng H, Hu D, Lin H, Tong Z, Xu J, Li, Y (2017) Somatic embryogenesis of immature *Cunninghamia lanceolata* (lamb.) hook zygotic embryos. Sci Rep 7:56

Ilah A, Mujib A, Junaid A, Samar F, Abdin MZ (2009) Somatic embryogenesis and two embryo specific proteins (38 and 33 kD) in *Catharanthus roseus*. Biologia 64(2):299–304

Jacobs DI, Gaspari M, van der Greef J, van der Heijden R, verpoorte R (2005) Proteome analysis of the medicinal plant *Catharanthus roseus*. Planta 221:690–704

Junaid A, Bhatt MA, Mujib A, Sharma MP (2006) Somatic embryo proliferation maturation and germination in *Catharanthus roseus*. Plant Cell Tiss Org Cult 84:325–332

Junaid A, Mujib A, Sharma MP (2014) Somatic embryos in *Catharanthus roseus*: a scanning electron microscopic study. Not Sci Biol 6(2):167–172

Kim Y-J, Lee OR, Kim K-T, Yang D-J (2012) High frequency of plant regeneration through cyclic secondary somatic embryogenesis in *Panax ginseng*. J Ginseng Res 36:442–448

Konar S, Karmakar J, Ray A, Adhikari S, Bandyopadhyay TK (2018) Regeneration of plantlets through somatic embryogenesis from root derived calli of *Hibiscus sabdariffa* L. (Roselle) and assessment of genetic stability by flow cytometry and ISSR analysis. PLoS One 13(8):e0202324. https://doi.org/10.1371/journal.pone.0202324

Krishna H, Alizadeh M, Singh D, Singh U, Chauhan N, Eftekhari M, Sadh RK (2016) Somaclonal variations and their applications in horticultural crops improvement. 3 Biotechnology 6:1–18

Kundu S, Salma U, Gantait S (2018) Cryopreservation of medicinal herbs: Major breakthroughs, hurdles and future. In: Kumar N (ed) Biotechnological approaches for medicinal and aromatic plants. Springer, pp 353–381

Landey RB, Cenci A, Guyot R (2015) Assessment of genetic and epigenetic changes during cell culture ageing and relations with somaclonal variation in *Coffea arabica*. Plant Cell Tiss Org Cult 122:517–531

Latif Z, Nasir IA, Riazuddin S (2007) Indigenous production of synthetic seeds in *Daucus carota*. Pak J Bot 39:849–855

Lelu-Walter M-A, Gautier F, Eliášová K, Sanchez L, Teyssier C, Lomenech A-M, Le Metté C, Hargreaves C, Trontin J-F, Reeves C (2017) High gellan gum concentration and secondary somatic embryogenesis: two key factors to improve somatic embryo development in *Pseudotsuga menziesii* [Mirb.]. Plant Cell Tiss Org Cult 132(1):137–55

Li K, Zhu W, Zheg K, Zhang Z, Ye J, Qu W, Rehman S, Heuer B, Chen S (2010) Proteome characterization of cassava (*Manihot esculenta* Crantz) somatic embryos, plantlets and tuberous roots. Proteome Sci 8:1–12

Lipaska H, Konradova H (2004) Somatic embryogenesis in conifers: the role of carbohydrate metabolism. In Vitro Cell Dev Biol Plant 40:23–30

Liu W, Liang Z, Shan C, Marsolais F, Tian L (2013) Genetic transformation and full recovery of alfalfa plants via secondary somatic embryogenesis. In Vitro Cell Dev Biol-Plant 49:17–23

Loureiro J, Travnıcek P, Rauchova J, Urfus T, Vit P, Stech M, Castro S, Suda J (2010) The use of flow cytometry in the biosystematics, ecology and population biology of homoploid plants. Preslia 82:3–21

Mahanom J, Norzulaani K, Rofina YO (2003) Plant regeneration from embryogenic suspension cultures of *Musa acuminata* cv. Mas (AA). Plant Cell Tiss Org Cult 75:209–214

Malik MQ, Mujib A, Gulzar B, Zafar N, Syeed R, Mamgain J, Ejaz B (2020) Genome size analysis of field grown and somatic embryo regenerated plants in *Allium sativum* L. J Appl Genet 61(1):25–35. https://doi.org/10.1007/s13353-019-00536-5

Maruyama E, Yoshihisa H, Katsuaki I (2007) Somatic embryogenesis and plant regeneration in yakutanegoyou, *Pinus armandii* Franch. var. amamiana (Koidz.) Hatusima, an endemic and endangered species in Japan. In Vitro Cell Dev Biol- Plant 43:28–34

Mehpara M, Mujib A, Khusrau M (2015) Preparation and low temperature short-term storage for synthetic seeds of *Caladium bicolor*. Not Sci Biol 7:90–95

Mehpara M, Mujib A, Siddiqui ZH (2012) Synthetic seed development and conversion to plantlet in *Catharanthus roseus* (L.) G. Don. Biotechnology 11:37–43

Mehpara M, Mujib A (2017) Yeast extract elicitation increases vinblastine and vincristine yield in protoplast derivedtissues and plantlets in Catharanthus roseus. Braz J Pharm 27: 549–556

Mendez-Hernandez HA, Ledezma-Rodriguez M, Avilez-Montalvo RN, Juarez-Gomez YL, Skeete A, Avilez-Montalvo J, De-la-Pena C, Loyola-Vargas VM (2019) Signaling overview of plant somatic embryogenesis. Front Plant Sci 10:1–15. https://doi.org/10.3389/fpls.2019.00077

Mikula A, Pozoga M, Tomiczak K, Rybczynski JJ (2015) Somatic embryogenesis in ferns: a new experimental system. Plant Cell Rep 34:783–794

Mostafa HHA, Wang H, Song J, Li X (2020) Effects of genotypes and explants on garlic callus production and endogenous hormones. Sci Rep 10:4867

Mujib A, Banerjee S, Maqsood M, Ghosh PD (2013) Somatic embryogenesis of some member ornamental generaof Amaryllidaceae and allied families: the similarities and differences. The Open Hort J 6: 9–18

Mujib A, Ilah A, Aslam J, Fatima S, Siddiqui ZH, Maqsood M (2012) Catharanthus roseus alkaloids: applicationof biotechnology for improving yield. Plant Growth Regul 68:111–127

Mujib A, Samaj J (2006) Somatic Embryogenesis. Springer-Verlag, Berlin, Heidelberg, New York, p 357

Mujib A, Das S, Dey S, Bhattacharya B (1995) Influence of agitation in *in vitro* cultivation of *Catharanthus roseus* (L). G. Don Multiple Shoot. Phytomorphology 45(3–4):239–245

Mujib A, Ali M, Isah T (2014) Somatic embryo mediated mass production of Catharanthus roseus in culture vessel (bioreactor) – A comparative study. Saudi J Biol Sci 21(5):442–449

Mujib A (2016) Somatic embryogenesis in ornamentals and its applications. Springer, Berlin, p 267. https://doi.org/10.1007/978-81-322-2683-3

Naing AH, Kim CK, Yun BJ, Jin JY, Lim KB (2013) Primary and secondary somatic embryogenesis in *Chrysanthemum* cv. Euro. Plant Cell Tiss Org Cult 112:361–368

Nandini BP, Sudarshana MS, Rajashekar N (2014) Plant regeneration through somatic embryogenesis and synthetic seed production in *Rumex vesicarius* L. - A potent medicinal herb. IOSR J Pharm Biol Sci 9:129–136

Nasim SA, Mujib A, Kapoor R, Fatima S, Aslam J, Mahmooduzzafar (2010) Somatic embryogenesis in *Allium sativum* L. (cv. Yamuna Safed 3): Improving embryo maturation and germination with PGRs and carbohydrates. Anales Biol 32:1–9

Nic-Can GI, Avilez-Montalvo JR, Avilez-Montalvo RN, Márquez-López RE, Mellado-Mojica E, Galaz-Ávalos RM, Loyola-Vargas VM (2016) The relationship between stress and somatic embryogenesis. In: Loyola-Vargas V, Ochoa-Alejo N (eds) Somatic embryogenesis: fundamental aspects and applications. Springer, Cham, pp 51–170. https://doi.org/10.1007/978-3-319-33705-0_9

Nitayadatpat R, Te-chato S (2005) Enhanced efficiency for propagation of anthurium by tissue culture technique. Songklanakarin J Sci Technol 27(5):1003–1008

Osuji C, Abubakar S, Mowobi G (2016) The place of somatic embryogenesis in crop improvement through genetic transformation. J Environ Life Sci 1:1–7

Palanyandy SR, Gantait S, Suranthran P, Sinniah UR, Subramaiam S, Aziz MA, Sarifa SRSA, Roowi SH (2015) Storage of encapsulated oil palm polyembryoids: Influence of temperature and duration. In Vitro Cell Dev Biol - Plant 5:118–124

Pan ZG, Liu CZ, Murch SJ, El-Demerdash M, Saxena PK (2003) Plant regeneration from mesophyll protoplasts of the Egyptian medicinal plants *Artmesia judaica* L. and *Echinops spinosissimus* Turra. Plant Sci 165:681–687

Pescador R, Kerbauy GB, Viviani D, Kraus JE (2008) Anomalous somatic embryos in *Acca sellowiana* (O. Berg) Burret (Myrtaceae). Braz J Bot 31:155–164

Popielarska M, Ślesak H, Goralski G (2006) Histological and SEM studies on organogenesis in endosperm-derived callus of kiwifruit (*Actinidia deliciosa* cv. Hayward). Acta Biol Cracov Sr Bot 48(2):97–104

Pulianmackal AJ, Kareem AV, Durgaprasad K, Trivedi ZB, Prasad K (2014) Competence and regulatory interactions during regeneration in plants. Front Plant Sci 5:142

Pullman GS, Skryabina A (2007) Liquid medium and liquid overlays improve embryogenic tissue initiation in conifers. Plant Cell Rep 26:873–887

Pullman GS, Johnson S, Bucalo K (2009) Douglas fir embryogenic tissue initiation. Plant Cell Tiss Org Cult 96:75–84

Raji MR, Lotfi M, Tohidfar M, Zahedi B, Carra A, Abbate L, Carimi F (2017) Somatic embryogenesis of muskmelon (*Cucumis melo* L.) and genetic stability assessment of regenerants using flow cytometry and ISSR markers. Protoplasma 255:873–883

Reed BM, Uchendu E (2008) Controlled rate cooling. In: Reed BM (ed) Plant cryopreservation: a practical guide. Springer, New York, pp 77–92

Reeves C, Hargreaves C, Trontin J-F, Lelu-Walter M-A (2017) Simple and efficient protocols for the initiation and proliferation of embryogenic tissue of Douglas-fir. Trees 32(1):175–190

Rihan HZ, Al-Issawi M, Al-Swedi F, Fuller MP (2012) The effect of using PPM (plant preservative mixture) on the development of cauliflower microshoots and the quality of artificial seed produced. Sci Hort 141:47–52

Robichaud RL, LesserVC MSA (2004) Treatments affecting maturation and germination of American chestnut somatic embryos. J Plant Physiol 161:957–969

Rowntree JK, Pressel S, Ramsay MM, Sabovljevic A, Sabovljevic M (2011) In vitro conservation of European bryophytes. In Vitro Cell Dev Biol- Plant 47:55–64

Saeed W, Naseem S, Gohar D, Ali Z (2019) Efficient and reproducible somatic embryogenesis and micropropagation in tomato via novel structures—rhizoid tubers. PLoS One 14(5):e0215929. https://doi.org/10.1371/journal.pone.0215929

Sangra A, Shahin L, Dhir SK (2019) Long-term maintainable somatic embryogenesis system in alfalfa (*Medicago sativa*) using leaf explants: embryogenic sustainability approach. Plants 8:278. https://doi.org/10.3390/plants8080278

Schlogl PS, Santos ALW, Vieira LN, Floh EIS, Guerra MP (2012) Gene expression during early somatic embryogenesis in Brazilian pine (*Araucaria angustifolia* (Bert) O. Ktze). Plant Cell Tiss Org Cult 108:173–180

Shrestha BR, Tokuhara K, Mii M (2007) Plant regeneration from cell suspension-derived protoplasts of *Phalaenopsis*. Plant Cell Rep 26:719–725

Silva de Almeida JA (2020) Observations on somatic embryogenesis in *Coffea arabica* L. IntechOpen. https://doi.org/10.5772/intechopenpp.1-20

Sushamakumari S, Asokan MP, Anthony P, Lowe KC, Power JB, Davey MR (2000) Plant regeneration from embryogenic cell suspension-derived protoplasts of rubber. Plant Cell Tiss Org Cult 61:81–85

Tahtamouni RW, Shibli AR, Al-Abdallat AM, Al-Qudah TS, Younis L, Al-Baba H, Al- Ruwaiei H (2015) In vitro conservation and cryopreservation of medicinal and aromatic plants: a review. Jord J Agr Sci 11 (1):147–167

Tchorbadjieva M (2016) Advances in proteomics of somatic embryogenesis. In: Mujib A (ed) Somatic embryogenesis in ornamentals and its applications. Springer, Berlin. https://doi.org/10.1007/978-81-322-2683-3_5

Te-Chato S, Hilae A, Moosikapala L (2005) Microcolony formation from embryogenic callus-derived protoplasts of oil palm. Songklanakarin. J Sci Technol 27:685–691

Thomas TD (2009) Isolation, callus formation and plantlet regeneration from mesophyll protoplasts of *Tylophora indica* (Burm. f.) Merrill: an important medicinal plant. In Vitro Cell Dev Biol- Plant 45:591–598

Thorpe T (2012) History of plant tissue culture. Methods Mol Biol 877:9–27

Tikkinen M, Varis S, Aronen T (2018) Development of somatic embryo maturation and growing techniques of norway spruce emblings towards large-scale field testing. Forests 9:325. https://doi.org/10.3390/f9060325

Tvorogova VE, Lutova LA (2018) Genetic regulation of zygotic embryogenesis in angiosperm plants. Russ J Plant Physiol 65:1–14. https://doi.org/10.1134/S1021443718010107

Ulisses C, Camara TR, Willadino L, Cavalcanti de Albuquerque C, Zoé de Brito J (2010) Early somatic embryogenesis in *Heliconia chartacea* Lane ex Barreiros cv. Sexy Pink ovary section explants, Braz Arch Biol Technol 53(1):11–18. https://doi.org/10.1590/S1516-89132010000100002

Vale EM, Heringer AS, Barroso T, Ferreira ATDS, DaCosta MN, Perales JEA, Catarina CS, Silveira V (2014) Comparative proteomic analysis of somatic embryo maturation in *Carica papaya* L. Proteome Sci 12:1–17

Vikrant JP (2018) Progress in understanding the regulation and expression of genes during plant somatic embryogenesis: a review. J Appl Biol Biotechnol 6 (05):49–56. https://doi.org/10.7324/JABB.2018.60508

von Arnold S, Sabala I, Bozhkov P, Dyachok J, Filonova L (2002) Developmental pathways of somatic embryogenesis. Plant Cell Tiss Org Cult 69:233–249

Wu G-Y, Wei X-L, Wang X, Wei Y (2020) Induction of somatic embryogenesis in different explants from *Ormosia henryi* Prain. Plant Cell Tiss Org Cult 142:229–240

Xiao W (2006) DNA methylation is critical for Arabidopsis embryogenesis and seed viability. Plant Cell 18:805–814

Yang X, Zhang X (2010) Regulation of somatic embryogenesis in higher plants. Crit Rev Plant Sci 29:36–57. https://doi.org/10.1080/07352680903436291

Yang X, Guo X, Zhang X, Nie Y, Jin S (2007) Plant regeneration from *Gossypium davidsonii* protoplasts *via* somatic embryogenesis. Biol Plant 51:533–537

Yang J, Wu S, Li, C (2013) High efficiency secondary somatic embryogenesis in *Hovenia dulcis* Thunb. through solid and liquid cultures. Sci World J 1–6

Yun-peng D, Wen-yuan L, Ming-fang Z, Heng-bin H, Gui-xia J (2012) The establishment of a slowgrowth conservation system *in vitro*for two wild lily species. Afr J Biotechnol 11(8):1981–1990

Zafar N, Mujib A, Ali M, Tonk D, Gulzar B, Malik M, Sayeed R, Mamgain J (2019) Genome size analysis of field grown and tissue culture regenerated *Rauvolfia serpentina* (L) by flow cytometry: Histology and scanning electron microscopic study for *in vitro* morphogenesis. Ind Crop Prod 128:545–555

Alkaloid Synthesis in In Vitro Cultures of *Catharanthus Roseus*: Potential and Limitations

5

Felipe A. Vázquez-Flota

Abstract

Catharanthus roseus (L.) G. Don, Madagascar periwinkle, was among the first plants to be explored for the in vitro production of fine chemicals, due to the high value of the cytotoxic dimer alkaloids that it accumulates in very low amounts. That goal has not been reached despite several decades, although without a doubt, cell culture technology has resulted in an extremely important tool in understanding the molecular basis that control alkaloid formation. This chapter presents an account of the effects of different media components and culture conditions, in the cellular mechanisms governing alkaloid formation and how this knowledge has contributed to make today the *Catharanthus* alkaloid biosynthetic pathway one of the better known. This level of understanding now allows the application of cutting-edge biotechnological approaches.

5.1 Introduction

The production of pharmaceutical compounds was one of the first biotechnological goals envisioned for the in vitro cell cultures. *Catharanthus roseus* (L.) G. Don., popularly called Madagascar periwinkle or simply periwinkle, is among the first species to be attempted (Van der Heijden et al. 1989; Moreno et al. 1995; Verpoorte et al. 2002). The first stable in vitro *C. roseus* cultures was obtained ca. 1945 from crown gall tumors derived from *Agrobacterium rhizogenes* stem infection (White1945), whereas stem long term callus cultures were first reported in 1947 (Ropp 1947). However, interest was then centered on morphogenic events and on the effects of different nutriments and culture conditions on them. After the serendipitous discovery of the powerful cytotoxic effects of the *Catharanthus* dimeric alkaloids (vinblastine, vincristine, vinleurosina, and vinrosidina), at the end of the decade of 1950 (Babcock and Carew 1962), the biosynthetic potential of cell cultures turned into a topic of intense research, with the occurrence of alkaloids being first described around 1962 (Harris et al. 1964). This was followed by a series of reports leading to the identification of nearly 20 different alkaloids by TLC criteria (Ritcher et al. 1965; Carew 1966; Farnsworth 1966). However, stable production of significant amounts of the dimeric alkaloids has been remaining elusive in undifferentiated cell

F. A. Vázquez-Flota (✉)
Unidad de Bioquímica y Biología Molecular de Plantas, Centro de Investigación Científica de Yucatán, Calle 43 No. 130, Chuburná de Hidalgo, Mérida, Yucatán 97205, México
e-mail: felipe@cicy.mx

© The Author(s), under exclusive license to Springer Nature Switzerland AG 2022
C. Kole (ed.), *The Catharanthus Genome*, Compendium of Plant Genomes,
https://doi.org/10.1007/978-3-030-89269-2_5

cultures, even when other alkaloids, such as ajmalicine, have been frequently reported (Das et al. 2020). Interestingly, catharanthine, one of the moieties required to conform the dimers, was also among the alkaloids detected in undifferentiated cell cultures and it was the lack of vindoline, the second unit, which remained as the main limitation for dimer formation in the in vitro cultures (Constabel et al. 1981). Genetic alterations hampering the formation of vindoline were discarded as, upon plant regeneration, this capacity was again acquired (Constabel et al. 1982). Instead, this inability was accounted to the transcriptional blockage of genes involved in the last reactions (Vázquez-Flota et al. 2002), which in turn was related to the specific tissue and cell requirement for the synthesis and storage of this alkaloid that was lost during the induction of cell cultures (St-Pierre et al. 1999; Vázquez-Flota et al. 2000).

However, despite disappointments in cell cultures as a system to produce the dimeric alkaloids, along with this nearly 70 years tour, in vitro culture provided an invaluable platform in understanding the complexity of the biosynthetic process and how it could be affected by different external stimuli. Moreover, *Catharanthus* cell cultures had played a critical role in the first drawings of the networks linking alkaloid biosynthetic pathways with those furnishing the starting units from primary metabolism (van der Fits and Memelink 2000). Lines below, a summary of the different *C. roseus* in vitro systems (Fig. 5.1) as well as their capacities and limitations are presented.

5.2 Callus and Cell Suspensions

Callus and cell suspension cultures represent an attractive experimental system for many reasons (Fig. 5.1a, b). They are easy to maintain, manipulate and propagate. They offer a controlled system for the application of different stimuli, such as variation of external conditions, nutrients, and other agents, although some facilities could be required, mainly culture rooms or chambers. To produce secondary metabolites, main limitations refer to a reduced biosynthetic capacity reflected as low accumulation levels and/or the loss of specific compounds (Vázquez-Flota and Loyola-Vargas 2003). As mentioned above, *C. roseus* cell cultures were among the first to be explored with that aim, in parallel with the development of technology for high volume cultures (Zhao and Verpoorte 2007). The lack of consistent results concerning the accumulation of the dimeric alkaloids prompted diverse approaches to surpass this constraint. Among them, obtaining elite cell lines, optimization of the culture media and conditions, supplying of biogenical precursors, and mimicking stress situations have been the most common.

5.2.1 Selection of Elite Lines, Media Formulation, and Culture Conditions

Selection of elite materials has been attempted by establishing cell cultures from outstanding individuals or tissues (Salma et al. 2018), by screening several previously established cell lines by automized methods (Scragg et al. 1988), or by tryptophan overproduction (Verma et al. 2013). Unfortunately, although some success in the accumulation of some alkaloids (such as ajmalicine, catharanthine, or serpentine) could result, this new capacity is rapidly lost due to the low genetic stability of suspension cultures (Whitmer et al. 2003). Major media components, such as carbon, nitrogen, and phosphorous have been assayed for their effects on alkaloid accumulation in *C. roseus* cell cultures. As a general rule, conditions promoting rapid cell growth have negative effects on alkaloid accumulation. Interestingly, designed alkaloid producing media (APM) is usually part of a biphasic process with a rapid growth stage before an alkaloid accumulation phase (Zenk et al. 1977; Knobloch and Berlin 1980). High doses of sucrose (6–8%), promoted the accumulation of ajmalicine, catharanthine and serpentine, but did not result in the presence of vindoline (Scragg et al. 1990). Medium nitrogen availability depends on total concentration, as well as on the proper balance

Fig. 5.1 *Catharanthus roseusin vitro* cultures. Leaf-derived callus (**a**); cell suspensions; (**b**) *A. rhizogenes* hairy root (**c**); and rootles shoot (**d**) cultures

between the oxidized (nitrate) and reduced (ammonium) forms. Again, nitrogen doses leading to rapid cell growth inhibited the accumulation of alkaloids (Knobloch and Berlin 1983). Regarding phosphate, usually low doses or total absence have promoting effects on alkaloid accumulation in different cell lines (Zenk et al. 1977; Knobloch and Berlin 1980). A similar trend has been observed for culture conditions such as temperature. Low temperatures (<20 °C) reduce growth and induce alkaloid accumulation (Curtois and Guern 1980). Light on the other hand, although plays a critical role in vindoline formation in organized tissues (Vázquez-Flota et al. 2000), in cell cultures only affected serpentine biosynthesis, promoting its formation from ajmalicine (Loyola-Vargas et al. 1992). Although in entire plants, plant growth regulators may affect alkaloid contents through improving effects on photosynthetic activity and nutrient uptake (Senbagalakshmiet al. 2017), in cell cultures, they exert negative effects in doses promoting growth (Van der Heijden et al. 1989; Moreno et al. 1995). Interestingly, abscisic acid (ABA) has positive effects on catharanthine accumulation, although it reduced cell expansion (Moreno et al. 1995; Senbagalakshmi et al. 2017).

From above, it could establish that in *C. roseus*, as in other species, cell culture biomass accumulation and alkaloid formation are incompatible and this is, at least partially, related to two issues: the role of secondary metabolites in plant environmental interactions and the specific cell organization requirements for alkaloid synthesis in *C. roseus*. Plant cells under stress reduce their growing rate due to nutrient limitation, whereas cell differentiation is preceded by cell divisions, which are then stopped once the mature and final stages are reached.

5.2.2 Elicitation and Stress Conditions

Elicitors are chemical or physical agents with the ability to induce a defensive response in plants, frequently involving synthesis of phytochemicals. In *C. roseus* in vitro cultures, the use of salt and water stress has resulted in limited increases in the accumulation of some alkaloids, such as ajmalicine and catharanthine. Heavy metals also display limited responses (Vázquez-Flota and Loyola-Vargas 2003; Senbagalakshmi et al. 2017).

In fact, interesting responses only have been observed with the use of octadecanoic acids, such as jasmonic acid and its methyl ester; methyl jasmonate (MeJa; Vázquez-Flota et al. 2020b). MeJa is a chemical mediator of the defensive reaction to external stimuli, such as infection or herbivory. For several years, it has been commonly used in in vitro cultures to induce the synthesis of secondary metabolites. Depending on the type of culture, both the biosynthetic genes and the final alkaloid accumulation could be differentially affected (Vázquez-Flota et al. 2009). The Octadecanoid-Responsive Catharanthus AP2-domain (ORCA1, ORCA2, and ORCA3) transcription factors, which play a critical role in the coordinate response of primary metabolism and alkaloid synthesis in response to Meja, were discovered in cell cultures (van der Fits and Memelink 2000) just as the transcriptional repressors ZCT.

5.3 Organ Culture

As in other species, alkaloid synthesis in *C. roseus* requires a complex organization involving different cell types, organelles, and transport systems (Murata and De Luca 2005; Oudin et al. 2007; Roepke et al. 2010). The lack of this structure in the undifferentiated cultures is considered as one of the main reasons for the poor alkaloid accumulation, especially for vindoline. Hence, the use or differentiated cultures, such as roots and shoots (Fig. 5.1c, d), has been proposed as an alternative (Hirata et al. 1990; Hernández-Domínguez et al. 2004). Although both shoot and root cultures display a sensitivity to similar stimuli found in cell cultures (Hirata et al. 1999; Campos-Tamayo et al. 2008; Vázquez-Flota et al. 2020a), important differences in the effects have been observed. For example, in cell cultures only ajmalicine was positively affected by MeJa exposure, with no effects on catharanthine, whereas in hairy roots, jasmonate affected both alkaloids and in shoot cultures, positive effects of jasmonate were only noticed on vindoline (Vázquez-Flota et al. 2009).

5.3.1 Shoot Cultures

Rootless shoot cultures display a similar alkaloid pattern to that of leaves; however, some important differences can also be noticed, mainly referring to those alkaloids found in roots of intact plants (Campos-Tamayo et al. 2008). Interestingly, vindoline was the main alkaloid (ca. 2 mg g DW^{-1}) found in rootless shoots, although low levels of ajmalicine were also found, compared to the rooted plantlets (Hernández-Domínguez et al. 2004). Moreover, during a four-week culture period, peaks of vindoline accumulation coincided with shoot formation suggesting a close relationship between morphogenic events and alkaloid formation in this species (Campos-Tamayo et al. 2008).

5.3.2 Roots Cultures

There are several reports of *C. roseus* root cultures (Vázquez-Flota and Loyola-Vargas 2003; Senbagalakshmi et al. 2017). In most cases, culture of *Agrobacterium rhizogenes* transformed tissues is preferred since they grow rapidly (Vázquez-Flota et al. 2009). However, the lack or low levels of vindoline accumulation limits its potential use as a suitable platform to produce dimeric alkaloids (Senbagalakshmi et al. 2017). In fact, *C. roseus* root cultures could be disaggregated into suspension cultures that reduce their capacity to accumulate alkaloids. However, after roots were differentiated again from these suspensions, the biosynthetic capacity (total alkaloid accumulation) was restored and new somaclonal variants were spontaneously detected (Vázquez-Flota et al. 2020b). This does not just confirm the requirement of tissue organization for an optimal alkaloid formation, but also suggests the possible use of this system to induce the appearance of new and interesting variants, which then could be selected as previously exposed.

5.4 Metabolic Engineering and Synthetic Biology Approaches

From the above, it can be noticed that cell and tissue organization exert a high hierarchical effect on the control of alkaloid synthesis in *C. roseus*. Specific cell types must be distributed following a well-defined tissue context. This is most evident in the case of vindoline, one moiety of the dimeric alkaloids. Therefore, conventional tissue culture strategies are fated to produce limited results in an attempt to increase their synthesis. Next generation technologies, together with genetic engineering methods, could help to surpass these limitations. However, although *C. roseus* is susceptible to genetic transformation *via Agrobacterium* infection and particle bombardment, and efficient procedures for plantlet regeneration are available (Verpoorte et al. 2002), no efficient methods for the recovery of transformed plants are still available. Hence, in vitro cultures remain a viable platform for the application of these approaches.

Metabolic engineering strategies attempt to modify metabolic pathways directing carbon flow towards the accumulation of specific products by either favoring its formation or avoiding intermediary diversion. This is accomplished by means of recombinant DNA technology and can be done using both structural and regulatory genes. Synthetic biology, on the other hand, designs and builds new biological functions, such as the introduction of new biosynthetic capacities. While metabolic engineering modifies the capacities already present, synthetic biology creates new capacities. Given of the complex intermediary traffic and membrane bound enzymes involved in alkaloid formation in *C. roseus* cells, the use of bacteria as a platform for their heterologous expression seems unlikely. However, remarkable results have been obtained using yeast and plant cells. The interested reader is referred to recent reviews (such as Sharma et al. 2020) for a more detailed description of those works, since this chapter is mainly dedicated to *C. roseus* in vitro cultures.

5.4.1 Modification of Alkaloid Biosynthesis in *C. roseus* by Metabolic Engineering

Most of the structural genes involved in the dimeric alkaloids have been isolated and characterized, making them available for application in metabolic engineering strategies (Stavrinides et al. 2015). Pushing forward strategies have been designed, both on the indole and terpenoid branches, by transforming *C. roseus* tissues with the tryptophan decarboxylase (TDC; Canel et al. 1998) and geraniol 10-hydroxylase genes (G10H; Collu et al. 2001). TDC transforms tryptophan in tryptamine, whereas G10H hydroxylates the monoterpene geraniol at C-10 that then converts in secologanin. Besides, the condensation of the indole and terpenoid moieties catalyzed by strictosidine synthase (STR; Canel et al. 1998), has also been tackled. Suspension cultures overexpressing TDC required the external addition of loganin and secologanin to achieve any important increase in alkaloid accumulation in strictosidine and catharanthine (Canel et al. 1998; Whitmer et al. 2003), contrasting to those transformed with STR, which presented a noteworthy increment in the accumulation of strictosidine, ajmalicine, catharanthine, serpentine and tabersonine (Canel et al. 1998). Although these results suggest that input from the indole branch would not be limiting for alkaloid synthesis, hairy roots expressing feedback resistant anthranilate synthase (AS), together with TDC, displayed a marked increase in serpentine accumulation (Hong et al. 2006). Hence, these results suggest that overexpressing TDC would not be sufficient due to a low supply of tryptophan, its substrate.

Engineering of the tabersonine to vindoline pathway has been attempted in hairy root cultures by expressing tabersonine 16-hydroxylase (T16H) and 16-hydroxytabersonine methyltransferase (16OMT). These two enzymes participate in the first set of reactions required in the transformation of tabersonine in vindoline (Sun et al. 2018). Although vindoline was not detected, new alkaloids, derivates from these methoxylated compounds, were observed coinciding with the reduction of the common root alkaloids (Sun et al. 2018). Interestingly, even the expression of the deacetylvindoline acetyltransferase (DAT), involved in the last reaction of vindoline synthesis, did not result in vindoline accumulation although other acetylated alkaloids were detected (Magnotta et al. 2007).

Besides structural genes, regulatory genes, corresponding to different transcriptional factors, have also been attempted. More than ten transcriptional factors (ORCA2, ORCA3, ORCA4, ORCA5, WRKY1, WRKY2, GATA, PIF, BIS1/BIS2, and ZCT1) have been identified, both acting as activators and repressors (Singh et al. 2020). Transcription factors regulate MIAs pathway genes by activating or repressing them. ORCA3 was identified in cell cultures as a master gene controlling the jasmonate induced alkaloid accumulation (Van der Fits and Memelink 2000). Interestingly, ORCA3 also fulfilled this function in rootless shoot cultures (Campos-Tamayo et al. 2008), although not in hairy roots (Peebles et al. 2009). However, once the terpenoid branch was "pumped out" by increasing the availability of strictosidine aglycone (by overexpression of strictosidine degluocosidase; SDG), a noteworthy increase in the alkaloid typical alkaloids was noticed (Sun and Peebles 2016). Similar results were obtained in root cultures transformed with G10H (Wang et al. 2010), although hairy roots overexpressing ORCA2 did not require any further intervention with structural genes to increase alkaloid contents (Liu et al. 2011). Positive effects on catharanthine, tabersonine, and serpentine accumulation were noticed in cell suspensions transformed with CrWRKY1 and CrWRKY2 (Suttipantaa et al. 2011). Operation of the iridoid pathway (leading from geraniol 10-hydroxyl to loganic acid) is under the control of two specific basic helix-loop-helix (bHLH) proteins named BIS1 and BIS2. However, overexpression of both, either alone or in combination with ORCA3, had no effects on vindoline biosynthetic genes (Schweizer et al. 2018). Attempts with other transcriptional factors had been pursued using transient transformation assays (see Sharma et al. 2020 for a review).

5.5 Conclusions

C. roseus in vitro cultures have remained as a current research topic in diverse laboratories around the world. Although production at commercial level has not been achieved, a better understanding of the complexity required for their formation has been acquired with valuable contributions coming for this technology. The main restriction, related to the tissue, cell, and subcellular organization could be surpassed by the use of organ culture. However, since each tissue presents a distinctive alkaloid profile, some limitations in diversity would occur in such systems. Besides, large volume organ cultures frequently present some tissue disruption. Moreover, genetic stability requires constant assessment to maintain a minimal guaranteed production level. Nevertheless, integrative (*omics*) approaches, combined with some of the traditional in vitro culture technologies, seem closer than ever to accomplishing that goal.

Acknowledgements This work was financed by CONACYT (National Council of Science and Technology, México; Grant CB-2016 285887).

References

Babcock PA, Carew DP (1962) Tissue culture of Apocynaceae 1. Culture requirements and alkaloid analysis. Lloydia 25(4):209–213

Campos-Tamayo F, Hernández-Domínguez E, Vázquez-Flota F (2008) Vindoline formation in shoot cultures of *Catharanthus roseus* is synchronously activated with morphogenesis through the last biosynthetic step. Ann Bot 102(3):409–415

Canel C, Lopes-Cardoso MI, Whitmer S, van der FitsL, Pasquali G, van der Heijden R, Hoge JHC, Verpoorte R (1998). Effects of over-expression of strictosidine synthase and tryptophan decarboxylase on alkaloid production by cell cultures of *Catharanthus roseus*. Planta 205(3):414–419

Carew DP (1966) Growth of callus tissue of *Catharanthus roseus* in suspension culture. J Pharm Sci 5(10):1153–1154

Collu G, Unver N, Peltenburg-Looman AM, van der Heijden R, Verpoorte R, Memelink J (2001) Geraniol 10-hydroxylase1, a cytochrome P450 enzyme involved in terpenoid indole alkaloid biosynthesis. FEBS Lett 508(2):215–220

Constabel F, Rambold S, Chatson KB, Kurz WGM, Kutney JP (1981) Alkaloid production in *Catharanthus roseus* (L.) G. Don. VI. Variation in alkaloid spectra of cell lines derived from one single leaf. Plant Cell Rep 1(1):3–5

Constabel F. Gaudet-LaPrairie P, Kurz WGM, Kutney JP (1982) Alkaloid production in *Catharanthus roseus* cell cultures. Plant Cell Rep 1(4):139–142. https://doi.org/10.1007/BF00269182

Courtois D, Guern J (1980) Temperature response of *Catharanthus roseus* cells cultivated in liquid medium. Plant Sci Lett 17(4):473–482

Das A, Sarkar S, Bhattacharyya S, Gantait S (2020) Biotechnological advancements in *Catharanthus roseus* (L.) G. Don. Appl Microbiol Biotechnol 104 (11):4811–4835. https://doi.org/10.1007/s00253-020-10592-1

De Ropp R (1947) The response of normal plant tissues and of crown-gall tumor tissues to synthetic growth hormones. Am J Bot 34(2):53–62

Farnsworth NR (1966) Biological and phytochemical screening of plants. J Pharm Sci 55(3):225–276

Harris AL, Nylund HB, Carew DP (1964) Tissue cultures studies of certain members of Apocynaceae. Lloydia 27(3):322–327

Hernández-Domínguez E, Campos-Tamayo F (2004) Vázquez-Flota F (2008) Vindoline synthesis in in vitro shoot cultures of *Catharanthus roseus*. Biotechnol Lett 26(8):671–674

Hirata K, Horiuchi M, Ando T, Miyamoto K, Miura Y (1990) Vindoline and catharanthine production in multiple shoot culture of *Catharanthus roseus*. J Ferm Bioeng 70(3):193–195

Hong SB Peebles CA, Shanks JV, San KY, Gibson SI (2006) Expression of the Arabidopsis feedback-insensitive anthranilate synthase holoenzyme and tryptophan decarboxylase genes in *Catharanthus roseus* hairy roots. J Biotechnol 122(1):28–38

Knobloch KH, Berlin J (1980) Influence of medium composition on the formation of secondary compounds in cell suspension cultures of *Catharanthus roseus* (L.) G. Don. Z Naturforsch c 35(7–8):551–556

Knobloch KH, Berlin J (1983) Influence of phosphate on the formation of the indole alkaloids and phenolic compounds in cell suspension cultures of *Catharanthus roseus* I. Comparison of enzyme activities and product accumulation. Plant Cell Tiss Org Cult 2 (4):333–340

Li CY, Leopold AL, Sander GW, Shanks JV, Zhao L, Gibson SI (2013) The ORCA2 transcription factor plays a key role in regulation of the terpenoid indole alkaloid pathway. BMC Plant Biol 13(1):1–17

Liu DH, Cui LJ, Zhang LD, Sun XF, Tang KX (2011) Enhanced accumulation of catharanthine and vindoline in *Catharanthus roseus* hairy roots by overexpression of transcriptional factor ORCA2. Afr J Biotechnol 10(17):3260–3268

Loyola-Vargas VM, Méndez-Zeel M, Monforte-González M, Miranda-Ham ML (1992) Serpentine accumulation during greening in normal and tumor tissues of *Catharanthus roseus*. J Plant Physiol 140(2):213–217

Magnotta M, Murata J, Chen J, De Luca V (2007) Expression of deacetylvindoline-4-O-acetyltransferase in *Catharanthus roseus* hairy roots. Phytochemistry 68(14):1922–1931

Moreno PR, Van der Heijden R, Verpoorte R (1995) Cell and tissue cultures of *Catharanthus roseus:* a literature survey. Plant Cell Tiss Org Cult 42(1):1–25

Murata J, De Luca V (2005) Localization of tabersonine 16-hydroxylase and 16-OH tabersonine-16-O-methyltransferase to leaf epidermal cells defines them as a major site of precursor biosynthesis in the vindoline pathway in *Catharanthus roseus*. Plant J 44 (4):581–594

Oudin A, Mahroug S, Courdavault V, Hervouet N, Zelwer C, Rodríguez-Concepción M, St-Pierre B, Burlat V (2007) Spatial distribution and hormonal regulation of gene products from methyl erythritol phosphate and monoterpene-secoiridoid pathways in *Catharanthus roseus*. Plant Mol Biol 65(1):13–30

Peebles CA, HughesEH SJV, San KY (2009) Transcriptional response of the terpenoid indole alkaloid pathway to the overexpression of ORCA3 along with jasmonic acid elicitation of *Catharanthus roseus* hairy roots over time. Metabol Eng 11(2):76–86

Richter I, Stolle K, Gröger D, Mothes K (1965) Überalkaloidbildung in gewebekulturen von *Catharanthus roseus* G. Don. Naturwissenschaften 52:305–306

Roepke J, Salim V, Wu M, Thamm AM, Murata J, Ploss K, Boland W, De Luca V (2010) Vinca drug components accumulate exclusively in leaf exudates of Madagascar periwinkle. Proc Nat Acad Sci USA 107(34):15287–15292

Salma U, Kundu S, Gantait S (2018) Conserving biodiversity of a potent anticancer plant, *Catharanthus roseus* through *in vitro* biotechnological intercessions: substantial progress and imminent prospects. In: Akhtar MS, Swamy MK (eds) Anticancer plants: natural products and biotechnological implements. Springer, Singapore, pp 83–107

Scragg AH, Cresswell R, Ashton S, York A, Bond P, Fowler MW (1988) Growth and secondary product formation of a selected *Catharanthus roseus* cell line. Enz Microb Tech 10(9):532–536

Scragg AH, Ashton S, York A, Bond P, Stepan-Sarkissian G, Grey D (1990) Growth of *Catharanthus roseus* suspensions for maximum biomass and alkaloid accumulation. Enz Microb Tech 12(4):292–298

Schweizer F, Colinas M, Pollier J, Van Moerkercke A, Bossche RV, De Clercq R, Goossens A (2018) An engineered combinatorial module of transcription factors boosts production of monoterpenoid indole alkaloids in *Catharanthus roseus*. Metabol Eng 48 (1):150–162

Senbagalakshmi P, Rao MV, Kumar TS (2017) In vitro studies, biosynthesis of secondary metabolites and pharmacological utility of *Catharanthus roseus* (L.) G. Don.: a review. In: Naeem M, Aftab T, Khan MMA (eds) *Catharanthus roseus*. Springer, Cham, Switzerland, pp 153–199

Singh SK, Patra B, Paul P, Liu Y, Pattanaik S, Yuan L (2020) Revisiting the ORCA gene cluster that regulates terpenoid indole alkaloid biosynthesis in *Catharanthus roseus*. Plant Sci 293 110408

Sharma A, Amin D, Sankaranarayanan A, Arora R, Mathur AK (2020) Present status of *Catharanthus roseus* monoterpenoid indole alkaloids engineering in homo- and hetero-logous systems. Biotechnol Lett 42 (1):11–23

St-Pierre B, Vazquez-Flota FA, De Luca V (1999) Multicellular compartmentation of *Catharanthus roseus* alkaloid biosynthesis predicts intercellular translocation of a pathway intermediate. Plant Cell 11 (5):887–900

Stavrinides A, Tatsis EC, Foureau E, Caputi L, Kellner F, Courdavault V, O'Connor SE (2015) Unlocking the diversity of alkaloids in *Catharanthus roseus*: nuclear localization suggests metabolic channeling in secondary metabolism. Chem Biol 22(3):336–341

Sun J, Zhao L, Shao Z, Shanks J, Peebles CA (2018) Expression of tabersonine 16-hydroxylase and 16-hydroxytabersonine-O-methyltransferase in *Catharanthus roseus* hairy roots. Biotechnol Bioeng 115 (3):673–683

Sun J, Peebles CAM (2016) Engineering overexpression of ORCA3 and strictosidine glucosidase in *Catharanthus roseus* hairy roots increases alkaloid production. Protoplasma 253:1255–1264

Suttipanta N, Pattanaik S, Kulshrestha M, Patra B, Singh SK, Yuan L (2011) The transcription factor CrWRKY1 positively regulates the terpenoid indole alkaloid biosynthesis in *Catharanthus roseus*. Plant Physiol 157(4):2081–2093

Van der Heijden R, Verpoorte R, Ten Hoopen HJ (1989) Cell and tissue cultures of *Catharanthus roseus* (L.) G. Don: A literature survey. Plant Cell Tiss Org Cult 18(3):231–280

Van der Fits L, Memelink J (2000) ORCA3, a jasmonate-responsive transcriptional regulator of plant primary and secondary metabolism. Science 289(5477):295–297

Vázquez-Flota, F, St-Pierre B, De Luca V (2000) Light activation of vindoline biosynthesis does not require cytomorphogenesis in *Catharanthus roseus* seedlings. Phytochemistry 55(6):531–536

Vázquez-Flota F, Loyola-Vargas VM (2003) In vitro Plant cell culture as the basis for the development of a research institute in Mexico: Centro de Investigación Científica de Yucatán. In vitro Cell Dev Biol-Plant 39 (2):250–258

Vázquez-Flota F, De Luca V, Carrillo-Pech M, Canto-Flick A, Miranda-Ham, ML (2002) Vindoline biosynthesis is transcriptionally blocked in *Catharanthus roseus* cell suspension cultures. Mol Biotechnol 22 (1):1–8

Vázquez-Flota F, E Hernández-Domínguez, ML Miranda-Ham, M Monforte-González (2009) A differential response to chemical elicitors in *Catharanthus roseus* in vitro cultures. Biotechnol Lett 31(4):591–595

Vázquez-Flota FA, Miranda-Ham ML, Navarrete-Loeza MP, Monforte-González M (2020a) A protocol for the selection of spontaneous variants from established *Catharanthus roseus* hairy root cultures. In: Srivastava V, Mehrotra S, Mishra S (eds) Hairy root cultures based applications. Rhizosphere Biology. Springer, Singapore, pp 65–76

Vázquez-Flota FA, Miranda-Ham ML, Coello-Coello J, Loyola-Vargas VM (2020b) Strategies for the optimization of culture conditions for increasing metabolite production through hairy root cultures: Monoterpene indole alkaloid production in *Catharanthusroseus*hairy roots. In: Srivastava V, Mehrotra S, Mishra S (eds) Hairy Root Cultures Based Applications. Rhizosphere Biology. Springer, Singapore, pp 77–86

Verma P, Mathur AK, Masood N, Luqman S, Shanker K (2013) Tryptophan over-producing cell suspensions of *Catharanthus roseus* (L) G. Don and their up-scaling in stirred tank bioreactor: detection of a phenolic compound with antioxidant potential. Protoplasma 250 (1):371–380

Verpoorte R, Contin A, Memelink J (2002) Biotechnology for the production of plant secondary metabolites. Phytochem Rev 1(1):13–25

Wang CT, Liu H, Gao XS, Zhang HX (2010) Overexpression of G10H and ORCA3 in the hairy roots of *Catharanthus roseus* improves catharanthine production. Plant Cell Rep 29(8):887–894

White P (1945) Metastatic (graft) tumors of bacteria-free crown-galls on *Vinca rosea*. Am J Bot 32(5): 237–241

Whitmer S, Canel C, Van der Heijden R, Verpoorte R (2003) Long-term instability of alkaloid production by stably transformed cell lines of *Catharanthus roseus*. Plant Cell Tiss Org Cult 74(1):73–80

Zenk MH, El-Shagi H, Arens H, Stöckigt J, Weiler EW, Deus B (1977) Formation of the Indole Alkaloids Serpentine and Ajmalicine in Cell Suspension Cultures *of Catharanthus roseus*. In: Barz W, Reinhard E, Zenk MH (eds) Plant Tissue Culture and its Biotechnological application. Proceedings in life sciences. Springer, Berlin, Heidelberg, pp 27–44

Zhao J, Verpoorte R (2007) Manipulating indole alkaloid production by *Catharanthus roseus* cell cultures in bioreactors: from biochemical processing to metabolic engineering. Phytochem Rev 6(2–3):435–457

Genetic Transformation in *Catharanthus roseus*

6

Lan Thi Ngoc Nguyen, Lien Thi Kim Vu, Quan Huu Nguyen, Ha Thi Bui, and Mau Hoang Chu

Abstract

Vincristine and vinblastine are two major terpenniod indole alkaloids (TIAs) produced naturally by *Catharanthus roseus*, which are a potential source for anticancer drugs. Vinblastine and vincristine are synthesized from vindoline via the catalysis of deacetylvindoline 4-O-acetyltransferase (DAT) in the last reaction of TIA pathway. In this chapter, we present the results of transformation for the purpose of creating new transgenic *C. roseus* lines carrying the *C. roseusDAT* (*CrDAT*) gene to increase vincristine and alkaloid biosynthesis in transgenic plants. The *CrDAT* gene (cDNA) isolated from the *C. roseus* cultivar TN1 has been cloned, and its nucleotide has been sequenced. The *CrDAT* gene has a size of 1320 bp, encoding 439 amino acids. Murashige and Skoog (MS) medium containing 30 g L^{-1} sucrose, 1.0 mg L^{-1} BAP, 0.6 mg L^{-1} IBA, 10 g L^{-1}agar, and coconut water (100 ml L^{-1}) at pH of 5.8 is suitable for shoot generation and growth from the stem segments (Murashige and Skoog 1962). Shoots induced from cotyledon nodes occurred in the similar MS medium with 0.5 mg L^{-1} BAP and 0.4 mg L^{-1} IBA. The efficiency of *GUS* gene transfer into the *C. roseus* via cotyledons was optimized at an *A. tumefaciens* density of OD 600 nm = 0.8, 100 µM acetosyringone and infection in 30 min, selected by 50 mg L^{-1} kanamycin. Structure 35S-CrDAT-cmyc was successfully transformed into rose-pink *C. roseus* TN with a genetic transformation efficiency of 1.02%. The integration of *CrDAT* transgene into the genome of *C. roseus* was confirmed to be successfully transferred and synthesized recombinant CrDAT (rCrDAT). Four T1 generation transgenic *C. roseus* lines, T1-1, T1-3, T1-6, and T1-7, obtained rCrDAT that the range of its content was 2.86–5.12 (µg mg^{-1}). Additionally, T1 generation lines had significant 1.63- and 2.48-fold increase in the vincristine content compared to that in

L. T. N. Nguyen · Q. H. Nguyen · M. H. Chu (✉)
School of Biology, Thai Nguyen University of Education, Thai Nguyen 250000, Viet Nam
e-mail: chuhoangmau@tnu.edu.vn

L. T. N. Nguyen
e-mail: lanntn@tnue.edu.vn

Q. H. Nguyen
e-mail: quannh@tnue.edu.vn

L. T. K. Vu
Institute of Theoretical and Applied Research, Duy Tan University, Hanoi 100000, Viet Nam
e-mail: vutkimlien@duytan.edu.vn

L. T. K. Vu
Faculty of Natural Sciences, Duy Tan University, Da Nang 550000, Viet Nam

H. T. Bui
Department of Biology, Thai Nguyen University of Medicine and Pharmacy, Thai Nguyen 250000, Viet Nam

© The Author(s), under exclusive license to Springer Nature Switzerland AG 2022

C. Kole (ed.), *The Catharanthus Genome*, Compendium of Plant Genomes,
https://doi.org/10.1007/978-3-030-89269-2_6

nontransgenic plants (WT), varying between 6.91 and 10.53 (μg g^{-1}fresh weight), in which the highest vincristine content belongs to T1-1. The total alkaloid content of the four transgenic lines was also higher than the total alkaloid content of WT plants (increased by 3.16- and 5.57-fold). Hence, overexpression of DAT enhanced the vincristine and total alkaloid content of transgenic *C. roseus* plants.

6.1 Introduction

Catharanthus roseus (L.) G. Don, commonly called as periwinkle, is a valuable medicinal plant capable of producing terpenoid indole alkaloids (TIAs). The pharmaceutical TIAs consist of vindoline, catharanthine, ajmalicine, ajmaline, serpentine, β-yohimbine, vinblastine, and vincristine, among which vinblastine and vincristine are the most important indole alkaloids. These compounds exhibit strong inhibition effect on cell division that leads to decrease cell proliferation. Thus, they are considered as clinically useful anticancer agents. TIAs are found in all parts of *C. roseus*, most commonly in leaves and roots. However, TIAs which have anticancer activities such as vinblastine and vincristine are found in very low concentrations in leaves (Pahwa 2009). Hence, there have been many research into effective ways to improve alkaloid production in the *C. roseus* (Van der Heijden et al. 2004). At the molecular level, the metabolic pathway for TIA synthesis in *C. roseus* is very complex, governed by many factors, with the participation of 30 enzymes and 35 intermediate reactions (Van der Heijden et al. 2004; Facchini and De Luca 2008; Zhao et al. 2013). TIA synthesis in *C. roseus* endures four pathways including methyler-ythritol phosphate, seco-iridoid, shikimate, and vindoline pathway. The vindoline pathway is well described, which starts with tabersonine undergoing six intermediate enzyme—catalyzed reactions to form deacetylvin doline. Firstly, tabersonine is changed to 16-hydroxytabersonine catalyzed by the enzyme tabersonine-16-hydroxylase and then methylated with 16-hydroxytabersonine-16-O-methyl transferase to make 16-methoxytabersonine (Levac et al. 2008). Then, oxidation occurs to change from 16-methoxy tabersonine to 16-methoxy-2, 3-dihydrotabersonine by the enzyme tabersonine 3-oxygenase. The next steps involve an enzyme N-methyltransferase to make deacetoxyvindoline (Liscombe et al. 2010). The final reactions are catalyzed by two enzymes, deacetoxyvindoline-4-hydroxylase, creating deacetylvindoline and the enzyme DAT, to form vindoline. Thus, DAT is the vital enzyme in the synthesis pathway of vindoline and catharanthine, two precursors for synthesis of vinblastine and vincristine. Overexpression of the *DAT* gene will increase intracellular DAT accumulation and secondary metabolic products, particularly vinblastine and vincristine. Therefore, the research direction of enhancing the expression of the gene encoding the DAT enzyme will improve the accumulation of vinblastine and vincristine in *C. roseus*.

To date, studies of alkaloid improvement in *C. roseus* have targeted to create transgenic *C. roseus* plants. Generation of genetically modified hairy root lines via *Agrobacterium rhizogens* and expression analysis of *DAT*, *DXS (1-deoxy-D-xylulose synthase)*, and *G10H (geraniol-10-hydroxylase)* genes in the hairy roots confirmed that the expression of the CrDAT protein altered the alkaloid metabolism and increased the accumulation of TIAs in *C. roseus* (Magnotta et al. 2007; Peebles et al. 2011; Zhou et al. 2012; Hanafy et al. 2016). Together with studying the direction of hairy root formation, Wang et al. (2010) commented that G10H and ORCA3 transgenic *C. roseus* hairy roots had 6.5 times higher catharanthine accumulation than non-transgenic hairy roots. Another study showed that overexpression of the *ORCA3* and *G10H* genes resulted a significant increase in vindoline, catharanthine, and ajmalicine accumulation in *C. roseus* plants (Pan et al. 2012). Research on creating transgenic *C. roseus* by *A. tumefaciens*-mediated transformation has been a concern of several authors (Wang et al. 2012; Verma et al. 2017; Vi et al. 2019). A regeneration system from hypocotyls and genetic transformation in *C. roseus* through *A. tumefaciens* was developed, and *DAT* gene overexpression was confirmed by real-time PCR analysis demonstrated overexpression via transcription (Wang et al. 2012) and was confirmed by analysis of overexpression of recombinant DAT protein (Pham et al. 2019). This chapter presents the result of the genetic transformation of *CrDAT* to increase the accumulation of TIAs in *C. roseus* plants.

6.2 Alkaloid in *C. roseus* Plant and Its Transgenic Vector

6.2.1 Biosynthesis of Alkaloid in *C. roseus* Plant

C. roseus is a herbaceous perennial plant, which often grows in thick bushes (Do et al. 2004). The commonly found flower colors in *C. roseus* are rose-pink, white, or white with a reddish eye, and the rose-pink *C. roseus* has the highest the amount of vincristine and vinblastine (Pahwa 2009). The seeds of rose-pink *C. roseus* variety TN supplied by the Thai Nguyen Center for Plant Variety, Vietnam, are used as transgenic material.

C. roseus constitutes a remarkable source of TIAs, which include vindoline, serpentine, catharanthine, vinblastine, and vincristine. These compounds from *C. roseus* plants are utilized for treating serious diseases as well as developing new drugs (Almagro et al. 2015; Liu et al. 2017). Vinblastine and vincristine are very important alkaloids in *C. roseus* plants that are used for the clinical treatment of various human cancers. Nevertheless, a very small amount of these secondary metabolites is naturally produced in *C. roseus* plants. To increase alkaloid production in *C. roseus* plants, several approaches have been explored to enhance the biomass of cell or hairy root (Wang et al. 2012; Iskandar and Iriawati 2016).

In *C. roseus*, the enzyme DAT is encoded by the *CrDAT* gene. The *CrDAT* isolated from *C. roseus* is 1320 bp in size, encoding 439 amino acids. The CrDAT protein has two regions, the PLN02481-omega-hydroxypalmitate O-feruloyl transferase region, from amino acids 8 to 438, and the transferase region from amino acids 11 to 434 which belongs to the transferase family. Anthranilate N-hydroxycinnamoyl/benzoyl transferase and DAT EC: 2.3.1.107 are also members of this family, which catalyzes the first step in phytoalex in biosynthesis and the last reaction of the vindoline biosynthesis, respectively. Protein deduced from the *CrDAT* gene isolated from the TN rose-pink rose *C. roseus* plants and the *CrDAT* sequence are used to design the PCR primer pair. The GenBank accession number AF053307 has 99.3% similarity, differing only in three amino acid positions: 427 (F - L), 428 (E - R), 429 (K - R). The spatial structure model of the DAT protein of the TN *C. roseus* plants and the DAT protein of the gene with the accession number AF053307, were established according to Kelley et al. (2015) and showed similarities (Fig. 6.1).

6.2.2 CrDAT Transgenic Vector

The DAT enzyme is involved in the reaction to create vindoline (Fig. 6.2). In our approach, the overexpression of the *CrDAT* gene is expected to enhance the accumulation of secondary metabolites and end-products of *C. roseus* in the TIA pathway. Thus, the content of vinblastine and vincristine in *C. roseus* plants is enriched by the *CrDAT* overexpression.

The *CrDAT* gene in the *pBT_CrDAT* vector was used to create a transgenic vector with a *CrDAT* gene expression structure in four steps (Fig. 6.3) (Bui et al. 2015a, b). The *CrDAT* gene was isolated from the cloning vector *pBT_CrDAT* (Fig. 6.3_ IA), the vector ring pRTRA7/3 was opened with *Nco I/Not I* (Fig. 6.3_II-C), and *CrDAT* was crossed into pRTRA7/3 to create the recombinant vector *pRTRA7/3_CrDAT* (Fig. 6.3_III-F). The structure *35S_CrDAT_cmyc_KDEL_polyA* was isolated from the recombinant vector *pRTRA7/3_CrDAT* (Fig. 6.3_IV-H), and the vector loop pBI121 was opened with *Hind*III (Fig. 6.3_IV-I). Grafting *35S_CrDAT_cmyc_KDEL_polyA* into pBI121 opened the ring to create a transgenic vector carrying a structure

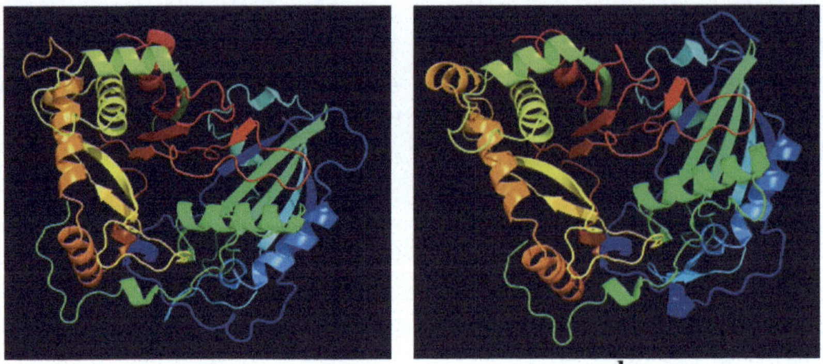

a b

Fig. 6.1 Structure model of CrDAT protein was established according to Kelley et al. (2015). **a** DAT protein was deduced from the *CrDAT* gene (accession number AF053307 in GenBank). Image colored by rainbow N → C terminus; Model dimensions (Å): X: 71.988 Y: 54.257 Z: 60.411. **b** DAT protein was deduced from the *CrDAT* gene isolated from TN periwinkle. Image colored by rainbow N → C terminus; Model dimensions (Å): X: 66.366 Y: 55.399 Z: 65.400

containing the *CrDAT* transgene (Fig. 6.3_IV-K). Each experiment was tested with agarose gel electrophoresis. The transgenic vector *pBI121_CrDAT* was transformed into *A. tumefaciens* and examined by colony-PCR.

6.3 Establishment of Genetic Transformation

6.3.1 Transformation Media

Investigating the effect of BAP only on shoot regeneration of *C. roseus* on five concentration scales 0, 0.5, 1.0, 1.5, and 2.0 mg L^{-1} after six weeks, the results showed that the number of shoots regenerated from the stem segment was 3.83 ± 0.26 (shoots/explant) at 1.0 mg L^{-1}BAP, and the number of shoots was 3.36 ± 0.21 (shoots/explant) from cotyledons at 0.5 mg L^{-1} BAP. Hence, the concentration of 1.0 and 0.5 mg L^{-1} BAP was selected to supplement in shoot produced media in combination with different concentrations of IBA (0.2, 0.4, 0.6, and 0.8 mg L^{-1}). After six weeks, the maximum number of shoot regeneration (7.41 ± 0.26 shoots/explant) from the stem segments were achieved on the medium supplemented with 1.0 mg L^{-1} BAP and 0.6 mg L^{-1} IBA. Similarly, the highest number of shoots (7.87 ± 0.37) was obtained on the medium containing 0.5 mg L^{-1} BAP and 0.4 mg L^{-1} IBA and ($p < 0.05$). The combination of 1.0 mg L^{-1} BAP and 0.6 mg L^{-1} IBA produced maximum number of shoots from the *C. roseus*stem segments. Alternatively, 0.5 mg/BAP combined with 0.4 mg L^{-1} IBA are also suitable for inducing shoots from *C. roseus* cotyledons. It was observed that shoot number per explant was similar from stem segments and cotyledons in each experiment formula (BAP alone or the combination of BAP and IBA).

6.3.2 Transformation Procedure

The *β—glucuronidase (GUS)* is the most important and common reporter gene in plant (Jefferson et al. 1987). Therefore, the transgenic vector containing *GUS* was used as an indicator to recognize the *GUS* expression in *C. roseus* plant and to establish transformation procedure for *CrDAT*. Many transgenic vectors such as pB1121 and pCAM301 are designed to carry the *GUS*, but the disadvantage of these vectors is that the intron segment was not designed at the beginning 5′ before the *GUS*. Thus, when X-gluc is stained, the *GUS* is expressed in both bacterial and plant cells, which makes it difficult to accurately recognize the presence of the *GUS* in plant tissues. To

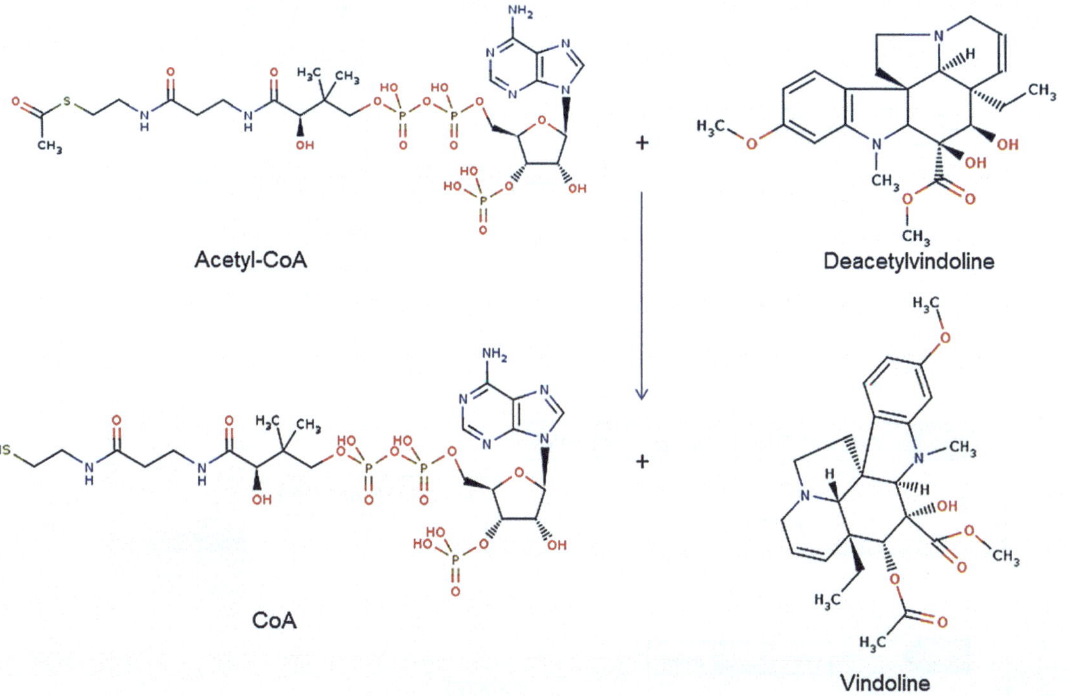

Fig. 6.2 Reaction schemes on vindoline generation with the participation of deacetylvindoline 4-O-acetyltransferase (DAT) DAT-*Nco*I_F/DAT–*Not*I_R primer pair for PCR to amplify the *CrDAT* gene from *C. roseus* variety TN was designed based on the *CrDAT* sequence of *C. roseus* published in GenBank with the accession number AF053307 (St-Pierre et al. 1998). Total RNA was extracted from *C. roseus* leaves (Saghai-Maroof et al. 1984), and cDNA was synthesized from total RNA by reverse transcription (Sambrook and Russell 2001). *CrDAT* was amplified from cDNA by PCR approximately 1.3 kb in size and cloned into the pBT vector and *E. coli* DH5α. The results of nucleotide sequencing showed that the cDNA segment isolated from the TN *C. roseus* sample had a size of 1320 bp. BLAST analysis in NCBI showed that the nucleotide sequence of the cDNA segment isolated from the TN *C. roseus* is the *DAT* gene segment of the *C. roseus* (*CrDAT*). The *CrDAT* gene isolated from the mRNA of TN *C. roseus* was published in GenBank with the accession number LN809930 (Bui et al. 2015b)

overcome this obstacle, the transgenic vector carrying the *GUS* was designed to add an intron at the 5′ end before the *GUS*. Therefore, only plant tissues having *GUS* activity are recognized by indigo dye deposits after staining with X-Gluc (Vancanneyt et al. 1990). The *A. tumefaciens* (strain C58/pGV2260) carrying vector pCB-gusplus was selected in this study.

The present study investigated factors affecting on *GUS* gene transformation efficiency including *A. tumefaciens* density (OD_{600nm} at 0.6, 0.8 and 1.0) concentration of acetosyringone (100 Mμ, 150 Mμ và 200 Mμ), infection time (10, 20 and 30 min), and kanamycin concentration (0, 25, 50, 75, 100 và 150 mg L^{-1}). The experiment was conducted to optimize conditions for transferring GUS into *C. roseus* plant. Thus, the transfer of the *GUS* via cotyledons mediated by *A. tumefaciens* with OD600 nm at 0.8, acetosyringone concentration 100 μM for 30 min and 50 mg L^{-1}kanamycin was used as selective agent. In this study, 84 explants (23.3%) generated shoots from 360 transformed cotyledons after two weeks. However, after four weeks, 34 shoots (9.45%) continued to grow, of which 21 shoots induced roots. After transfer to greenhouse conditions, 15 plantlets successfully grew up; their leaves and roots were treated with

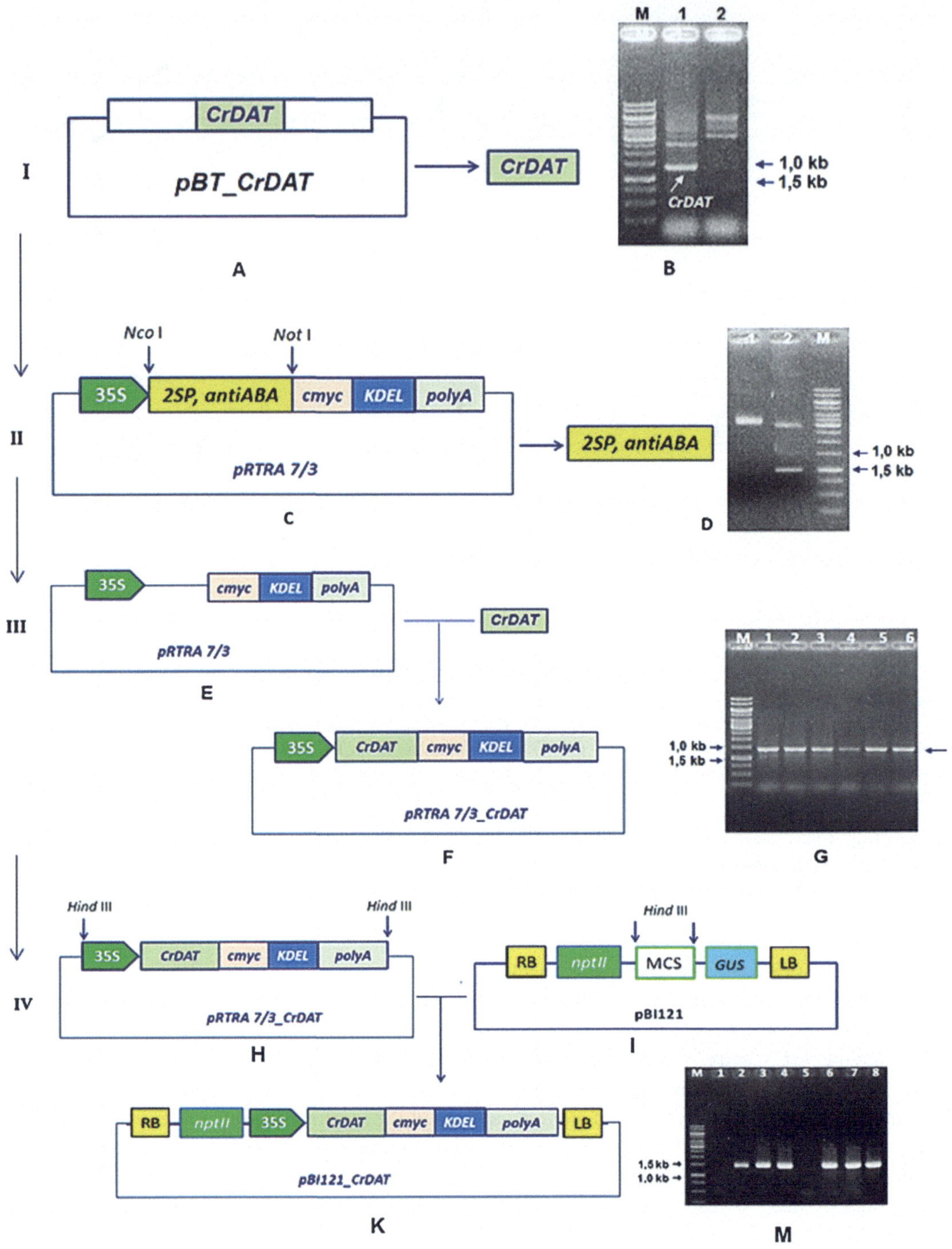
M 1 2
CrDAT
pBT_CrDAT
CrDAT
1,0 kb
1,5 kb
CrDAT
I
A
B
Nco I
Not I
35S
2SP, antiABA
cmyc
KDEL
polyA
pRTRA 7/3
2SP, antiABA
1 2 M
1,0 kb
1,5 kb
II
C
D
35S
cmyc
KDEL
polyA
pRTRA 7/3
CrDAT
III
E
M 1 2 3 4 5 6
35S
CrDAT
cmyc
KDEL
polyA
pRTRA 7/3_CrDAT
1,0 kb
1,5 kb
F
G
Hind III
Hind III
35S
CrDAT
cmyc
KDEL
polyA
pRTRA 7/3_CrDAT
Hind III
RB
nptII
MCS
GUS
LB
pBI121
IV
H
I
RB
nptII
35S
CrDAT
cmyc
KDEL
polyA
LB
pBI121_CrDAT
M 1 2 3 4 5 6 7 8
1,5 kb
1,0 kb
K
M

◀ **Fig. 6.3** Diagram of creating transgenic vector and electrophoresis to check the results of process to create the *pBI121_CrDAT* transgenic vector. **I-**Cut the vector *pBT_CrDAT* by *Nco I/Not I*. A: *pBT_CrDAT*; B: Results of electrophoresis to check cutting products *pBT_CrDAT*; M: DNA 1 kb; 1: *pBT_CrDAT* cut; 2: *pBT_CrDAT* uncut. **II-**Open the vector ring pRTRA7/3. C: Cut vector pRTRA7/3 by *Not I/Nco I* to create pRTRA7/3 opened ring and segment 2SP, antiABA. D: Results of electrophoresis test for cutting products pRTRA7/3; M: DNA 1 kb; 1: pRTRA7/3 uncut; 2: pRTRA7/3 cut. **III-**Results of creating recombinant vector pRTRA7/3_CrDAT. E: pRTRA7/3 opened round; F: pRTRA7/3_CrDAT; G: Results of electrophoresis checking colony-PCR products of colony lines containing pRTRA7/3_CrDAT. M: 1 kb DNA size marker; 1–6: Colony lines. **IV-**Creating the transgenic vector pBI121_CrDAT. H: Cut vector pRTRA7/3_CrDAT by *Hind* III to create structure *35S_CrDAT_cmyc_KDEL_polyA*; I: Open the ring pBI121 with *Hind* III; K: hybridization *35S_CrDAT_c-myc_KDEL_polyA* into pBI121 to create transgenic vector pBI12_CrDAT; M: Results of colony-PCR tests for *A. tumefaciens* lines containing transgenic vector pBI121_CrDAT

X-gluc to determine *GUS* activity in transformed *C. roseus*. The leaves and roots of 15 plants were stained blue (Fig. 6.4), and the transgenic efficiency was 4.17%.

6.4 The *CrDAT* Gene Transformation

C. roseus is a dicotyledonous plant, and the application of genetic transformation in dicots for the purpose of gene overexpression to increase the protein content, which is a product of the same gene, has been of interest in the research of many authors (Dao et al. 2015; Lo et al. 2015; Vi et al. 2017, 2019; Vu et al. 2018; Nguyen et al. 2019, 2020). For dicotyledonous plants, *Agrobacterium*-mediated genetic transformation was used effectively in bacterial infection via cotyledon nodes (Dao et al. 2015; Lo et al. 2015; Nguyen et al. 2019, 2020). The *C. roseus* seed is small, and the cotyledon size is also very small, making it more difficult to create transgenic material in the *C. roseus* than in other dicotyledonous plants, whose seeds are as large as legume seeds. Therefore, more skillful techniques might be required not to damage the cotyledon axillary system. Injured cotyledons were soaked in *A. tumefaciens* bacteria, cocultured, and then transferred to a shoot regeneration medium to produce transgenic *C. roseus* roots (Fig. 6.4).

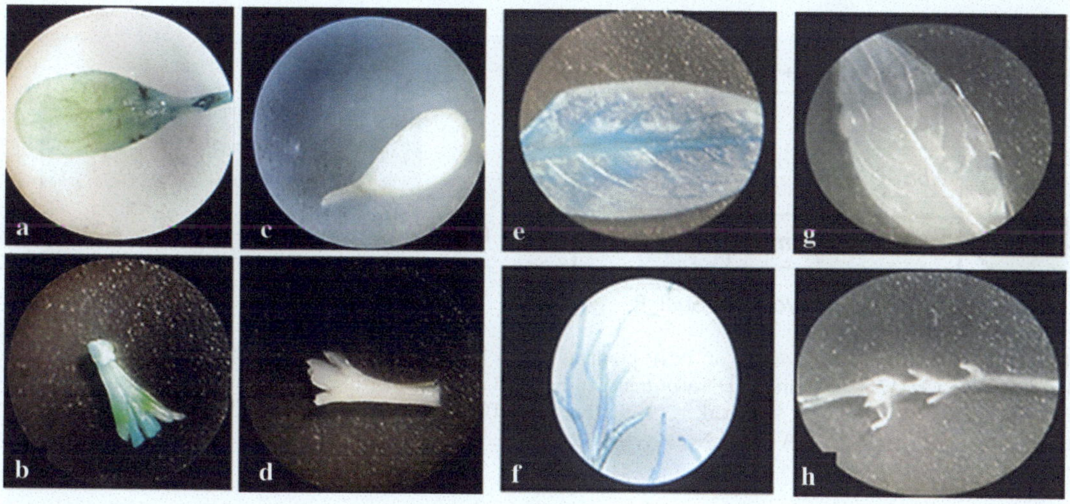

Fig. 6.4 Results of *GUS* gene expression in transgenic *C. roseus* plant. Temporary expression *GUS* in transgenic cotyledon (**a**) and stem segment (**b**); **c, d**: The cotyledon (**c**) and the stem segment (**d**) of WT. **e** *GUS* transgenic *C. roseus* leaves; **f** *GUS* transgenic *C. roseus* root; **g** The WT leaves; **h** Roots of WT

Cotyledons obtained from seeds of germinating *C. roseus* were used as transformed explants. The structure 35S_CrDAT_cmyc carrying the *CrDAT* was transferred into *C. roseus* through *A. tumefaciens* and infected through the cotyledon nodes (Olhoft et al. 2006). The transformation process consisted of the stage of culturing and selecting, regenerating the plant, and creating transgenic *C. roseus*.

Three hundred ninety transformed samples were created as the result of transformation of the *CrDAT* with repetition of three times, among which there were 233 shoot-forming samples. Through two antibiotic selections in the multi-shoot regeneration and shoot growth stages, 164 survival shoots were cultured on root inducing medium. After that, 120 rooted shoots were selected as putative transgenic plantlets. Among which, 70 plantlets with normal stem, leaf, and root morphology were transferred to pots filled with substrates, and 19 plants continuously grew under greenhouse conditions. Hence, out of 390 transformed explants, only 19 putative *CrDAT* transgenic plants (4.87%) were able to grow under greenhouse conditions. Only 7 of 19 transgenic plants (T0 generation) were well developed in the greenhouse. The seven T0 generation plants were named as T0-1, T0-2, T0-3, T0-4, T0-5, T06, and T0-7.

6.5 Detection of the Presence of the *CrDAT* Transgene

6.5.1 PCR and Southern Blot Analysis

Seven transgenic plants obtained from transformation process were analyzed for the presence of the *CrDAT* transgene using PCR. Total DNA was extracted from young leaves of T0 transgenic plants, and PCR was performed using the pair of primer *XbaI-DAT_F/cmyc-KDEL-Sac I_R*. Electrophoresis results of the PCR product on agarose gel showed the presence of a single DNA band that is about 1.3 kb in size, which is similar to the *CrDAT* transgene (Fig. 6.5).

The endogenous *CrDAT* gene of *C. roseus* without cmyc-tagged and KDEL is 1320 bp in size; *CrDAT* transgene was tagged cmyc and KDEL with a size of 1383 bp, containing the cutoff point of restriction enzymes *Xba*I and *Sac*I. Distinguishing point between endogenous *CrDAT* gene and *CrDAT* transgene expressed in PCR result with the pair of primer *Xba*I- DAT-F/cmyc-KDEL-*Sac*I-R. PCR amplified only the *CrDAT-cmyc-KDEL* structure and contained cmyc-tagged and KDEL (1383 bp). Thus, all T0 transgenic *C. roseus* plants were confirmed for the presence of the *CrDAT* transgene by PCR. However, it is necessary to continue to

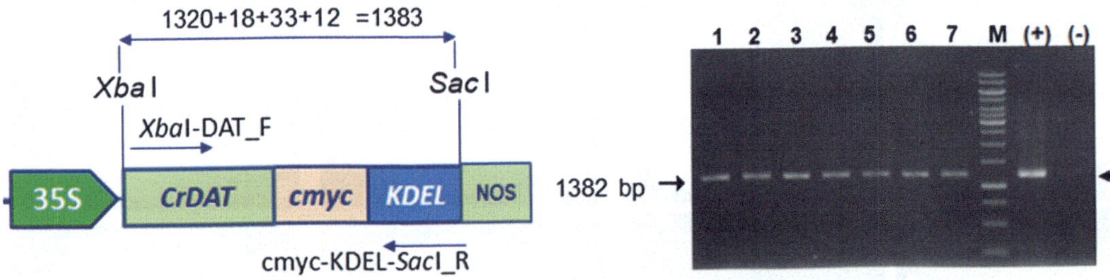

Fig. 6.5 Results of electrophoresis of PCR products amplifying the *CrDAT* gene from the DNA of transgenic *C. roseus* plants. M: 1 kb DNA size marker; (+) positive control was plasmid pBT_CrDAT; (−) Negative control was PCR result of cloning *CrDAT* gene from WT; 1, 2, 3, 4, 5, 6, 7 lines of *C. roseus* transgenic *CrDAT* (T0-1, T0-2, T0-3, T0-4, T0-5, T0-6, T0-7)

demonstrate by other molecular techniques to identify whether *CrDAT* transgene is integrated into the genome of the host cells.

The integration of the *CrDAT* transgene in the T0 generation was verified by Southern blot (Southern 1975). Seven PCR-positive transgenic *C. roseus* plants and control plants were used for Southern blot analysis. Genomic DNA extraction from the C. roseus leaves of the transgenic and control plants were purified and treated with *Sac*I. The results of cutting the total DNA of transgenic *C. roseus* lines were DNA fragments containing the *npt*II gene of different sizes (Fig. 6.5a). The result of Southern blot revealed that the presence of DNA band appeared in transgenic lines T0-1, T0-2, T0-4, and T0-7, whereas transgenic lines T0-3, T0-5, T0-6, and nontransgenic plants did not exhibit any DNA band. T0-2 and T0-4 plants had two copies; T0-1 and T0-7 had a unique copy. Southern blot-positive transgenic lines were further evaluated for growth, development, and seed collection for the analysis of transgenic plants in the T1 genome. Thus, the PCR and Southern blot results confirmed that the *CrDAT* transgene was successfully transferred and integrated into the *C. roseus* genome.

6.5.2 Western Blot and ELISA Analysis

T0-1, T0-2, T0-4, and T0-7 lines produced seeds that were planted to create T1 transgenic plants named T1-1, T1-2, T1-4, and T1-7, respectively. Protein extracted from leaves of these plants was used for SDS-PAGE (sodium dedocyl sulfate–polyacrylamide gel electrophoresis) protein electrophoresis analysis on a polyacrylamide gel and Western blot analysis. The rCrDAT protein distinguished from endogenous CrDAT protein contained the cmyc tag at the C end of the polypeptide chain. Therefore, Western blot allowed the detection of proteins on nitrocellulose hybrid membranes that carried cmyc tags when binding specific antibodies and for a color reaction with substrates (Fig. 6.5b).

Western blot analysis results in Fig. 6.5b show that on the nitrocellulose hybrid membrane, a color band appeared at approximately 51 kDa in all four T1 transgenic lines, which are similar to the molecular size of the CrDAT protein, and the *CrDAT* transgene in four T1 transgenic *C. roseus* lines translated into rCrDAT protein. The results of Western blot analysis revealed that the inheritance of *CrDAT* transgene from T0 transgenic plants that was successfully done and expressed as the CrDAT protein in T1 generation. The transgenic efficiency at the translation stage was 1.02% (4/390).

The content of rCrDAT from four transgenic lines in the T1 generation was analyzed using enzyme-linked immunosorbent assay (ELISA) assay. The total protein extracted from *C. roseus* leaves was denatured and analyzed by electrophoresis on a 10% SDS-PAGE gel (Laemmli 1970) and then transferred to a nitrocellulose membrane. The membranes were blocked with 5% skim milk in PBS-T overnight. The protein-containing films were incubated with cmyc antibodies for 3 h, shaken at room temperature, and incubated with secondary antibodies for 2 h. The results were determined using TMB (3, 3′, 5, 5′-tetramethyl benzidine). The content of the recombinant CrDAT (rCrDAT) protein was determined by ELISA (Sun et al. 2006).The results showed that four transgenic *C. roseus* lines had recombinant protein content varying between 2.87 and 5.12 (μg mg^{-1}). The T1-1 had the highest rCrDAT protein (5.12 μg mg^{-1}), and the T1-2 had the lowest rCrDAT protein (2.87 μg mg^{-1}). The rCrDAT protein content in T1 plants was placed in rank order from low to high as T1-2 < T1-7 < T1-4 < T1-1. Thus, the *CrDAT* gene in T1 generation was enhanced, which led to increase in the CrDAT content in transgenic plants compared to the nontransgenic plants.

6.5.3 Alterations in Total Alkaloid and Vincristine Content

Four T1 transgenic *C. roseus* plants exhibited normal morphological characteristics as WT plants. Analysis of the total alkaloid content of four T1 transgenic *C. roseus* lines and WT plants showed that T1 plants had total alkaloid content ranging from 0.38 g100 g^{-1} to 1.82 g100 g^{-1} (dry weight). The alkaloid content, which was highest in T1-1 line (1.82 g100 g^{-1}) was reduced in the T1-4 (1.52 g100 g^{-1}) and T1-7 (0.64 g100 g^{-1}). The T1-2 line had the lowest alkaloid content (0.38 g100 g^{-1}); meanwhile, the non-transgenic plants had an alkaloid content of 0.12 g100 g^{-1}. Thus, compared to the WT plants, the total alkaloid content of the *CrDAT* transgenic *C. roseus* lines increased from 3.16 to 15.17 (times); this result demonstrated that overexpression of *CrDAT* in *C. roseus* increased CrDAT content and led to enhance CrDAT enzyme activity. The implementation of *CrDAT* transgenic technique can improve the alkaloid content in *C. roseus* accordingly.

In the TIA biosynthesis pathway of *C. roseus*, the final product is vincristine that is involved with activity of *CrDAT*. Vincristine content of the four T1 transgenic lines and WT was found at the range of 4.25 to 10.53 (μg g^{-1} fresh weight). Out of four T1 plants and WT, the vincristine content in the T1-1 line ranked number one with a 2.48-fold increase compared to WT, the vincristine content of the T1-6 and T1-7 plants significantly decreased with 7.18 μg g^{-1} and 7.02 μg g^{-1}, respectively. The T1-3 plant had the lowest content of vincristine (6.91 μg g^{-1}) in comparison to other T1 generation lines, but nevertheless it was approximately 1.63-fold increase as compared with WT. Therefore, the transgenic *C. roseus* plants had a higher content of vincristine based on the rCrDAT overexpression.

A. rhizogene-mediated transformation with the *DAT* involving the production of hairy roots in *C. roseus* was reported by Magnotta et al. (2007). DAT enzyme activity in hairy root lines was confirmed to be significantly higher than DAT enzyme activity in nontransgenic hairy root lines. In addition, previous research of Li et al. (2015) indicated that the overexpression of *CrBPF1* leads to elevated gene expression level associated in both indole and terpenoid biosynthesis; thus, precursors for TIA have been produced as a result. In addition, the *CrBPF1* overexpression plays a crucial role in adapting the transcriptional levels for 11 out of 13 genes, which were predicted to regulate TIA transcriptional repressors. According to Wang et al. (2012), *DAT* transgene expression was analyzed with RT-PCR and Western blot analysis to confirm integration of the *DAT* transgene into the *C. roseus* genome. The results showed that the amount of vindoline in transgenic *C. roseus* plants ranged between 1.42 and 2.72 (μg mg^{-1} dry weight), whereas the amount of vindoline in WT (1.15 μg mg^{-1}) was remarkably lower than T1 transgenic plants. This result suggested that mRNA degradation or nontranslation of *DAT* transgene might be occurred due to inhibition at the posttranscriptional level following RNA interference in cells. In our work, the results of Western blot and ELISA assay confirmed the overexpression of the *CrDAT* transgene and the presence of the rCrDAT protein in *C. roseus* plants (Fig. 6.5b). The alkaloid content in the T1-1, T1-2, T1-4, and T1-7 plants was threefold higher than the alkaloid content in the WT plants. As depicted in Fig. 6.5a, the integration of the *CrDAT* transgene into different loci of the host genome might lead to improved rCrDAT protein content in the T1 generation lines, which can be ranked in order from high to low values of rCrDAT protein content as T1-1 > T1-4 > T1-7 > T1-2 > WT. The TIA biosynthetic pathways in *C. roseus* showed overexpression of the rCrDAT protein that increases the content of vindoline, vinblastine, and vincristine. Vincristine, the final product of TIA biosynthesis related to the expression of *CrDAT*, has been estimated to be naturally low. Hence, this study aims to improve the expression of the *CrDAT* gene. The amount of vincristine in the T1 transgenic *C. roseus* plants and WT was performed using an Agilent HPLC system. Four T1 transgenic lines had an increase in vincristine content compared to WT, from 162.59% to 247.76%.

6.6 Conclusions

The *CrDAT* gene isolated from the TN *C. roseus* variety has been cloned, and nucleotides have been sequenced. The *CrDAT* (cDNA) gene has a size of 1320 bp, encoding 439 amino acids. The MS medium contained 30 g L^{-1} sucrose, 10 g L^{-1} agar, 1.0 mg L^{-1}BAP, 0.6 mg L^{-1} IBA, and 100 ml L^{-1} coconut water, which was adjusted to pH 5.8 and was suitable for shoot generation from stem segments. Cotyledons were cultured on the same MS medium with BAP (0.5 mg $L^{-1)}$ and IBA (0.4 mg L^{-1}) performed the best in shoot generation and growth. The efficiency of *GUS* gene transfer into the *C. roseus* via cotyledons was optimized at an *A. tumefaciens*density with OD600 nm of 0.8, 100 μM acetosyringone and 30 min infection period, selected by 50 mg L^{-1} kanamycin. Structure *35S-CrDAT-cmyc* was successfully transformed into rose-pink *C. roseus* TN with a genetic transformation efficiency of 1.02%. rCrDAT protein was expressed in four T1 transgenic *C. roseus* lines with a molecular weight of approximately 51 kDa, and the recombinant protein content ranged from 2.87 to 5.12 (μg mg^{-1}). The *CrDAT* transgenic *C. roseus* lines in the T1 generation had increased total alkaloid content compared with the nontransgenic control from 3.16 to 15.17 (times).

References

Almagro L, Fernández-Pérez F, Pedreño MA (2015) Indole alkaloids from *Catharanthus roseus*: bioproduction and their effect on human health. Molecules 20:2973–3000

Bui HT, Ho MT, Hoang PH, Le VS, Nguyen TT, Chu HM (2015a) Molecular cloning and designing transgenic vector carrying DAT gene isolated from *Catharanthus roseus* (L.) G. Don). Acad J Biol 37 (2):236–243. https://doi.org/10.15625/0866-7160/v37n2.6835

Bui HT, Hoang HP, Nguyen TT, Chu MH (2015b) *Catharanthus roseus* mRNA for DAT enzyme (DAT gene), isolate TN1 (Pink-purple flower). GenBank: LN8099301. https://wwwncbinlmnihgov/nuccore/LN809930. Accessed November 5, 2020

Dao XT, Ho MT, Vu TTT, Le VS, Chu HM (2015) Cloning and overexpression of GmDREB2 gene from a Vietnamese drought-resistant soybean variety. Braz Arch Biol Technol 58:651–657. https://doi.org/10.1590/S1516-89132015050170

Do HB, Dang QT, Bui XC, Nguyen TD, Do TD, Vu NL, Pham DM, Pham KM, Nguyen T, Tran T (2004) Medicinal plants and animals in Vietnam, vol 1. Hanoi Science & Technology Publishing House, Vietnam

Facchini PJ, De Luca V (2008) Opium poppy and Madagascar periwinkle: model non-model systems to investigate alkaloid biosynthesis in plants. Plant J 54:763–784

Hanafy MS, Matter MA, Asker MS, Rady MR (2016) Production of indole alkaloids in hairy root cultures of *Catharanthus roseus* L. and their antimicrobial activity. S Afr J Bot 105:9–18. https://doi.org/10.1016/j.sajb.2016.01.004

Iskandar NN, Iriawati I (2016) Vinblastine and vincristine production on Madagascar periwinkle (*Catharanthus roseus* (L.) G. Don) callus culture treated with polethylene glycol. Makara J Sci 20(1):7–16. https://doi.org/10.7454/mss.v20i1.5656

Jefferson RA, Kavanagh TA, Bevan MW (1987) GUS fusion: β-glucuronidase as a sensitive and versatile gen fusion marker in higher plants. EMBO J 6:3901–3907

Kelley LA, Mezulis S, Yates CM, Wass MN, Sternberg MJE (2015) The Phyre2 web portal for protein modeling, prediction and analysis. Nat Protoc 10:845–858. https://doi.org/10.1038/nprot.2015.053

Laemmli UK (1970) Cleavage of structural protein during the assembly of the head of bacteriophage T4. Nature 227:680–685

Levac D, Murata J, Kim WS, De Luca V (2008) Application of carborundum abrasion for investigating leaf epidermis: molecular cloning of *Catharanthus roseus* 16-hydroxy-tabersonine-16-Omethyltransferase. Plant J 53:225–236

Li CY, Leopold AL, Sander GM, Shanks JV, Zhao L, Gibson SI (2015) CrBPF1 overexpression alters transcript levels of terpenoid indole alkaloid biosynthetic and regulatory genes. Front Plant Sci 6:818. https://doi.org/10.3389/fpls.2015.00818

Liscombe DK, Usera AR, O'Connor SE (2010) Homologue of tocopherol C methyltransferases catalyses N methylation in anticancer alkaloid biosynthesis. Proc Natl Acad Sci USA 107(44):18793–18798. https://doi.org/10.1073/pnas.1009003107

Liu J, Cai J, Wang R, Yang S (2017) Transcriptional regulation and transport of terpenoid indole alkaloid in *Catharanthus roseus*: Exploration of new research directions. Intl J Mol Sci 18(1):53. https://doi.org/10.3390/ijms18010053

Lo TS, Le HD, Nguyen VTT, Chu HH, Le VS, Chu HM (2015) Overexpression of a soybean expansin gen, GmEXP1, improves drought tolerance in transgenic tobacco. Turk J Bot 39(6):988–995. https://doi.org/10.3906/bot-1502-40

Magnotta M, Murata J, Chen J, De Luca V (2007) Expression of deacetylvindoline-4-O-acetyltransferase in *Catharanthus roseus* hairy roots. Phytochemistry

68(14):1922–1931. https://doi.org/10.1016/j.phytochem.2007.04.037

Murashige T, Skoog F (1962) Arevised medium for rapid growth and bioassays with tobacco cultures. Physiol Plant 15:473–479. https://doi.org/10.1111/j.1399-3054.1962.tb08052.x

Nguyen HQ, Le THT, Nguyen TNL, Nguyen TG, Sy DT, Tu QT, Vu TTT, Le VS, Chu HM, Vu TKL (2020) Overexpressing GmCHI1A increases the isoflavone content of transgenic soybean (*Glycine max* (L.) Merr.) seeds. In Vitro Cell Dev Biol-Plant. https://doi.org/10.1007/s11627-020-10076-x

Nguyen HQ, Vu TKL, Nguyen TNL, Pham TTN, Nguyen THY, Le VS, HM C (2019) Overexpression of the GmDREB6 gene enhances proline accumulation and salt tolerance in genetically modified soybean plants. Sci Rep.https://doi.org/10.1038/s41598-019-55895-0

Olhoft PM, Donovan CM, Somers DA (2006) Soybean (*Glycine max)* transformation using mature cotyledonary node explants. Methods Mol Biol 343:385–396. https://doi.org/10.1385/1-59745-130-4:385

Pahwa D (2009). Catharanthus alkaloids. Scribd. 6:1–9

Pan Q, Wang Q, Yuan F, Xing S, Zhao J, Choi YH, Verpoorte R, Tian Y, Wang G, Tang K (2012) Overexpression of ORCA3 and G10H in *Catharanthus roseus* plants regulated alkaloid biosynthesis and metabolism revealed by NMR-Metabolomics. PLoS One 7(8):e43038. https://doi.org/10.1371/journal.pone.0043038

Peebles CA, Sander GW, Hughes EH, Peacock R, Shanks JV, San KY (2011) The expression of 1-deoxy-D-xylulose synthase and geraniol-10-hydroxylase or anthranilate synthase increases terpenoid indole alkaloid accumulation in *Catharanthus roseus* hairy roots. Metab Eng 13(2):234–240. https://doi.org/10.1016/j.ymben.2010.11.005

Pham TTN, Nguyen TNL, Bui TH, Nguyen HQ, Nguyen TT, Le VS, Chu HM (2019) Agrobacterium-mediated transformation of the CrDAT gene and selection of transgenic periwinkle lines with a high vincristine accumulation. J Hortic Sci Biotechnol 94(5):591–598. https://doi.org/10.1080/14620316.2019.1607572

Saghai-Maroof MA, Soliman KM, Jorgensen RA, Allard RW (1984) Ribosomal DNA spacer-length polymorphisms in barley: mendelian inheritance, chromosomal location and population dymnamics. Proc Natl Acad Sci USA 81(24):8014–8018. https://doi.org/10.1073/pnas.81.24.8014

Sambrook JF, Russell DW (2001) Molecular cloning: a laboratory manual. Cold Sping Harbor Laboratory Press, New York

Southern EM (1975) Detection of specific sequences among DNA fragments separated by gel electrophoresis. J Mol Biol 98(3):503–517. https://doi.org/10.1016/s0022-2836(75)80083-0

St-Pierre B, Laflamme P, Alarco AM, De Luca V (1998) The terminal oacetyltransferase involved in vindoline biosynthesis defines a new class of proteins responsible for coenzyme A dependent acyl transfer. Plant J 14(6):703–713. https://doi.org/10.1046/j.1365-313x.1998.00174.x

Sun HJ, Cui ML, Ma B, Ezura H (2006) Functional expression of the tastemodifying protein, miraculin, in transgenic lettuce. FEBS Lett 580(2):620–626. https://doi.org/10.1016/j.febslet.2005.12.080

Van der Heijden R, Jabos DJ, Snoeijer W, Hallard D, Verpoorte R (2004) The Catharanthus alkaloids: pharmacognosy and biotechnology. Curr Med Chem 11(5):607–628. https://doi.org/10.2174/0929867043455846

Vancanneyt GRS, Connor-Sanchez AO, Willmitzer L, Sosa MR (1990) Construction of an intron-containing marker gene: splicing of the intron in transgenic plants and its use in monitoring early events in *Agrobacterium*-mediated plant transformation. Mol Gen Genet 220(2):245–250. https://doi.org/10.1007/BF00260489

Verma P, Mathur AK, Khan SA, Verma N, Sharma A (2017) Transgenic studies for modulating terpenoid indole alkaloids pathway in *Catharanthus roseus*: present status and future options. Phytochem Rev 16:19–54. https://doi.org/10.1007/s11101-015-9447-8

Vi TXT, Le HD, Nguyen VTT, Le VS, Chu HM (2017) Expression of the ZmDEF1 gene and α-amylase inhibitory activity of recombinant defensin against maize weevils. Turk J Biol 41:98–104. https://doi.org/10.3906/biy-1512-64

Vi TXT, Nguyen TNL, Pham TTN, Nguyen HQ, Nguyen THY, Tu QT, Le VS, Chu HM (2019) Overexpression of the ZmDEF1 gene increases the resistance to weevil larvae in transgenic maize seeds. Mol Biol Rep 46(2):2177–2185. https://doi.org/10.1007/s11033-019-04670-5

Vu TNT, Le THT, Hoang PH, Sy DT, Vu TTT, Chu HM (2018) Overexpression of the Glycine max chalcone isomerase (GmCHI) gene in transgenic *Talinum paniculatum* plants. Turk J Bot 42(5):551–558. https://doi.org/10.3906/bot-1801-22

Wang CT, Liu H, Gao XS, Zhang HX (2010) Overexpression of G10H and ORCA3 in the hairy roots of *Catharanthus roseus* improves catharanthine production. Plant Cell Rep 29(8):887–894. https://doi.org/10.1007/s00299-010-0874-0

Wang Q, Xing S, Pan Q, Yuan F, Zhao J, Tian Y, Chen Y, Wang G, Tang K (2012) Development of efficient Catharanthus roseus regeneration and transformation system using *Agrobacterium tumefaciens* and hypocotyls as explants. BMC Biotechnol 12:34. https://doi.org/10.1186/1472-6750-12-34

Zhao L, Sander GW, Shanks JV (2013) Perspectives of the metabolic engineering of terpenoid indole alkaloids in Catharanthus roseus hairy roots. Adv Biochem Eng Biotechnol 134:23–54. https://doi.org/10.1007/10_2013_182

Zhou ML, Zhu XM, Shao JR, Wu YM, Tang YX (2012) An protocol for genetic transformation of *Catharanthus roseus* by *Agrobacterium rhizogenes* A4. Appl Biochem Biotechnol 166(7):1674–1684. https://doi.org/10.1007/s12010-012-9568-0

Molecular Markers and Descriptor's Suitable for Quantitative Trait Loci and Genome Mapping in *Catharanthus roseus*

7

Shashi Pandey Rai, Sanjay Kumar Rai, Apoorva, Krishna Kumar Rai, and Sushil Kumar

Abstract

Most investigated perennial herb *Catharanthus roseus* is the only natural source for durable bioactive compounds vinblastine (VBL) and vincristine (VCR) as anticancer drugs. The VBL and VCR are the two main dimeric downstream products of the branched terpenoid indole alkaloids (TIAs) pathway. The *in-planta* high TIA production and storage abilities are essential for reducing the production cost and high availability of VBL and VCR. A large number of publications has been already made in the dissection of TIA pathway including various transcription factors, regulatory enzymes and genes. The derivation of *Catharanthus roseus* genotypes for high *in-planta* yields of pharmaceutical alkaloids is being carried out either by overexpression or underexpression of regulatory genes, steps/routes of the TIAs pathway. Selection of elite genotypes has been done through mutation, molecular breeding and transgenic approaches. The molecular mapping of quantitative trait loci (QTL) associated with different levels of individual metabolites (mQTL) combined with the mapping of morphological traits linked with TIAs are essential tools to investigate the genetic basis of variation/covariation in metabolites and plant phenology in C. *roseus* population. It is shown that the development of large number of molecular markers, genetics related to TIA pathway and molecular breeding approaches necessitates pyramidation of quantitative traits that contribute toward high alkaloid yield. The saturated genetic linkage map construction on molecular markers along with alkaloid yield-related QTLs in relation to DNA is a prerequisite for tracking QTLs. By the use of molecular DNA markers, observations recorded on a genetic resource collection, the quantitative and morphological traits and accessions possessing their contrasting manifestations have been already identified in *C. roseus*. The life cycle features and morphological descriptors of *C. roseus* are referred here in this chapter which render it favorable for speedy development of the desired QTL map. The knowledge generated through the scaffolded genome sequence of *C. roseus* it is possible now to understand the genes and their functions by integrating multidimensional

S. P. Rai (✉) · Apoorva · K. K. Rai
Department of Botany, Center of Advance Study in Botany, Institute of Science, Banaras Hindu University (BHU), Varanasi 221005, India
e-mail: pshashi@bhu.ac.in

S. K. Rai
Dr. Rajendra Prasad Central Agricultural University, PUSA, Samastipur, Bihar, India

S. Kumar
SKA Institute for Research, Education and Development (SKAIRED), 4/11 Sarv Priya Vihar, New Delhi, India

© The Author(s), under exclusive license to Springer Nature Switzerland AG 2022
C. Kole (ed.), *The Catharanthus Genome*, Compendium of Plant Genomes,
https://doi.org/10.1007/978-3-030-89269-2_7

informations. The genome sequence data is essential to map all possible QTLs related to the trait of interest which can be further analyzed to identify the superior alleles for TIAs' synthesis in plant/organ by targeted genetic engineering approaches.

7.1 Introduction

Plant secondary metabolites are natural bioactive compounds like alkaloids, flavonoids, terpenoids, phenolics, glucosinolates, etc. that are commonly used in traditional medicine and have wide range of commercial applications. The interest in enhancing the production of these natural bioactive compounds is mainly focused on obtaining high yields suitable for commercial exploitation but these contents are controlled by genetic, environmental and agronomic factors. These bioactive phytocompounds are products of secondary metabolism and are serving as essential lifesaving drugs. Currently, scientists are focused on sustainable development through natural potential reconsideration for characterization and designing of new pharmacological agents. Increasing the yield of pharmaceutically important metabolites in the plants that synthesize them is an important area of research for production of phenolics and alkaloids, which possess wide range of biological properties including anticancer, antidiabetic, antihypertensive and antimicrobial activities (Memelink 2005).

C. roseus, commonly called as periwinkle, serves as a remarkable factory for anticancerous bioactive compounds, and it is the leading single plant species reported to produce such a wide array of complex alkaloids (Misra and Kumar 2000). For the commercial use of periwinkle alkaloids, plants are harvested about three months after germination of seeds when they begin to flower. Traditionally, this plant is seed propagated, which limits uniform yield of dimeric alkaloids due to genetic segregation (Aslam et al. 2010); therefore, large number of breeding programs have been carried out in this crop to modify its terpenoid indole alkaloid (TIAs) content (Van der Heijden et al. 2004). The pharmaceutically important TIAs occur in low levels in plant; therefore, commercial VBL and VCR are synthesized from their natural precursors vindoline (V) and catharanthine (C) which occur in plant parts in relatively higher concentrations. Similarly, ajmalicine (A) is semi-synthesized from serpentine (S) which is relatively higher in plant parts. Since chemical synthesis of these TIAs (A, S, V, C, VBL and VCR) are cost effective, it is now important to work on *Catharanthus roseus* to increase the concentration of pharmaceutically important alkaloids or their precursors in different plant parts/organs. To enhance the production of these precursors, studies leading to identification, cloning, localization, and gene manipulations that direct synthesis and storage of TIAs in different organs are needed and characters that determine the development, growth and primary metabolism with specific cellular functions in organs on the other hand are required to achieve the high alkaloid yield *in planta*.

The molecular genetics of commercially important secondary metabolites such as terpenoid indole alkaloids are polygenic in nature. The small undetectable environmental effects on individual minor/regulatory genes and lack of defined segregation pattern make it difficult to monitor through traditional breeding approaches for improvement in the amount of these compounds in different plant parts. However, efficient development and use of molecular markers have largely facilitated construction of genetic linkage maps and the detection of quantitative trait loci (QTLs) associated with yield-related traits. The DNA or molecular markers closely linked with the major identified QTLs having positive and additive effects can also be pyramided to achieve higher yield of TIAs. However, till date, very few molecular linkage maps have been reported in pharmaceutically important medicinal plants. Molecular mapping and detection of therapeutic QTLs have been already reported in periwinkle (Gupta et al. 2007). Utilizing molecular markers including simple sequence repeat (SSR), restriction fragment length polymorphism) RFLP and random amplified polymorphic

DNA (RAPD) flanking the QTLs can be used for breeding widely adaptive varieties with improved phytomedicine contents (Kole and Abbott 2008). The recombinant inbred lines (RILs) being developed now will be used to construct a genetic linkage map and locating QTLs controlling contents of other phytomedicines (Kole et al. 2010). The concepts and strategies employed for development of TIAs can serve as a model for genetic improvement of medicinal plants and nutraceutical crops. The QTL mapping requires (1) selection of germplasms and evaluation of genetic diversity in the accessions/mutants; (2) construction of genetic linkage maps utilizing molecular markers; (3) mapping of QTLs controlling the contents of secondary metabolites; (4) marker-assisted selection of genotypes; and (5) gene pyramiding of traits linked with high content of bioactive secondary metabolites.

7.2 *Catharanthus roseus* as a Genetic Model for Secondary Metabolism

After decades of continuous research, it was observed that *C. roseus* could serve as a perfect model system that could be used in exploring the monoterpene indole alkaloid metabolism. As a model organism, *C. roseus* has several characteristics as it takes a short duration to complete its life cycle (two to four generations/year) due to absence of dormancy and has a small diploid complement of chromosomes ($2n = 2x = 16$). The herbaceous, small, erect habit allows propagation of large populations both under in vitro and *ex-vitro* conditions. Due to pereniality, individual plants can be also maintained for several years. The flower size and structure is suitable for large-scale controlled pollinations as the species is largely self-pollinated (50–100%). The seeds are highly amenable for physical and chemical mutagenesis (Pandey-Rai et al. 2003). The *C. roseus*, model plant of apocynaceae family, had genome size of 522.65 Mb, and about 33,829 protein-coding genes are known till date. Nonetheless, numerous types of proteins (1287) and genes (134) have been also reported and characterized. The large morphological diversity makes this plant species suitable for hybridization for producing new crop varieties. With the release of the three genome assemblies (one full and two partial) of *C. roseus, ASM94934V1(522,653,741)*; *BACT16H(115,520)*; *BACSGD(114,931)*, it is now necessary to locate the sequences on chromosomes and annotate the gene functions at genome level. Recently, 53 RNA-seq datasets were also made available in public for different plant parts such as seedling, leaf, flower, shoot and root. The deep transcriptome sequencing of *C. roseus* was carried out to identify genes that are involved in synthesis of bioactive compounds. About 65% transcripts of *C. roseus* showed homology with available public repositories sequences along with35% unigenes. Recently, 1820 transcription factor-encoding genes and 11,620 genic SSRs have also been identified through in silico analysis. All these advances along with high-throughput sequencing and genome editing can serve as better tool in more precise high TIA producing cultivar development in *C. roseus.*

The genetic resources and genetic variability are the basic requirement for plant breeding. These are the primary sources of important gene (s) associated with yield enhancement and quality improvement. In addition, DNA markers are useful for screening and isolation of alleles in breeding populations, construction of linked functional marker (FM) haplotypes, and pyramiding of FM alleles associated with TIAs' biosynthesis which is a complex trait. The DNA marker technology is also helpful in identifying QTLs which can be further used in genetic mapping.

Although identifying genes, molecular and functional markers (FMs) associated with plant phenotypic variation is very difficult, still QTL identification and linkage mapping is a powerful approach to tag desired alleles/genes of traits of interest with DNA markers for allele introgression associated with QTLs by marker-assisted selection (MAS) breeding. Reproductive behavior of *C. roseus* with control on pollination allows development of mapping populations. Permanent mapping populations such as RILs have also been developed after self-pollination in the progenies of hybrids in *C. roseus* up to F_6 generations. These mapping populations are helpful in understanding the relationship with QTLs and genetic elucidation of new cultivars or varieties with deep exploration of their primary and secondary metabolism are essential for achieving a high quantity and quality output of TIAs such as VBL and VCR.

7.3 Genetic Resources and Genetic Variability in *Catharanthus roseus*

In plants, the yield of each TIA is represented by a product of weight of the organ or plant and its accumulation in specific organs. The natural variability present in different accessions and induced variability/mutants serve as the reservoir for verities of alleles/genes that govern TIA content in the respective organs. These genetic variations of *C. roseus* for different traits including plant height, flower color, leaf shape, color and size, TIA content in various plant parts or organs present in natural population along with stable mutants can be utilized in the breeding programs as improvement strategies for TIA yield (De Luca et al. 1986; Kulkarni et al. 2008). The observed phenotypic heritable variations linked with accumulation of TIAs can be tracked for polymorphism detection among the known genes of interest through homologous or comparative genomics approaches. About 10,8472 nucleotide sequences are presently available on the databases for *C. rosues*. The DNA sequences (loci) linked with QTL alleles should be identified, characterized and mapped on the different linkage groups. With regard to the already defined genes, the DNA array hybridizations and/or cloning should be followed by sequencing across a panel of germplasm for prevalent polymorphism detection. Depending upon the allelic diversity at a QTL region, the quantitative traits can have varying phenotypic range and FMs associated with these QTLs are helpful in establishing the importance of QTL analysis. Thus, polygenic traits such as TIAs' concentrations that are not easily detectable through classical breeding methods could easily be tagged with DNA markers. Fine mapping of QTL, followed by expression analysis will allow identification of differences at the nucleotide level among the parents of mapping population. The role of variant alleles in the manifestation of TIA yield can be confirmed by (1) determining function of different alleles for organ-wise TIA yields, (2) possibilities of transgenic complementation or allele replacement by homologous recombination, (3) understanding the role of alleles in regulation of protein or enzyme product and (4) knockout/knockdown of the gene affecting the trait of interest (TIA yield). The above criterias will be helpful in establishing genetic correlation between variation among alleles and yield-related traits.

7.4 Importance of Genetic Resource Collection and Variability in QTL Mapping

The variability in periwinkle has been found to be very narrow in the Asian region. Therefore, the germplasm from tropical and subtropical regions is being utilized by the workers in the breeding programs along with mutation breeding approaches. Further, double-mutant recombinants through gene pyramiding approach have been also carried out for enhancing the alkaloid contents in this plant to discover allelic variation in mutagenized populations. Further, the targeting induced local lesions in genomes (TILLING) approach could be adopted utilizing information's from mutant libraries. Identification and characterization of the stable phenotypes from the available genetic resources and construction of RILs is the primary step toward genetic linkage mapping and marker-assisted breeding for high TIAs yields from *C. roseus*. The genetic resources of *C. roseus* from Indian subcontinent can help and allow identification of the marker traits useful in map construction such as (1) quantitative traits affecting the yield of TIAs; (B) organ specific quantitative traits linked with TIA yield; (C) morphological traits affecting high alkaloid yield; and (D) other morphological traits closely linked with QTLs.

The variable traits and mQTL mapping allow clear understanding between the relationship of QTL numbers contributing to the trait variation in a population, additive effects of QTL, dominant and epistatic nature of QTL and genetic correlation between different traits. It can also demonstrate pleiotropy and/or close linkage along with QTL interaction with environments. The genetic architecture of quantitative traits intimately related to various applications, such as marker-assisted prediction, selection and gene introgression. The first requirement for efficient fine-scale QTL mapping is the large segregating bi-parental mapping population such as F_2 population, RILs and double haploids (DHs) (Fig. 7.1). It is very common that each of the RILs selfed for several generations ($F_6 \sim F_9$ populations) for QTL analyses and functional genomics to answer tight linkages associated with TIAs.

7.4.1 Genetic Improvement Strategies for High TIA Yields

In periwinkle, the plant organs/parts are the principal source of A, S, V and C for the large-scale production of VBL and VCR. To make its production more economic, the quantities of S, V and C must be increased in different plant parts. The efficient deployment of molecular genetics-based plant breeding techniques, transgenosis and their combinations, can serve in improving *Catharanthus roseus* cultivars for desired TIAs and their yield (Dwivedi et al. 2000; Mishra et al. 2001). Transgenosis approach can permit transfer of selected designed gene cassettes for the improvement of TIA biosynthesis and plant morphogenesis for organ development. On the other hand, the classical/molecular breeding will allow pyramiding of already identified high alkaloid yield determining QTLs or alleles. Various biotechnological techniques and molecular breeding strategies could be adopted for the enhancement of TIA yield in different cultivars of *C. roseus*. The agronomic performance and genetic characterization of different cultivars/mutants with consideration of their

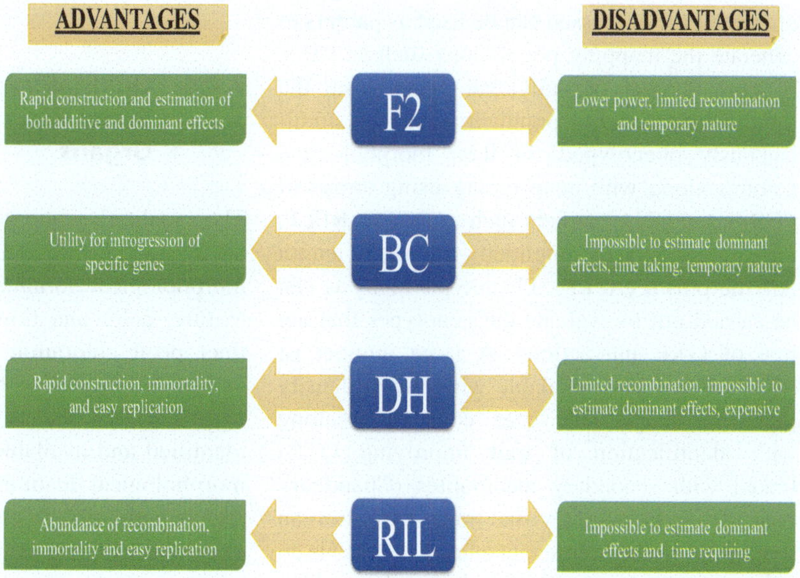

Fig. 7.1 Advantages and disadvantages of different mapping populations required in QTL mapping

physiological behavior along with secondary metabolism (TIA production) are prerequisite for obtaining a high quality and quantity output of terpenoid indole alkaloids. There is large number of reports related to transfer of TIA pathway genes successfully into cells under in vitro conditions (Whitmer et al. 1998), but whole plant transgenics related to TIA regulatory genes are not reported till date. Although, efficient recovery of transgenics in this crop is reported, application of molecular breeding approaches in *C. roseus* improvement is desirable for improvement in TIAs' yield.

7.4.2 Mapping of TIA Yield Associated Quantitative Trait Loci

The organ size such as root and leaf in *C. roseus* is directly linked with the contents of TIAs along with many principal quantitative traits that affect TIAs' yield. These types of quantitative traits that are linked with leaf and root organogenesis/ development need to be identified/converted into inherited easy to assay DNA markers that can help in molecular breeding. To achieve this linkage of the traits needs to be determined and accessions of *C. roseus* in which contrasting trait of interest are expressed can be used as parents to generate the mapping populations such as RILs. A large number of individual plants from the segregating/ mapping populations can be simultaneously phenotyped for their morphological features along with geno typing using arrays of PCR-based DNA markers such as RAPD, RFLP, microsatellite and amplified fragment length polymorphism (AFLP). Phenotypic analysis can be carried out to evaluate the genotypes that are free of GXE interactions. A large number of examples are now available where high density molecular maps have been constructed along with identification of trait improving QTLs linked with secondary metabolites (Chaudhary et al. 2013). In case of visual traits such as organ shape/size, bulked segregant analysis (BSA) is an efficient method to establish linkage relationships with molecular markers.

7.4.3 Quantitative Traits Directly Related to Yield of TIA

The pharmaceutical TIAs in *C. roseus* are largely present in root, stem and leaves of the plant. The sizes of the organ and alkaloid contents present in root, stem and leaves are the principal alkaloid yield determining quantitative traits. Different accessions of Asian region in which sizes of the root, aerial part (stem and leaf organ) are contrastingly large and small are listed in Table 7.1. There are few accessions in which organ yields differ by two or threefold and TIAs by 2- to 8-fold due to difference in organ size. These variations can be helpful for marker-assisted breeding (MAS) for high yields of TIAs in *C. roseus*. The sizes of shoot organs (stem and leaves) of accessions can be measured in terms of fresh/dry weight, along with number of branches. The contrasting visual observations on these characters could be measured quantitatively and can be grouped into various phenological categories such as: normal height/tallness; normal height/dwarfness; thin stem/thick stem; many branches/fewer branches; many leaves/fewer leaves; and large leaf/small leaf. Table 7.1 gives contrasting expressions of the shoot/leaf organ sizes in different accessions of *C. roseus*.

7.4.4 Qualitative Morphological Traits Related to TIA Rich Organs

The qualitative morphological characters also have importance that are related to stem and leaf morphological variations with respect to shape, texture, color and flower morphology with distinct petal coloration (Fig. 7.2). The different accessions/ varieties harboring the contrasting manifestations for such characters are already identified and available (Table 7.1). Some other morphological features related to inflorescence architecture, flower color, size, shape/structure, orientation of flower organs, siliqua shape/size, structure and color of seeds have been also identified.

Table 7.1 Range of variation in different morphological qualitative and quantitative traits affecting alkaloid yield in genetic resources of *Catharanthus roseus*

	Trait/characters	Range of variations
(A)	Shoot-related attributes	
1.	Plant height	Dwarf (5–20 cm), tall (100–150 cm)
2.	Main stem diameter	Thin (14–45 mm), thick (100–150 mm)
3.	Branching	Scanty (10–35), profuse (36–62)
4.	Leaf node number	Low ($\leq$250), high (500–750)
5.	Leaf size/area	Small ($\leq$1 cm^2), large ($\geq$20 cm^2)
(B)	Stem-related attributes	
1.	Stem width	Narrow, medium, broad
2.	Stem color	Green *(137A, 138B), dark purplish red (N77C, N78B, C)
3.	Pattern of stem branching	Basal, erect
4.	Petiole/ stem angle	30°, 60°
(C)	Leaf-related attributes	
1.	Leaf shape	Linear, oblong, elliptic, ovate, subovate
2.	Leaf color	Dark green *(137 B, 139 A/B), yellow green (144A,146B/C/D, 147B)
3.	Leaf margin	Entire, undulate
4.	Leaf variegation	Present, absent
5.	Leaf apex	Acute, acuminate, mucronate, obtuse, cuspidate, emarginated
6.	Leaf texture	Brittle, malleable
7.	Leaf surface	Smooth, rough
8.	Stomata size	Small ($\leq$150 μm^2), large ($\geq$600 μm^2)
(D)	Inflorescence and flower-related attributes	
1.	Inflorescence	Terminal racemes, axillary cymes
2.	Flower size	Small ($\leq$3 cm), large ($\geq$7 cm)
3.	Flower petal color	Pink *(49A/B/C, 65C,58, B, N57C, 63C), purple *(75C, N78C,75B,78A,75C,76A/B), red *(47A, 45A,62A,63B,51A,50A), white *(N157C, 155A/B/C/D)
4.	Flower orifice color	Pink *(63C, N66C), purple (N57, 58, 61A, 62A, 72A, N74, 75A, 77A, N78, 71C, 71A), red *(47A, 45A, 63A/B, 64A, 58B,52A), white *(16A/D,159A-D,155A-D), yellow *(158A,160A,6A,8C,8A,1B,1D,2B)
5.	Corolla tube diameter	Thin ($\leq$0.2 cm), thick ($\geq$0.7 cm)
6.	Corolla tube length	Small ($\leq$2.1 cm), long ($\geq$4.5 cm)
7.	Style length	Short ($\leq$1.8 cm), long ($\geq$4.2 cm)
8.	Attachment of anthers to filament	Basifixed, dorsifixed
9.	Pollen size	Small ($\leq$45 μm^2), large ($\geq$60 μm^2)
10.	Pollen surface	Smooth, colporate
(E)	Fruit-related attributes	
1.	Siliqua color	Green *(137B,138B,139A/B), purple *(58, 61A, 62A, 72A, N74)
2.	Siliqua tip	Pointed, rounded

(continued)

Table 7.1 (continued)

	Trait/characters	Range of variations
3.	Siliqua orientation	Parallel, divergent
4.	Siliqua size	Small ($\leq$0.15), large ($\geq$0.31)
5.	Seed number per siliqua	Low ($\leq$6), high ($\geq$35)
6.	Seed size	Small ($\leq$0.25 mm), bold ($\geq$0.70 mm)
7.	Seed shape	Oval, oblong

*Given in parenthesis are the color codes of Royal Horticulture Society (RHS) color chart

Fig. 7.2 Flower shape, color and leaf size variation in the different accessions/varieties harboring the contrasting manifestations that are already identified and available in *Catharanthus rosues*

Milo et al. (1985) studied the inheritance of flower color and the pigmentation in petals of periwinkle. These identified contrasting qualitative morphological traits can be utilized in QTL mapping. In 35 traits (qualitative and quantitative) which are listed in Table 7.2, nine quantitative traits show direct linkage with the yield of one/more alkaloids in plant parts/organs. Six traits are

Table 7.2 List of different DNA markers identified and used in linkage/QTL mapping of *C. roseus*

S. No	Markers	Parameters/characteristics	Number/values
1	Expressed sequence tags (ESTs)	Total number of ESTs	19,899
		Total sequence analyzed (in bp)	10,196,333
		Total masking (in bp)	100,087
		EST after vector and Poly A/T removal	19,868
		Total number of singletons	5646
		Total number of sequences examined	2692
		Total size of examined sequences (in bp)	1,911,399
2	Simple sequence repeats (SSRs)	Total number of SSR loci located	338
		Mononucleotide	163
		Dinucleotide	84
		Trinucleotide	90
		Hexanucleotide	1
		Mapped SSRs	39
3	EST-SSRs	Frequency of ESTs-SSR loci	1 per 9.66 kb
		Total EST-SSRs	2034
		Dinucleotide	872
		Trinucleotide	853
		Tetranucleotide	139
		Pentanucleotide	96
		Hexanucleotide	74
		Mapped SSRs	39
4	Plastome SSRs	Mapped ESTs-SSRs	72
		Lineage specific SSRs	14
5	Sequence tagged microsatellite (STMS) marker	Mapped	28
3	Randomly amplified polymorphic DNA (RAPD) marker	Mapped	360
4	Amplified fragment length polymorphism (AFLP) marker	Mapped	103
5	Restriction fragment length polymorphism RFLP marker	Identified	25
6	Inter-simple sequence repeat (ISSRs) marker	Mapped	11
7	Gene targeted markers (GTM)	Mapped	64
8	miRNA	Mapped	08
9	QTLs detected	Mapped	23

associated with content of alkaloids in organs, and three are related to biomass (FW/DW) of the plant parts. Large number of traits can be treated as qualitative traits showing multigenic behavior but sometimes they may behave as mono/oligogenic. These contrasting traits all together have potential in mapping and can add several scores of markers to the linkage map of *C. roseus*. Further, the identification of QTLs for identified/selected traits, TIA yield-related quantitative traits such as total alkaloids (TIA) contents (%) in roots (TR), leaves (TL) and stems (TS) have strong correlation with the alkaloid yield. Various reports suggest that the TL shows strong correlation with the contents of VL, CL and VBL along with their yields. Further, the biomass (dry weight) of stems leaves and roots was also found linked directly or indirectly with other traits.

7.5 Importance of Mutants in QTL Mapping of *C. roseus*

Mutation breeding is an efficient tool for creation of variability and improvement of crops that have narrow genetic base. The role of mutation breeding approaches for increasing the extent of genetic variability in various medicinal plants has already been proved beyond doubt. The induced mutagenesis through chemical and physical mutagens are the promising approaches for induction of new genetic variability having desired attributes (Fig. 7.3). The induced mutagenesis is widely used in *C. roseus* also, and large number of mutants have been already reported in this crop having high content of economically important metabolites (Baskaran et al. 2013). Nevertheless, periwinkle has a variety of induced mutants with altered traits including contents of alkaloids in different plant parts. Macro- and micro-mutants having altered phenotypes have been isolated and evaluated for their alkaloid yield. The variable morphological features in mutants are mainly height of plant, leaf size and shape, flowering and reproductive traits along with change in abiotic stress (salt, heat and water stresses), tolerance behavior. Among height architecture mutants (tall, dwarf, semi-dwarf and bushy), the dwarf and semi-dwarf characters were monogenic recessive in nature, whereas bushy character was governed by non-allelic recessive gene (Kulkarni et al. 2016). The double-mutant recombinations of dwarf and

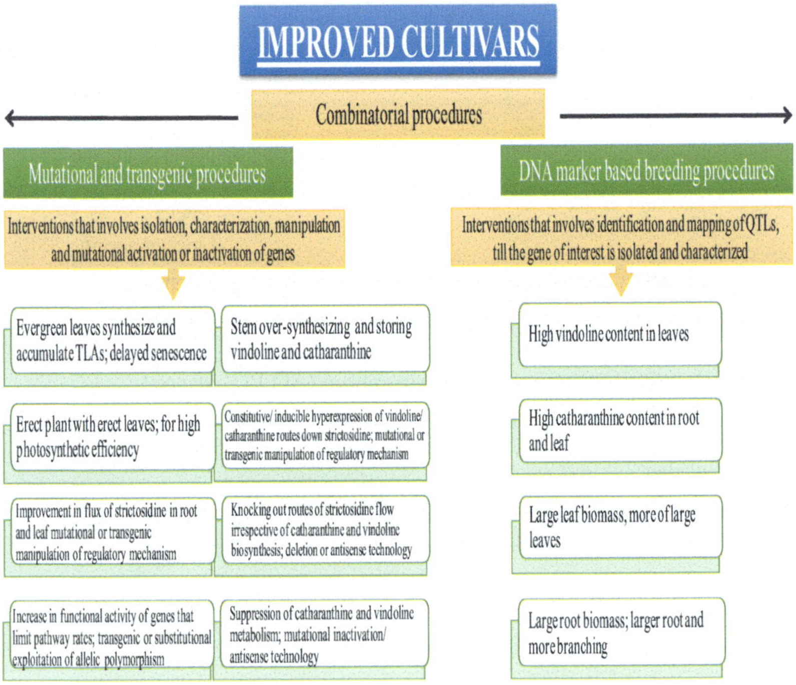

Fig. 7.3 Conventional and molecular breeding strategies for genetic improvement and enhanced yield of pharmaceutically important terpenoid indole alkaloids in *Catharanthus roseus*

bushy mutants exhibited 20% higher content of root alkaloids compared to its parental types (Kulkarni and Baskaran 2014). Leaf mutants having wavy leaf margin, necrotic leaf and nerium like leaf showed higher contents of leaf TIAs. Male sterile mutants and mutations affecting the gynoecium have also been reported in periwinkle for commercial exploitation of hybrid seeds. Recessive mutant producing single cotyledon and heterocarpous flowers, with one, two and three carpels with high fertility has been reported by Pandey-Rai and Kumar (2000, 2001). An altered inflorescence mutant 'leafless inflorescence' has been also reported and the concerning locus '*lli*' has been mapped by utilizing molecular markers (Sharma et al. 2011). High alkaloid accumulation in mutants exhibiting tolerance to abiotic stresses has been characterized by several workers. Eight glycophytic salinity tolerance mutants that are tolerant to high-temperature (45 °C) stress and salinity (250 mM NaCl) stress have been isolated and fully characterized (Pandey-Rai et al. 2001, 2003). The higher TIA contents in two of these *gsr* mutants (*gsr3* and *gsr6*) were also validated with the expression of strictosidine synthase, desacetoxyvindoline 4-hydroxylase and deacdetyl vindoline 4-O-acetyl transferase (key biosynthetic pathway genes). Few of these gsr mutations were assumed to be in loci/genes that have wide regulatory roles in signaling network of *C. roseus* for metabolism of TIAs, morphological development and adaptation to adverse environmental conditions. These mutants showed epigenetic regulation through hypomethylaion at repeat regions and controls the up-regulation and down-regulation of transcripts for several genes causing pleiotropic alterations for several traits affecting TIAs yield per plant (Chaudhary et al. 2013; Kumari et al. 2013a, b).

The genetic or epigenetic changes along with dominant mutations have also been reported in plants derived from seeds exposed to space flight (Ou et al. 2010). All these reports related to mutation breeding approach can be further exploited for enhanced yield of TIAs *in planta*. It can be easily achieved through site-targeted mutagenesis or by generating vindoline (V) and catharanthine (C)-rich mutants by knockout of serpentine route gene functions (Singh et al. 2008). To increase the efficient use of mutation breeding, mutation screening through high-throughput technologies can be efficient tool in periwinkle such as TILLING and ECOTILLING and high-resolution melt analysis (HRM). For identification of allelic variation in natural plants or mutants and TIA pathway genes, the reverse genetics techniques utilizing recombinant DNA technology and virus-induced silencing approach to analyze gene function *in-planta* have already been utilized in periwinkle (Liscombe and O'Connor 2011).

7.6 Genomic Resources, Molecular Markers, Genetic Linkage Maps and QTLs

Availability of genome sequence can enable insights into the physiology, biochemistry, function and evolution of various genes responsible for TIA biosynthesis. Two different cDNA libraries (leaf base and root tips) were created by Murata et al. (2006) which was further sequenced to generate 5023 non-redundant expressed sequence tags (ESTs) that served as the first genomic platform to harvest the molecular markers (microsatellite). The whole genome sequence of Sunstorm Apricot variety of *C. roseus* using shotgun sequencing was described by Kellner et al. (2015). An assembly of 522,653,759 Mb having GC content 33.6% was generated with an N50 scaffold size of 26.249 kbp, L50 scaffold size of 5808 kbp, representing 147,992 number of contigs and the total 33,829 protein-coding genes were annotated. Number of studies have been now reported and the utility of annotation, gene function and genes network analysis for inferring the specific biological processes is being presently carried out utilizing genomic information's. Availability of sequence data from whole genome sequencing along with bioinformatics tools can maximize the identification and characterization of available sequences that can be further utilized in generating molecular markers. For example to search open reading frames (ORFs) in the sequences

Various tools such as, Metagene Annotator, Prodigal and GeneMark can be utilized along with databases GENBANK and UniProt for identification of their function which can be further mapped to different metabolic pathways using a KEGG database. Further, the advances in genomic technologies are helpful in generating expressed sequence tags (ESTs), thereby open a new direction toward development of EST-SSR markers by data mining. With increasing emphasis on structural and functional genomics of *C. roseus*, large numbers of ESTs are being developed in a cost-effective manner (Mishra et al. 2011). Currently, 215 GEO datasets, about32,867 ESTs, 114,657 genomic sequences, 139 full gene sequences, 118 GSS and 46 transcriptomes of *Catharanthus* are available in public domain along with plastome sequence containing 14 lineages of specific SSRs. These informations are now being utilized for development of SSRs, sequence tagged microsatellite sites (STMS), ESTs, single nucleotide polymorphisms (SNPs) and candidate gene markers (Shokeen et al. 2005, 2007; Jhang et al. 2012). From genomic-enriched libraries, STMS and microsatellite marker resources (423) have been also developed. It was investigated that out of 350 unigene characterized microsatellite markers 80 were cotransferable in apocynaceous species too. About 2520 SSR markers have been generated from the transcriptome data of seedlings with a subset of 48 validated markers (Kumar et al. 2014). The complete plastid sequence of this plant is also available and large number of EST-SSRs have been also analyzed (Ku et al. 2013). These available DNA marker data can be further analyzed through a mixed linear model-based association analysis (MLMA) approach (Fig. 7.4) for related traits of yield of TIAs to identify potential genes in different linkage groups along with associated SNPs that might be responsible for yield performing traits.

Large number of cultivar-specific DNA markers were also detected among them about 100 DNA markers are associated with flower characteristics. The maximum number of flowering trait-related markers was obtained during amplified fragment length polymorphism (AFLP) analysis (65), while the minimum in number during random amplified polymorphic DNA (RAPD) analysis (14) (Chaudhary et al. 2011). Some of these markers can be utilized to analyze their linkage with the levels of TIAs. Large number of ESTs (19,899) were analyzed for generation of SSR (2034) markers having 42.8% di, 41.9% tri, 6.8% tetra, 4.7% penta and 3.6% hexanucleotide repeat regions. The high reproducibility, transferability, scorable features and polymorphism in these EST-SSR markers make them highly suitable for their exploitation in diversity analysis, molecular mapping, analysis of QTL and comparative mapping in *C. roseus* (Table 7.2).

The analysis of available data set of ESTs is a rapid and cost-effective method to unravel the transcriptome by utilizing several computational methods related to assembly and annotation. Verma et al. (2014) optimized the de novo assembly for 59,220 unique transcripts of 1284 bp in average and functional annotation along with comprehensive gene ontology (GO) analysis revealed molecular functions of several genes having role in different gene functions and biological processes (Fig. 7.5). Analysis of sequences that are available in various public repositories showed 65% of homology with C. *roseus* transcripts and35% *C. rosues* specific unigenes. Further, 11,620 genic SSRs and 1,820 transcription factor (TFs) encoding genes were also analyzed through in silico approaches which can be utilized as an excellent resource for better understanding toward plant-secondary metabolism and in production of commercially important pharmaceuticals *in planta*.

Gupta et al. (2007) have constructed the first genetic linkage map of *C. roseus* with the help of six morphological traits as markers,125 DNA markers utilizing 79 RAPD, 2 EST-SSR, 7 ISSR, 37 other DNA markers and mapped them in 14 linkage groups of 1,131.9 cM map length (Figs. 7.6 and 7.7). The second molecular linkage map was published by Shokeen et al. (2011) utilizing 423 codominant markers and 111 F_2 individuals (CrN1 (Nirmal) × CrN82 (Kew), and a total of 114 markers were mapped on eight

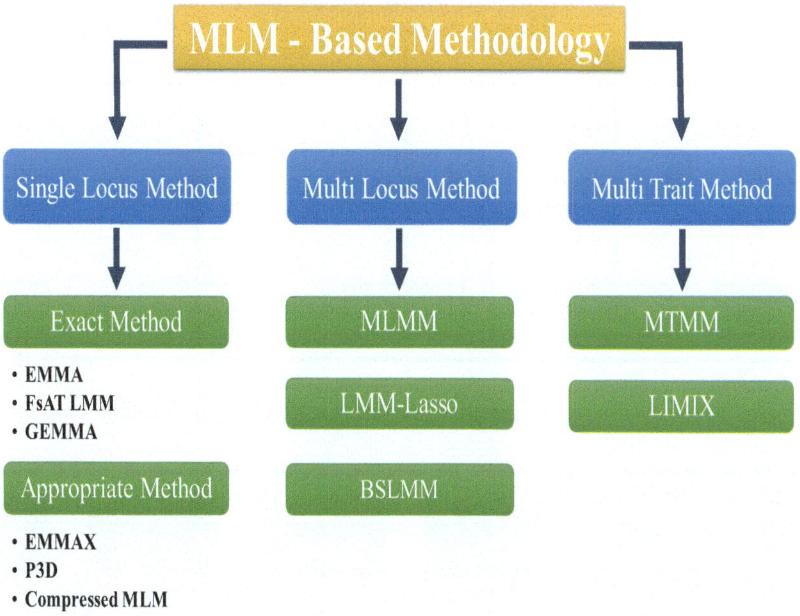

Fig. 7.4 Mixed linear model-based statistical methods suitable for single locus, multi-locus and multi-trait mapping in *Catharanthus roseus*. (BSLMM: Bayesian sparse linear mixed model; EMMA: efficient mixed model association; EMMAX: efficient mixed model association eXpedite; FaST-LMM: factored spectrally transformed linear mixed models; GEMMA: genome-wide efficient mixed model; LMM-Lasso: linear mixed model-Lasso; P3D: population parameters previously determined; MTMM: multi-trait mixed model; MLMM: multi-locus mixed model

linkage groups with average 5.55 cM marker distance spanning 632.7 cM region of the genome.

A linkage map utilizing one of the mutant leafless inflorescence trait linked with 172 DNA markers was also constructed by Chaudhary et al. (2011). Twenty QTLs for alkaloid yield-related traits such as contents of alkaloids in different organs and biomass (total dry matter weight) were identified associated with five linkage groups (LG1, LG2, LG3, LG4 and LG6).

This study further elaborated the utility of MAS for TIAs-related traits in periwinkle. The identified QTLs were further validated for their efficient use in marker-assisted breeding which expressed 16 morphological traits linked with the yield vindoline, vinblastine and catharanthine from leaves, stem and roots having favorable alleles of 13 QTLs. As QTLs for quantitative traits are strongly influenced by environmental conditions and genetic background (GB), the identification of GB independent QTLs along with environmental stability is prerequisite for efficient marker-assisted breeding. However, these reported QTLs in *C. roseus* may not show the same effects in other genetic backgrounds and also affected by environmental conditions; therefore, its further validation is required. This can be achieved by the use of diverse recipient parents to assess the performance of a QTL in one population and the consistency of the QTL effects in the GBs became less obvious by the use of several recipient parents which is a major concern for marker-assisted breeding of complex traits such as TIAs in *C. roseus*. Further, QTL × environment interaction effects can also enhance or counteract phenotype/ traits along with the main influence of QTL.

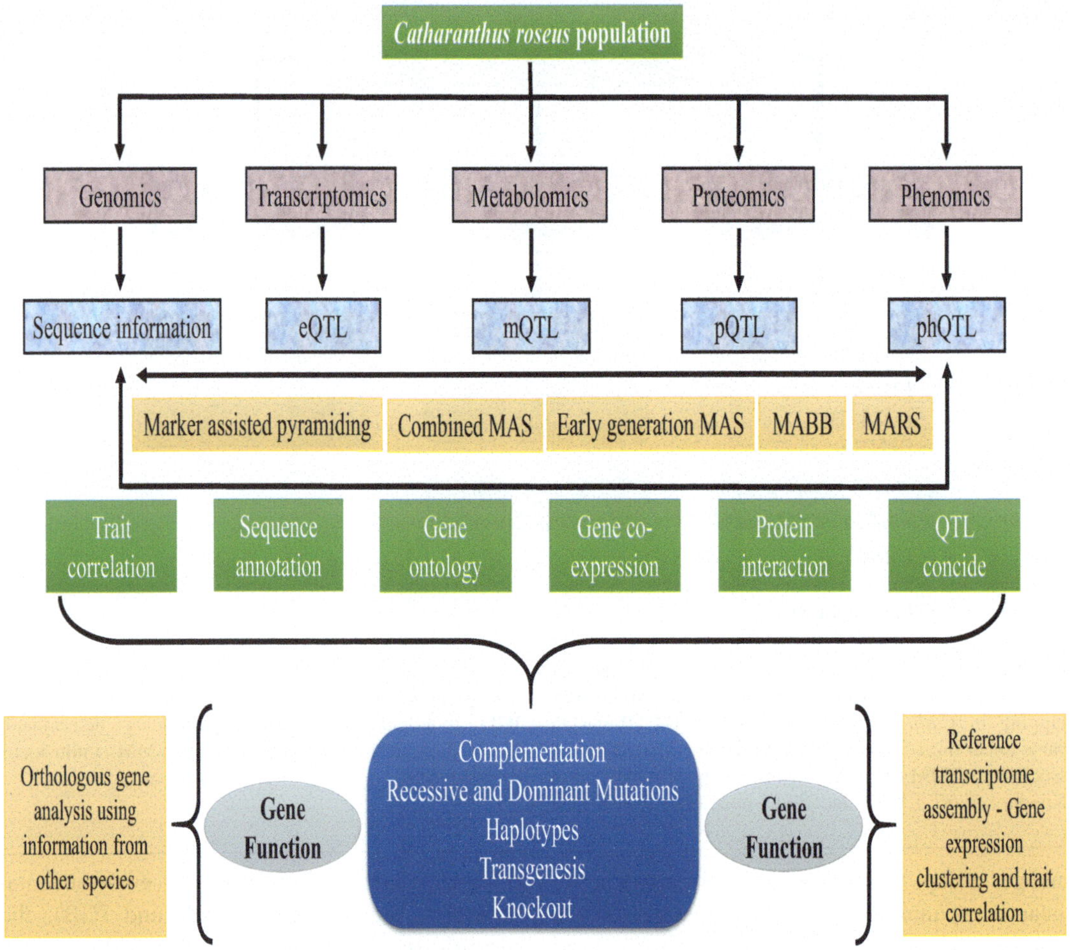

Fig. 7.5 Different omics strategies suitable for generation of QTL_S and gene function analysis in *Catharanthus rosues*

7.7 Gene-To-Metabolite Relationship for TIAs in *C. roseus*

In understanding the interaction of QTLs to environmental factors related to TIA metabolism, there is a strong need to integrate transcriptomic data (gene expression) with metabolomics data (metabolite profile) for decipheringgene-to-metabolite networks related to secondary metabolism. There are in vitro studies showing correlation between transcripts and metabolites in periwinkle cells which can help in understanding gene-to-gene and gene-to-metabolite networks that are involved in TIA biosynthesis. The inducible short-term environmental effects (transcript expression and accumulation of metabolite also successfully demonstrated by several workers in other plant species. The exploitation of large number of periwinkle ESTs undoubtedly can provide us opportunities to understand relationship between biosynthetic pathways and its associated signaling cascades along with genetic mapping. *C. roseus* transcriptome data can offer large number of sequence resources to decipher better insights toward gene diversity and its function. Further, next-generation sequencing has also being utilized for high-throughput complete genes/transcripts that can be used as

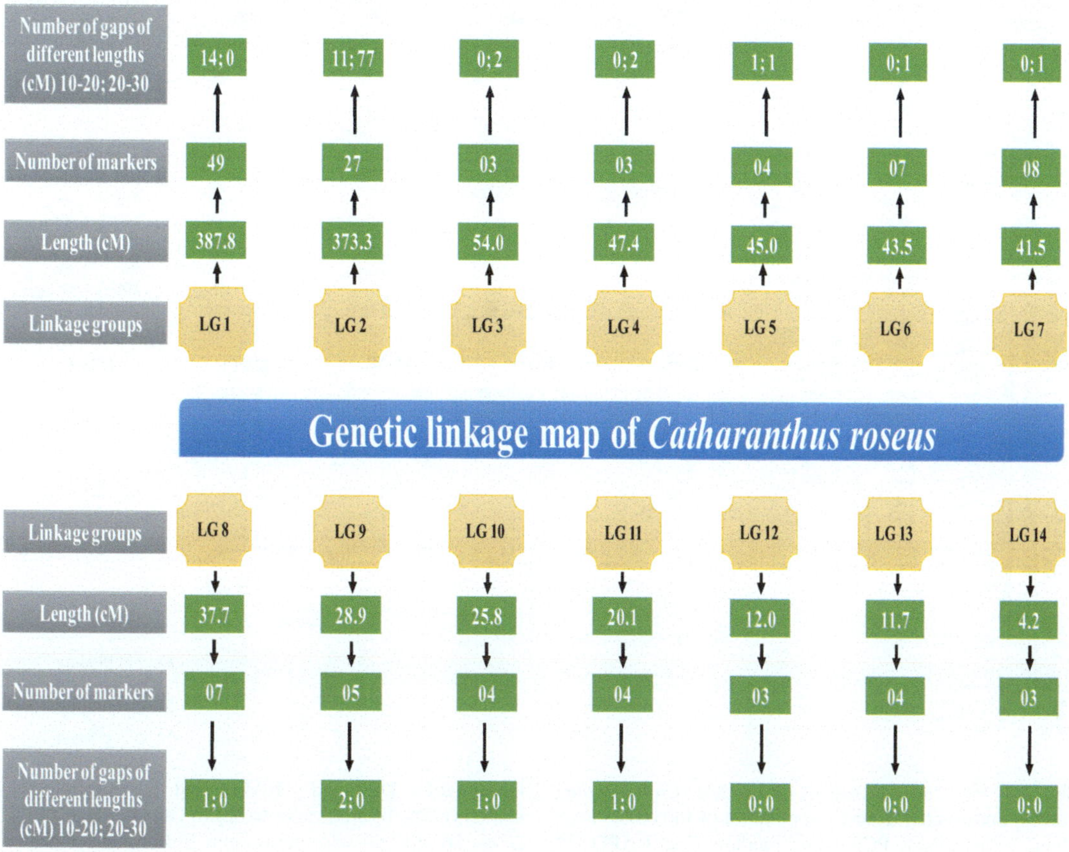

Fig. 7.6 Diagrammatic representation of first genetic linkage map of *Catharanthus roseus*

efficient molecular markers. About 7687 unigenes were found to have cluster of orthologous groups associated with genes/genome pathways of which 81 known TF families resulted in identification of 11,004 SSRs through in silico studies in *C. roseus*. In periwinkle, the most common was the trinucleotide repeats (AAG/CTT) compared to dinucleotide repeats (AG/CT). For understanding the organogenesis, development of plant and TIAs biosynthesis, these resources are helpful in SSR/novel gene discovery for utilization in marker-assisted selective breeding in *C. roseus*. The understanding of candidate genes involved in QTL-controlling networks of transcripts/metabolites is essential toward deciphering regulatory factors influencing TIA metabolism. A systematic metabolomic profiling and its association with mQTLs for all known gene/enzymatic loci is the prerequisite to test heritability/epistatic interaction related to metabolomic networks in *C. roseus*.

7.7.1 Concluding Remarks

The TIA pathway of *Catharanthus roseus* is very complex, and the transcription factors/genes that determine its ramifications are yet to be dissected for optimization of the essential/dispensable elements of the TIA biosynthesis and its regulation. There is strong need of pyramiding the alleles/genes responsible for high content of TIAs utilizing plant breeding approach by the help of morphological, DNA markers closely linked with alkaloid content determining traits on the genetic linkage map of *C. roseus*. These DNA markers can permit identification of

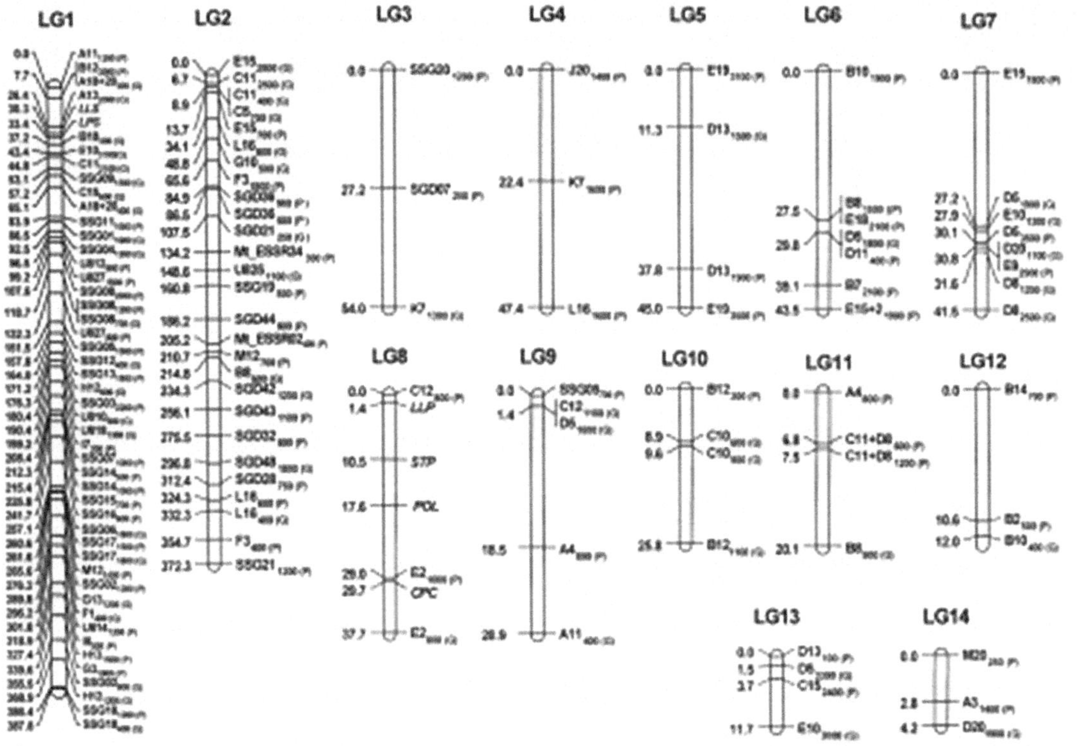

Fig. 7.7 First linkage map of periwinkle (*Catharanthus roseus*) constructed from F_2 population of the cross 'Pink Delhi' × gsr8 using PCR-based markers (260 RAPD, 24 RAPD combination 42 ISSR 42 EST-SSR 49 Designed,55 Designed + RAPD and 6 Morphological markers). The figure is adopted from Gupta et al. (2007) *J Genet* 86, pp 266 with permission from Springer Nature publishing group

rare segregants/recombinants when they exist in close proximity of the yield of TIAs determining mQTLs. The QTLs and variations recorded by different workers earlier in the genetic resources of *C. roseus* have allowed identification of quantitative and qualitative traits linked with alkaloid yield. Additionally, large number of accessions have also been described having contrasting phenotypes. The life cycle characters of *C. roseus* make it highly amenable for achieving the goal toward genomic linkage and QTL mapping. Recently, the plant has emerged as a model plant for the study of TIAs biosynthesis in plants. Marker-assisted breeding for high alkaloid content by identification of QTLs for yield-related traits has been already initiated by several workers in *C. roseus*. Further, identification/validation mQTLs in different genetic backgrounds are now essential for simultaneous improvement in alkaloid contents for commercial production in the near future. However, the genetic analyses of background of all parents and RILs have not been fully explored which can be achieved by considering the following points (i) to evaluate morphological appearance traits and their correlations with TIAs, (ii) identification of the mQTLs through composite interval mapping method, and (iii) validation of the QTL-linked molecular markers with previously reported QTLs. Furthermore, an additional challenge is to the strike prevalence of gene regulatory mechanisms underlying association with multifactorial trait that regulate cellular/temporal effects on target expression of TIAs. Modulation of transcription factors at transcriptional/post-transcriptional levels can surely facilitate molecular breeding of this plant for high TIA yield by maintaining the level of metabolite flux. However, it is well known that the whole TIA biosynthetic pathway of *C. rosues*, is influenced by the many regulatory networks consisting of multiple transcription factors. The interactomics, such as protein–protein interactions, subcellular trafficking, protein nucleo-cytoplasmic partitioning, are crucial areas to understand the role of transcription factors in secondary metabolism of TIAs. The knowledge generated by these researches can enhance the understanding of TIA metabolism and would be beneficial to change/modify the morphological/genetic architecture to get high TIA yielding varieties in *C. rosues*.

References

Aslam J, Khan SH, Siddiqui ZH, Fatima Z, Maqsood M, Bhat MA, et al. (2010) *Catharanthus roseus* (L.) G. Don. An important drug: it's applications and production. Pharmacie Globale (IJCP) 4(12):1–16

Baskaran K, Srinivas KVNS, Kulkarni RN (2013) Two induced macro-mutants of periwinkle with enhanced contents of leaf and root alkaloids and their inheritance. Ind Crops Prod 43:701–703

Chaudhary S, Pandey R, Sharma V, Tripathi BN, Kumar S (2013) Detection and mapping of QTLs affecting contents of pharmaceutical alkaloids in leaf and root of *Catharanthus roseus*. Agric Res 2:9–23

Chaudhary S, Sharma V, Prasad M, Bhatia S, Tripathi BN, Yadava G, Kumar S (2011) Characterization and genetic linkage mapping of the horticulturally important mutation leafless inflorescence (*lli*) in periwinkle *Catharanthus roseus*. Sci Hort 129:142–153

De Luca V, Balsevich J, Tyler RT, Eilert U, Panchuk BD, Kurz WGW (1986) Biosynthesis of indole alkaloids: developmental regulation of the biosynthetic pathway from tabersonine to vindoline in *Catharanthus roseus*. J Plant Physiol 125:147–156

Dwivedi S, Singh M, Singh AP, Sharma S, Uniyal GC, Kumar S (2000) Assessment of genetic divergence for its purposeful exploitation in periwinkle *Catharanthus roseus* (Apocynaceae). J Genet Breed 54:95–99

Gupta S, PandeyRai S, Srivastava S, Naithani SC, Prasad M, Kumar S (2007) Construction of genetic linkage map of the medicinal and ornamental plant *Catharanthus roseus*. J Genet 86:259–268

Jhang T, Gautam TP, Shukla S, Fatayal D, Annula, Kulkarni RN (2012) Development of microsatellite marker resource for genome analysis of *Catharanthus roseus*. In: Proceedings of international plant conference on "Molecular Mapping and Marker Assisted Selection", University of Veterinary Medicine, Veterinärplatz 1, 1210 Vienna, Austria, 8–11th February 2012, International Plant Conference Association, p 39

Kellner F, Kim J, Clavijo BJ, Hamilton JP, Childs KL, Vaillancourt B, Cepela J, Habermann M, Steuernagel B, Clissold L, McLay K, Buell CR, O'Connor SE (2015) Genome-guided investigation of plant natural product biosynthesis. Plant J 82(4):680–92. https://doi.org/10.1111/tpj.12827. Epub 2015 Apr 11. Erratum in: Plant J 2019 May; 98(4):760. PMID: 25759247

Kole C, Abbott AG (eds) (2008) Principles and practices of plant genomics, vol 1. Genome Mapping. Science Publishers, Enfield, NH, USA

Kole C, Olukolu BA, Kole P, Abbott AG (2010) Association mapping of fruit traits and phytomedicine contents in a structured population of bitter melon (*Momordica charantia* L.). In: Thies JA, Kousik S, Levi A (eds) Cucurbitaceae 2010 proceedings. American Society of Horticultural Science, Alexandria, VA, pp 42–45

Ku C, Chung W-C, Chen L-L,Kuo C-H (2013) The Complete plastid genome sequence of Madagascar periwinkle *Catharanthus roseus* (L.) G. Don: plastid genome evolution, molecular marker identification, and phylogenetic implications in Asterids. PLoS One 8:e68518. https://doi.org/10.1371/journal.pone.0068518

Kulkarni RN, Baskaran K, Sreevalli Y (2008) Genetics of corolla colour in periwinkle: relationship between genes determining violet, orange-red and magenta corolla. J Appl Hort 10:20–23.

Kulkarni RN, Baskaran K (2014) Increasing total leaf alkaloid concentrations in periwinkle (*Catharanthus roseus*) by combining the macro-mutant traits of two induced leaf mutants ('necrotic leaf' and 'nerium leaf'). J Hort Sci Biotechnol 89:513–518

Kulkarni R, Baskaran K, Jhang T (2016) Breeding medicinal plant, periwinkle [Catharanthus roseus (L) G. Don]: a review. Plant Genet Resour 14 (4):283–302. https://doi.org/10.1017/S1479262116000150

Kumar S, Shah S, Garg V,Bhatia S (2014) Large scale in-silico identification and characterization of simple sequence repeats (SSRs) from *de novo* assembled transcriptome of *Catharanthus roseus* (L.) G. Don. Plant Cell Rep 33:905–918.

Kumari R, Sharma V, Sharma V,Kumar S (2013a) Pleiotropic phenotypes of the salt-tolerant and cytosine hypomethylated leafless inflorescence, evergreen dwarf and irregular leaf lamina mutants of *Catharanthus roseus* possessing Mendelian inheritance. J Genet 92:369–394

Kumari R, Yadav G, Sharma V, Sharma V,Kumar S (2013b) Cytosinehypomethylation at CHG and CHH sites in the pleiotropic mutants of Mendelian inheritance in *Catharanthus roseus* . J Genet 92:499–511

Liscombe DK, O'Connor SE (2011) A virus-induced gene silencing approach to understanding alkaloid metabolism in *Catharanthus roseus*. Phytochemistry 72(16):1969–1977

Memelink J (2005) The use of genetics to dissect plant secondary pathways. Curr Opin Plant Biol 8:230–235

Milo J, Levy A, Akavia N, Ashri A, Palevitch D (1985) Inheritance of corolla colour and anthocyanin pigments in periwinkle *Catharanthus roseus* (L) G. Don. Z Pflanzenzücht 95:352–360

Mishra P, Uniyal GC, Sharma S, Kumar S (2001) Pattern of diversity for morphological and alkaloid yield-related traits among the periwinkle *Catharanthus roseus* accessions collected from in and around Indian subcontinent. Genet Resour Crop Evol 48:273–286

Mishra RK, Gangadhar BH, Yu JW, Kim DH, Park SP (2011) Development and characterization of EST based SSR markers in Madagascar periwinkle (*Catharanthus roseus*) and their transferability in other medicinal plants. Plant Omics J 4:154–162

Misra P, Kumar S (2000) Emergence of periwinkle (*Catharanthus roseus*) as a model system for molecular biology of alkaloid: phytochemistry, pharmacology, plant biology and *in vivo* and *in vitro* cultivation. J Med Arom Plant Sci 22:306–337

Murata J, Bienzle D, Brandle JE, Sensen CW, De Luca V (2006) Expressed sequence tags from Madagascar periwinkle (*Catharanthus roseus*). FEBS Lett 580:4501–4507

Ou X, Long L, Wu Y, Yu Y, Lin X, Qi X, Liu B (2010) Spaceflight-induced genetic and epigenetic changes in the rice (*Oryza sativa* L.) genome are independent of each other. Genome 53:524–532

Pandey-Rai S, Kumar S (2000) Induced mutation to monocotyledony in periwinkle, *Catharanthus roseus*, and suppression of mutant phenotype by kinetin. J Genet 79:1–18

Pandey-Rai S, Kumar S (2001) Heterocarpous flowers resulting from a recessive mutation in periwinkle, *Catharanthus roseus*. Curr Sci 80:1581–1584

Pandey-Rai S, Luthra R, Gupta MM, Kumar S (2001) Pleiotropic morphological and abiotic stress resistance phenotypes ofthe hyper-abscisic acid producing Abo™ mutant in the periwinkle *Catharanthus roseus*. J Biosci *26*(1):57–70

Pandey-Rai S, Luthra R, Kumar S (2003) Salt-tolerant mutants in glycophytic salinity response (GSR) genes in *Catharanthus roseus*. Theor Appl Genet 106:221–230

Sharma V, Chaudhary S, Srivastava S, Pandey R, Kumar S (2011) characterization of variation and quantitative trait loci related to terpenoid indole alkaloid yield in a recombinant inbred line mapping population of *catharanthus roseus*. J Genet 91(1):49–69

Shokeen B, Sethy NK, Choudhary S, Bhatia S (2005) Development of STMS markers from the medicinal plant Madagascar periwinkle [*Catharanthus roseus* (L.) G. Don.]. Mol Biol Notes 5:818–820

Shokeen B, Sethy NK, Kumar S, Bhatia S (2007) Isolation and characterization of microsatellite markers for analysis of molecular variation in the medicinal plant Madagascar periwinkle [*Catharanthus roseus* (L.) G. Don.]. Plant Sci 172:441–451

Shokeen B, Choudhary S, Sethy NK, Bhatia S (2011) Development of SSR and gene-targeted markers for construction of a framework linkage map of *Catharanthus roseus*. Ann Bot 108(2):321–336

Singh DV, Rai SK, Pandey-Rai S, Srivastava S, Mishra RK, Chaudhary S, Sharma S, Kumar S (2008) Predominance of the serpentine route in monoterpenoid indole alkaloid pathway of *Catharanthus roseus*. Proc Indian Natl Sci Acad 74:97–109

Van der Heijden R, Jacobs DI, Snoeijer W, Hallard D, Verpoorte R (2004) The Catharanthus alkaloids: pharmacognosy and biotechnology. Curr Med Chem 11:607–628

Verma M, Ghangal R, Sharma R, Sinha AK, Jain M (2014) Transcriptome analysis of *Catharanthus roseus* for gene discovery and expression profiling. PLoS One 9(7):e103583

Whitmer S, Canel C, Hallard D, Goncalves C, Verpoorte R (1998) Influence of precursor availability on alkaloid accumulation by transgenic cell line of *Catharanthus roseus*. Plant Physiol 116:853–857

Sequencing the Genome of *Catharanthus roseus*: Status and Prospects

8

Saumya Shah, Divya Vashisth, and Ajit Kumar Shasany

Abstract

Catharanthus roseus is a highly valuable plant often used for its medicinal applications. This plant synthesizes potent antitumor drugs known as vinblastine and vincristine which are produced via monoterpene indole alkaloids (MIAs) biosynthesis pathway. MIA biosynthesis is a complicated pathway that involves about 30 coordinated steps and produces at least 35 intermediates in *C. roseus*. A combination of recent 'omics' studies has created a new means to obtain biologically relevant information about the MIA biosynthesis in *C. roseus*. However, the availability of the draft genome of *C. roseus* has expanded our knowledge in understanding the MIA biosynthesis to the next level. The accessibility of genome sequence of *C. roseus* has also revealed the biochemical processes involved, their regulation and phylogeny of genes responsible for MIA biosynthesis. This chapter presents a summary of the status and future trends in sequencing the genome of *C. roseus*.

S. Shah · D. Vashisth
Biotechnology Division, CSIR-Central Institute of Medicinal and Aromatic Plants, Lucknow, Uttar Pradesh 226015, India

A. K. Shasany (✉)
ICAR-National Institute for Plant Biotechnology (NIPB), Pusa Campus, New Delhi 110012, India
e-mail: ajit.shasany@icar.gov.in

8.1 Introduction

The latest advancements in genomic DNA sequencing technologies have made genome studies one of the most influential and cutting-edge fields in life sciences (Hu et al. 2017). Genome sequencing techniques have provided many avenues for identification of the novel as well as therapeutically important genes, development of molecular markers in non-model plants, detection of single nucleotide sequences (SNPs) for the comparison between different cultivars, etc. (Edwards and Batley 2010). The information acquired from genome data is not restricted only to plant biology but also are of great importance in agriculture, evolutionary biology, computational biology, combinatorial chemistry, functional genomics, pharmacology, molecular medicine, etc. (Kaul et al. 2000). The data obtained after genome sequencing could also be utilized to find out the effect of environmental responses on plant growth and development and allow to access and understand the structure and dynamics of plant genomes. One of the best utilizations of high-end genomic techniques could be seen in sequencing the genome of medicinal plants. Medicinal plants produce many specialized secondary metabolites having great medicinal importance. These secondary metabolites are highly complex structures, synthesized via complicated biosynthetic pathways (Moore et al. 2014). Moreover, the large size of

© The Author(s), under exclusive license to Springer Nature Switzerland AG 2022
C. Kole (ed.), *The Catharanthus Genome*, Compendium of Plant Genomes,
https://doi.org/10.1007/978-3-030-89269-2_8

genomes, redundancy in genes, cellular localization and multidimensional regulatory processes make medicinal plant research elusive and the relative progress at a slow pace. However, recent development of sequencing strategies and the low cost of genomic data sequencing have facilitated the completion of medicinal plant genome projects dramatically. Accessibility of medicinal plant genome sequence data has made the analysis of unknown genes and biochemical pathways involved in their biosynthesis, very simple. Further, the genome sequences of medicinal plants have provided an opportunity to expand the traditional medicinal plants breeding into living reservoirs of phytomedicines (Hao and Xiao 2015). The complete genome sequences of medicinal plants also provide a deep understanding of genetic similarities and diversity across the species as well as among different genera. There are many medicinal plants whose genome has been sequenced recently, for example, *Ocimum tenuiflorum* (Rastogi et al. 2015; Upadhyay et al. 2015), *Salvia miltiorrhiza* (Xu et al. 2016), *Artemisia annua* (Shen et al. 2018), *Polygonum cuspidatum* (Li et al. 2019), *Catharanthus roseus* (Ku et al. 2013; Kellner et al. 2015; She et al. 2019), etc. The accessibility of processed data of the genome sequences of these medicinal plants has helped researchers in studying functional genomics and making new developments in major biological pathways (She et al. 2019). Genomic sequences of these medicinal plants have encouraged plant biologists in identifying the specialized metabolites of therapeutic importance for their large production. Furthermore, the characterization of large gene families participating in the production of specific compounds which are pharmacologically active would aid in the validation of the remedial aspect of medicinal plants. Hence, considering the recent discoveries and developments in the genomic studies of medicinal plants, this chapter will include the status and prospects of *C. roseus* genome sequencing.

8.2 Genome Sequencing of *C. roseus*

C. roseus belonging to the order Gentianales and family Apocynaceae is a well-known medicinal plant. The plant obtains its name from two Latin words katharos (pure) and anthos (flower; Kaushik et al. 2017). Madagascar periwinkle, rosy periwinkle, sadabahar (in Hindi), Chang Chung Hua (in Chinese) are some common names of this plant. In homeopathy, the plant is considered as 'Mother tincture' describing its restorative properties {https://Catharanthus Roseus Herb Uses, Benefits, Cures, Side Effects, Nutrients (herbpathy.com)}. The plant has been used to cure many diseases since ancient times. Because of its immense therapeutic properties, *C. roseus* has been widely used in different traditional systems of medicines such as Chinese and Ayurveda for treating diabetes, eye disease, throat infection, and lung disease (Kaushik et al. 2017). The plant sequesters hundreds of specialized natural products and phytocompounds (Kellner et al. 2015). However, the presence of remarkably valuable monoterpene indole alkaloids (MIAs) makes *C. roseus* an exceptional medicinal plant (O'Connor and Maresh 2006). MIAs are produced from secologanin (an iridoid class of monoterpenes) and tryptophan (an amino acid). These precursors of MIA synthesis are generated from 2-C-methyl-D-erythritol 4-phosphate (MEP) and mevalonate (MVA) pathways operated in plastids and cytoplasm, respectively. Despite widespread research carried out to characterize the genes of MIA biosynthesis in *C. roseus*, many key genes involved in their biosynthesis are still missing. The plant synthesizes *bis*-indole MIAs, viz. vinblastine and vincristine which are potent and valuable anticancerous agents. Like vinblastine and vincristine, *C. roseus* also synthesizes several other MIAs such as yohimbine, raubasine (ajmalicine), alstonine, serpentine, etc. (Fig. 8.1).

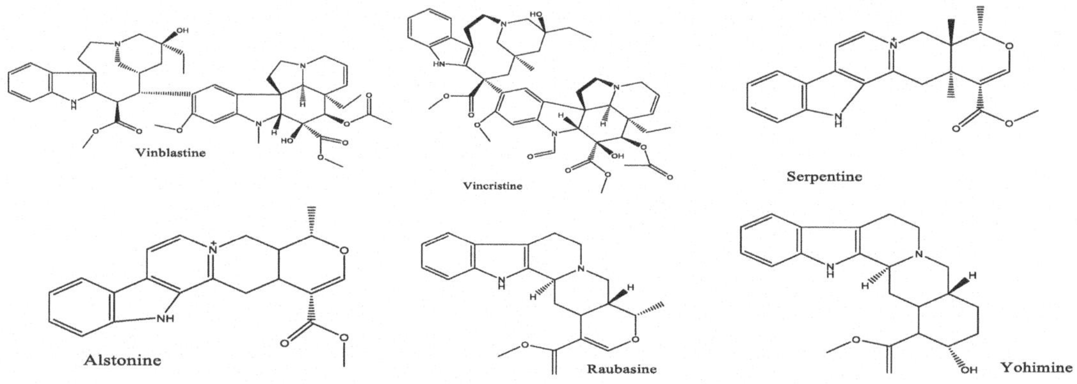

Fig. 8.1 Major alkaloid biosynthesized from the *C. roseus*

These phytocompounds are also used in treating many diseases. India ranks third in manufacturing vinblastine and vincristine in the world and exports these compounds to Western countries (Aslam et al. 2010). However, the demand for these phytocompounds in the world market could not be fulfilled only by relying on the plant. Also, the synthetically produced pharmacological compounds have many side effects that may cause other health problems (Sen et al. 2011). Hence, increasing the production of these compounds using applied research in *C. roseus* is required. Moreover, this plant may also be possibly involved in the biosynthesis of several other economically and health-benefiting important compounds except for vinblastine, vincristine, yohimbine, raubasine, alstonine, serpentine, etc. To identify the medicinal potentials as well as metabolic networks involved in the synthesis of specialized metabolites, whole genome, and genome-wide studies in this plant are required.

The genome size of *C. roseus* is 738 Mbp which makes it an exceptional model plant for genome sequencing (Guimaraes et al. 2012). Recently, a genome-directed analysis of specialized metabolite biosynthesis of *C. roseus* was published by Kellner et al. (2015) which highlighted the importance of draft genome sequence accessibility in promoting the investigation of valuable metabolites in *C. roseus* (Kellner et al. 2015). The draft genome of *C. roseus* (variety Sunstorm Apricot) was carried out using whole-genome shotgun sequencing approach. This resulted in an assembly of 523 Mb with an N50 scaffold size of 26.2 kbp. A total of 33,829 genes were found in annotation. Although many RNA and protein sequences are available for *C. roseus* (Murata et al. 2008; Champagne et al. 2012; Góngora-Castillo et al. 2012; Van Moerkercke et al. 2013; Verma et al. 2014), the access of genome sequence of *C. roseus* has unlocked a new way for understanding the evolution of genes, physicochemical processes, and gene regulations involved in the MIA biosynthesis.

C. roseus is a diploid plant with 16 chromosomes (Kaushik et al. 2017). However, substantial genetic variation has been detected in the *C. roseus* which resulted from crossing, induced mutagenesis, and polyploidy (Kaushik et al. 2017). The genome of *C. roseus* has shown the genomic background of crucial points within the MIA biosynthetic pathway, including physically grouped genes, segmental duplication, expression subfunctionalization, and putative neofunctionalization, using a near-complete representation of the genic space. This data has also revealed the hi-res co-expression analyses, highlighting three different groups of co-expression among the MIA pathway's components. In addition, the genome sequence has also uncovered the presence of enzyme-rich genic regions and transporters of biosynthetic enzymes further signifying the application of genome sequencing in studying secondary metabolite biosynthesis. Thus, the information gained from the genome sequence and annotation of *C. roseus* provides an important source to accelerate the research on secondary metabolite biosynthesis, breeding, and molecular evolution of *C. roseus*.

8.3 Plastid Genome Sequencing in *C. roseus*

Plastid genome sequencing is a valuable tool used for surveying molecular diversity, phylogenetic and evolutionary studies. It could also be used in the development of barcode regions in *C. roseus* (Sonnante 2019). Comparatively stable structure and uniparental inheritance among higher plants make the plastid genome an efficient source for understanding the phylogenetic relationships at different taxonomical stages (Ravi et al. 2008). An increased number of whole plastome sequencing has made the plastome-based analyses feasible. The plastid genome is effectively used for reducing the sampling errors and in resolving the previously unresolved evolutionary relationships (Jansen et al. 2007; Moore et al. 2010). Not long ago, a plastid genome of the *C. roseus* was published. This was the second whole plastid genome sequenced among the clade asterid order Gentianales (Ku et al. 2013) after the plastid genome of *Coffea arabica* (Samson et al. 2007). A reference-assisted approach was utilized for *C. roseus* plastome assembly. For this, 14 nearly complete plastome sequences of Asclepias and whole plastome sequences from *Coffea arabica* and four species of *Nicotiana* were used as reference (Ku et al. 2013). The length of the complete plastome of *C. roseus* was obtained to be 154,950 bp with 85,765 bp large single copy (LSC) region, 17,997 bp small single copy (SSC) region and 25,594 bp a pair of inverted repeats (IRa and IRb). The gene content of the *C. roseus* plastome included 86 protein-coding genes, 37 tRNA genes and 8 rRNA genes. Overall, the complete plastome of *C. roseus* highlighted the variation of Apocynaceous plastomes, with variations in the sizes of intronic regions (e.g., rps2-rpoC2 and IRb-ndhF) and the presence or absence of *accD*, *clpP* and *ycf1* genes. However, the organization and gene content of *C. roseus* was found to be very similar to its ancestral Apocynaceous plastome. Further, the intergenic regions (IGR) and simple sequence repeats (SSRs) of the *C. roseus* plastome were identified to find the potential locus of efficacy for the characterization and phylogenetic analyses of *Catharanthus* species and cultivars. The *C. roseus* plastome contains a total of 56 SSRs with 41 lineage-specific SSRs and 15 shared with *Asclepias syriaca*. The analysis of SSRs further emphasizes the plastome diversity within Apocynaceae as the number of SSRs in *A. syriaca* (103) plastome was twice than the *C. roseus* (56) plastome. This study also provided insight into the status of Gentianales in asterid plastome phylogenies.

8.4 Application of *C. roseus* Genome Sequence in Understanding MIA Biosynthesis

MIAs are a large group of alkaloids found in plants. There are more than 3000 alkaloids known to date. Several MIAs having therapeutic importance such as quinine (an antimalarial drug), camptothecin (the antineoplastic drug), strychnine (the rat poison and tonic), reserpine (the antihypertensive and tranquilizer), vinblastine and vincristine (an antitumor agent) are solely isolated from *Cinchona officinalis, Strychnos nux-vomica, Rauvolfia* species, *C. roseus,* respectively (Kutchan 1995). Since *C. roseus* is the sole source for vinblastine and vincristine synthesis, this plant is broadly used as a model medicinal plant. Biosynthesis of vinblastine and vincristine comprises complex, multienzymatic networks that are tightly controlled by developmental and environmental factors (Pan et al. 2016). This plant also synthesizes ajmalicine (an antihypertensive agent) and serpentine (sedative agent) in small quantities. Therefore, to understand its in-depth medicinal potentials, many researchers from all over the world have put in comprehensive efforts and conducted thorough research on MIA biosynthesis in *C. roseus* over the past few decades (Pan et al. 2016). Several genes encoding structural biosynthetic enzymes, regulatory proteins and transporters of the MIA biosynthesis pathway of *C. roseus* have been identified, functionally characterized and utilized to increase MIA production. However, there are still many genes, enzymes, intermediates and their regulatory mechanisms in the MIA biosynthesis that are yet to be discovered. Mining of the whole-genome sequence of *C. roseus* using several computation tools would help to unravel the uncharacterized parts of the MIA biosynthesis, transport, signaling and regulation. For example, Kellner et al. (2015) reported four near-identical secologanin synthase (*SLS*) genes involved in the iridoid pathway from four different scaffolds of leaf tissue. Among which SLS_4 got physically clustered next to reticuline oxidase-like protein and alcohol dehydrogenase (tetrahydroalstonine synthases, *THAS*). Recent studies suggested that the alcohol dehydrogenase *THAS* converts strictosidine aglycone to a monomeric MIA of the heteroyohimbine structural class MIA tetrahydroalstonine (Stavrinides et al. 2015).

Similarly, the genome sequence of *C. roseus* could be utilized for retrieving unidentified genes of the MIA biosynthesis pathway or novel biochemical pathways. The scaffolding details obtained from the *C. roseus* genome sequence would increase the length of contigs which in turn uncover more gene families and lead to the finding of previously unknown genes of the pathway encoding new biochemical enzymes. Likewise, genome sequences could also be utilized to understand gene duplication and sub/neofunctionalization to examine the evolution among the genes, which otherwise cannot be identified from short-read derived transcriptome data alone. The draft genome of *C. roseus* revealed the existence of multiple paralogs of many MIA genes which cleared the presence of lineage-specific evolution of MIA biosynthetic pathway genes. The presence of functionally distinct paralogs of T16H having 88% (nt) identity was found to be tandemly duplicated on a single 32 kbp scaffold. In the same way, four paralogs of *SLS* genes having 94–98% nt sequence identity were found on four different scaffolds that expressed in leaf tissue. Additionally, phylogenetic studies revealed six strictosidine synthase (*STR*) family genes grouped with the previously analyzed STR from the MIA pathway.

Furthermore, to understand MIA biosynthesis, it is very important to understand the regulatory role of many transcription factors (TFs) involved in the MIA biosynthesis pathway. TFs are DNA-binding proteins that binds with the upstream region of targeted genes and control their expression for normal growth and physiological function (Gantet and Memelink 2002). Jasmonate-responsive ORCAs (octadecanoid-responsive Catharanthus AP2-domain proteins) belonging to AP2/ERF (APETALA2/ethylene-responsive factor) family are well-known regulatory TFs of *C. roseus* MIA biosynthesis pathway (Pan et al. 2016). The regulatory role of *ORCA2* and *ORCA3* in MIA biosynthesis under MeJA treatment is well established in *C. roseus* (van der Fits and Memelink 2000). These two TFs were identified from genome data which is coherent with earlier investigations (Menke et al. 1999; van der Fits and Memelink 2000). Besides these TFs, two additional *ORCAs* were identified from the *C. roseus* whole-genome sequence further suggesting the presence of tandem duplication. The draft genome also revealed the presence of several other TFs physically co-localized on MIA gene scaffolds. While several TFs involved in the regulation of MIA pathway have been identified, the accessibility of the *C. roseus* draft genome would lead to the detection of uncharacterized TFs that participated in the regulation of important genes involved in its branches, such as the vindoline, iridoid and *bis*-indole alkaloids pathways. Figure 8.2 summarizes the utilization of *C. roseus* genome sequence in understanding MIA biosynthesis in *C. roseus*.

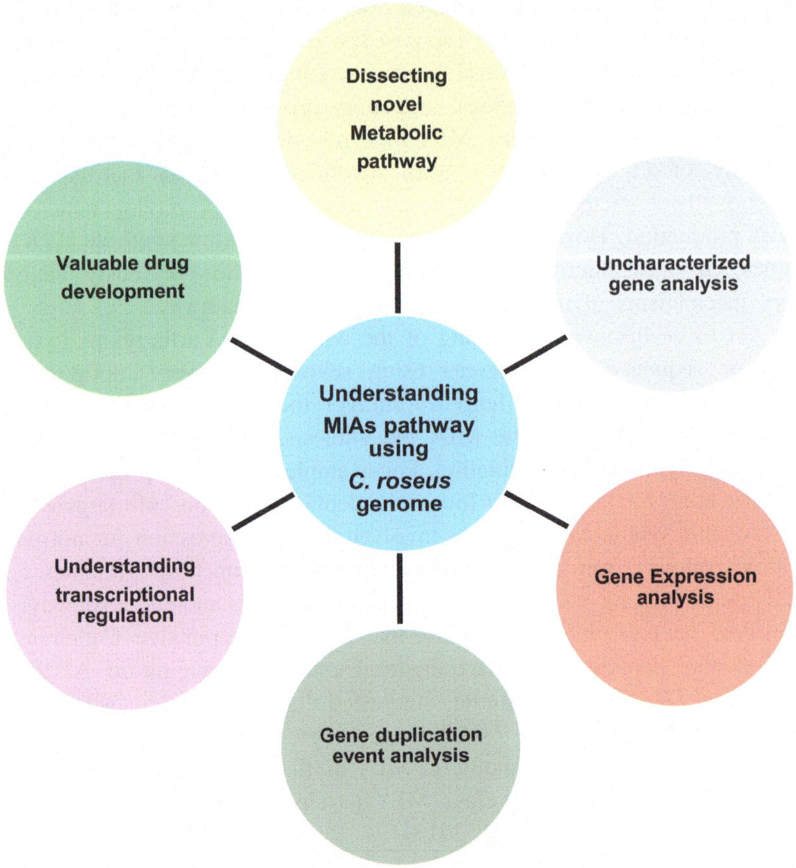

Fig. 8.2 Application of *C. roseus* genome in understanding the MIA biosynthesis

8.5 Functional Genomics Research in *C. roseus*

Although many genes, TFs and transporters involved in MIA biosynthesis have been identified and utilized in enhancing the production of MIA using metabolic engineering approaches, still research at the genetic level is required to identify the metabolic networks concerning the complete metabolic pathway (Pan et al. 2016). Recent studies involving the integration of 'multi-omics' (genomics, transcriptomics, proteomics and metabolomics) analysis could be a scientific tool for exploring the biological mechanisms implicated in the secondary metabolism of *C. roseus* (Fig. 8.3).

Table 8.1 reports different types of genes characterized for enhancing the production of MIA biosynthesis using a natural product targeted metabolic approach. However, with the release of the *C. roseus* draft genome, sequence analysis may aid in the detection of minor substitution in the amino acid at the substrate-binding sites of secondary metabolite biosynthesis pathway genes, resulting in the exploration of specific therapeutic properties of medicinal herbs. Besides, in the postgenomic era, metabolomics has turned out to be an effective method for functional genomic studies (Sumner et al. 2003). Metabolomics includes metabolite profiling, fingerprinting, footprinting and metabolic flux analysis. Metabolite profiling targets quantitative analysis of a specific class of compounds

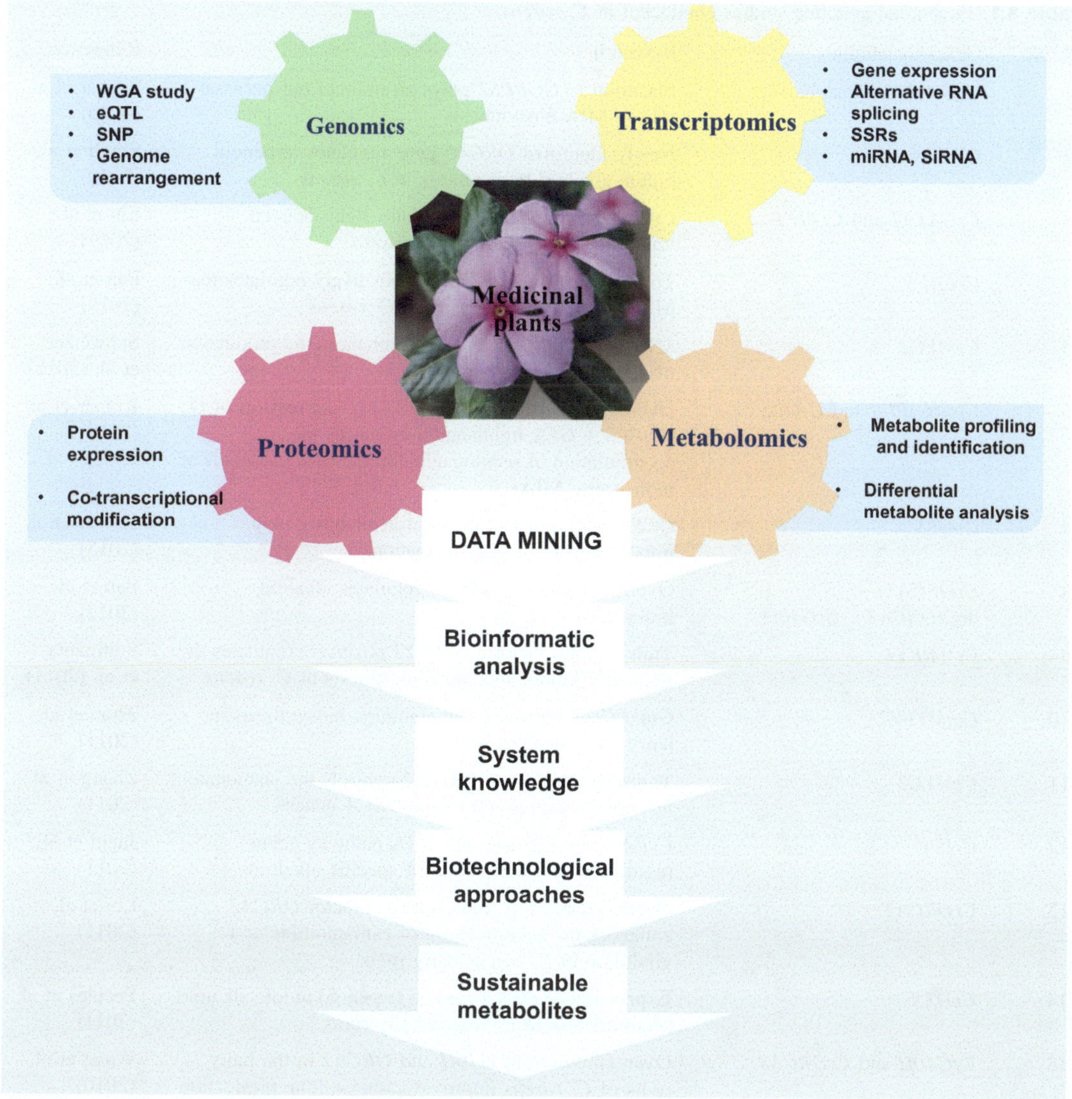

Fig. 8.3 Diagrammatic illustration of workflow showing the utility of integrated 'omics' analysis to obtain sustainable sources for medicinally important metabolites

in a particular biochemical pathway. Fingerprinting utilizes high-throughput qualitative assessment of the metabolic constituents for sample comparison and differential analysis in an organism or tissue. In contrast, footprinting is the qualitative and differential analysis of compounds that are excreted by cells into the culture medium. Whereas metabolic flux analysis (MFA) includes the quantification of all intracellular fluxes in an organism. MFA is typically carried out using gas chromatography (GC), liquid chromatography (LC), mass spectroscopy (MS) and nuclear magnetic resonance spectroscopy (NMR). NMR has been used in plant metabolomics studies of *C. roseus* for the detection of new compounds, the identification of metabolic pathways, classification and characterization of metabolic flux. Table 8.2 reports the secondary metabolites identified in *C. roseus* using the nuclear magnetic resonance

Table 8.1 Functional genomic studies conducted in *C. roseus*

S. No	Genes/metabolites	Research	Reference
1	*G(G)PPS2* (Plastidial)	Plastidial *CrGGPPS2* plays an indirect but necessary role in MIA biosynthesis	Kumar et al. (2020)
2	ORCA6	Newly identified *ORCA6* gene regulates terpenoid indole alkaloid biosynthesis in *C. roseus*	Singh et al. (2020)
3	*CrGATA1* and *CrPIF1*	*CrGATA1* and *CrPIF1* regulates light-induced vindoline biosynthesis in *C. roseus*	Liu et al. (2019)
4	*CrERF5*	Transcription factor *CrERF5* positively regulates the MIA biosynthesis pathway in *C. roseus*	Pan et al. (2019)
5	*CrMYC2*	*MYC2* with *ORCA3* and *BIS1* enhances the production of atleast 23 MIAs	Schweizer et al. (2018)
6	*CrG(G)PPS* and *CrGES*	Overexpression of *G(G)PPS* and co-expression of *G (G)PPS* + *GES* significantly enhances the accumulation of secologanin and elevates the levels of monomeric MIAs	Kumar et al. (2018)
7	*CrGES*	*GES* expression and geraniol availability play important role in MIA formation	Kumar et al. (2015)
8	*CrORCA3* or*CrORCA3* + *CrG10H*	Overexpression of *ORCA3* regulates alkaloid biosynthesis	Pan et al. (2012)
9	*CrWRKY1*	Transcription factor *CrWRKY1 positively* regulates the terpenoid indole alkaloid biosynthesis in *C. roseus*	Suttipanta et al. (2011)
10	*GmMYBZ2*	*GmMYBZ2* represses catharanthine biosynthesis in hairy roots of *C. roseus*	Zhou et al. (2011)
11	*CrMYC2*	Transcription factor *CrMYC2* controls the jasmonate-responsive expression of the *ORCA* genes	Zhang et al. (2011)
12	*CrPrx*	*CrPrx* gene regulates the MIA pathway genes, regulators, and production of specific alkaloids	Jaggi et al. (2011)
13	*CrORCA2*	Overexpression of transcriptional factor *ORCA2* enhances the accumulation of catharanthine and vindoline in *C. roseus* hairy roots	Liu et al. (2011)
14	*CrDXS*	Expression of *DXS* increases terpenoid indole alkaloid accumulation in *C. roseus* hairy roots	Peebles et al. (2011)
15	*CrG10H* and *CrORCA3*	Overexpression of *G10H* and *ORCA3* in the hairy roots of *C. roseus* improves catharanthine production	Wang et al. (2010)
16	*CjMDR1*	MDR1, an ABC transpoter from *Coptis japonica* improves accumulation of ajmalicine and tetrahydroalstonine	Pomahačová et al. (2009)
17	*CrORCA3*	Transcriptional response of the terpenoid indole alkaloid pathway to the overexpression of *ORCA3* along with jasmonic acid elicitation of *C. roseus* hairy roots over time	Peebles et al. (2009)
18	*Cr*ASαβ	Effect of the engineered indole pathway on accumulation of phenolic compounds in *C. roseus* hairy roots	Chung et al. (2007)
19	*CrDAT* (deacetylvindoline-4-O-acetyltransferase)	Expression of deacetylvindoline-4-*O*-acetyltransferase in *C. roseus* hairy roots	Magnotta et al. (2007)

(continued)

Table 8.1 (continued)

S. No	Genes/metabolites	Research	Reference
20	*CrASα* + *CrASβ* (anthranilate synthase)	Expression of the Arabidopsis feedback-insensitive anthranilate synthase holoenzyme and tryptophan decarboxylase genes in *C. roseus* hairy roots	Hong et al. (2006)
21	ASα(anthranilate synthase)	Expression of a feedback-resistant anthranilate synthase in *C. roseus* hairy roots provides evidence for tight regulation of terpenoid indole alkaloid levels	Hughes et al. (2004a)
22	*CrTDC*(tryptophan decarboxylase)	Metabolic engineering of the indole pathway in *C. roseus* hairy roots and increased accumulation of tryptamine and serpentine	Hughes et al. (2004b)
23	*CrHMGR* (hamster 3-hydroxy-3-methylglutaryl-CoA reductase)	Overexpression in *C. roseus* hairy roots of a truncated hamster 3-hydroxy-3-methylglutaryl-CoA reductase gene	Ayora-Talavera et al. (2002)
24	*CrTDC*	Effect of precursor feeding on alkaloid accumulation by a tryptophan decarboxylase over-expressing transgenic cell line T22 of *C. roseus*	Whitmer et al. (2002a)
25	*CrSTR*	Effect of precursor feeding on alkaloid accumulation by a strictosidine synthase over-expressing transgenic cell line S1 of *C. roseus*	Whitmer et al. (2002b)
26	*CrG10H1* (Geraniol 10-hydroxylase1), a cytochrome P450	Involvement of Geraniol 10-hydroxylase, cytochrome P450 in terpenoid indole alkaloid biosynthesis	Collu et al. (2001)
27	*CrORCA3*	A *C. roseus* BPF-1 homologue interacts with an elicitor-responsive region of the secondary metabolite biosynthetic gene Str and is induced by elicitor via a JA-independent signal transduction pathway	Van der Fits et al. (2000)

(NMR) technique. The recent advancement in the robust flux analysis methods for the *C. roseus* would be useful in determining the metabolic effects in response to genetic or environmental modulation.

8.6 *C. roseus* Functional Genomics Database (croFGD)

Comparative co-expression network study across the organisms is a powerful tool for improving gene annotations and their functional analysis (Pathania et al. 2016). By the integration of genomic and transcriptomic data, a comparative gene function analysis at a multidimensional grid and epigenome level is possible (You et al. 2016). There is currently an ample of transcriptomic data accessible from the public domain, which serve as the basis for the *C. roseus* study (Kellner et al. 2015). Utilizing such 53 RNA-seq datasets of various tissues (seedling, leaf, flower, shoot and root) and treatments {Methyl jasmonate (MeJA), peanut witches' broom (PnWB) infection and yeast elicitor) available on public platform, an online *C. roseus* functional genomics database (croFGD; http://bioinformatics.cau.edu.cn/croFGD/) was constructed to provide an online server for gene annotation and functional analysis (She et al. 2019). croFGD was designed to assist the society to study functional genomics and provide new information about molecular regulatory pathways. The database comprises a co-expression network using an in-house data processing pipeline and a combination of algorithms to investigate the

Table 8.2 Secondary metabolites identified from *C. roseus* using NMR analysis

Secondary metabolites	Chemical shift	References
4-*O*-Caffeoyl quinic acid	7.67 (d), 7.17 (d), 7.08 (dd), 6.44 (d), 2.09 (m)	Pan et al. (2014)
3-*O*-Dimethylcyanidin	8.80 (s), 8.42 (m), 8.17 (br-s), 7.39 (br-s), 7.21 (m), 6.86 (s), 4.02 (s), 3.98 (s)	Toki et al. (2008)
7-*O*-Methylcyanidin	8.74 (s), 8.25 (br-d), 8.22 (br-s), 7.28 (s), 7.12 (d), 6.87 (s), 4.00 (s)	Toki et al. (2008)
Phenylpropanoids	6.51 (d), 7.02 (s)	Mustafa et al. (2009)
Anthocyanidin-1	8.92 (s), 8.39 (dd), 8.26 (d), 7.42 (d), 7.15 (d), 6.78 (d), 4.04 (s), 3.99 (s)	Toki et al. (2008)
Anthocyanidin-2	8.96 (s), 8.33 (dd), 8.14 (d), 7.17 (s), 7.04 (d), 6.70 (s), 4.04 (s)	Toki et al. (2008)
Catharanthine	1.10 (t), 7.36 (d), 7.55 (d)	Pan et al. (2014)
Chlorogenic acid	7.61 (d), 7.14 (d), 7.05 (dd), 6.36 (d)	Pan et al. (2014)
Flavonoid glucosides 1	7.15 (d), 7.12 (d)	Chung et al. (2009)
Flavonoid glucosides 2	7.05 (br-s)	Chung et al. (2009)
Flavonoid glucosides 3	7.33 (d), 7.16 (m), 7.03 (m), 5.56 (dd), 5.42 (d), 2.39 (d)	Chung et al. (2009)
Flavonoid glucosides 4	7.39 (dd), 7.37 (d), 7.33 (d), 5.32 (dd), 3.19 (d), 3.11 (m)	Chung et al. (2009)
Gentisic acid glucoside	7.56 (d), 7.17 (dd), 6.83 (d), 4.88 (d)	Mustafa et al. (2009)
Kaempferol 1	6.30 (d), 6.51 (d), 7.00 (d), 8.04 (dd)	Pan et al. (2014)
Kaempferol 2	6.30 (d), 6.51 (d), 6.98 (d), 8.08 (dd)	Pan et al. (2014)
Quercetin-3-O-glucoside	7.87 (d), 7.65 (dd), 6.99 (d), 6.31 (d), 6.50 (d), 5.33 (d)	Pan et al. (2014)
Quinic acid	1.88 (dd), 1.93 (m)	Pan et al. (2014)
Salicylic acid	7.81 (d), 7.37 (t), 6.90 (t), 6.88 (d)	Mustafa et al. (2009)
Secologanin	7.44 (dd), 7.56 (d), 9.65 (d)	Pan et al. (2014)
Serpentine	8.31 (d), 8.35 (d), 8.46 (d)	Pan et al. (2014)
Strictosidine	7.80 (s)	Pan et al. (2014)
Syringetin-3-o-robinobioside	7.69 (s), 6.23 (d), 3.95 (s), 2.46 (d)	Brun et al. (1999)
Vindoline	0.51 (t), 2.00 (s), 2.66 (s), 5.93 (m), 6.22 (d), 7.11 (d)	Pan et al. (2014)
Vindolinine	1.08 (d), 5.92 (m), 6.51 (dd), 6.82 (d), 7.24 (d)	Pan et al. (2014)

global, tissue-specific and treatment-response network with expression insight. The database contains almost 90% of the genes of *C. roseus* and identified over 6000 functional modules covering the annotation of vindoline and vinblastine biosynthesis, hormone response, MeJA response and pathogen resistance, etc. The database is supported by many bioinformatics tools through which several important genes involved in crucial biosynthetic processes could be identified and contributes to a deeper understanding of the molecular mechanism involved in the metabolic pathways. Despite its utility and importance in functional genomics and gene annotation, the croFGD needs some refinements with the publication of the chromosome-level genome of *C. roseus*. Additionally, integration of transcriptomic data of other tissues and treatments, microarray and draft genome data would also improve this database.

8.7 Conclusion and Prospects

Until recently, genomic research in medicinal plants lagged behind model and crop plants. However, modern-day genomic studies and cost-effective high-throughput sequencing have enabled plant biologists to explore the metabolic pathways especially that are implicated in the regulation and synthesis of specialized metabolites. Combination of statistical and computational/bioinformatics tools with the high-dimensional intricated omics dataset has driven the understanding of biological systems and construction of metabolic models for medicinal plants. Even though the health system is improving gradually, the side effects of modern medicines have made medicinal plants an alternative source for drugs used for treating fatal diseases. The in-depth knowledge of metabolic pathways in medicinal plants like *C. roseus* could be widely utilized to enhance the specific metabolites. The information acquired from the specialized metabolic pathway studies could be the most dependable approach to predict the gene function. Genome editing tools such as CRISPR/Cas9 could also be employed for generating transgenic lines with desired traits and studying secondary metabolites. However, the lack of effective protocol for plant transformation and time taking regeneration techniques are two major limitations in studying MIA biosynthesis in *C. roseus*. Refinement of secondary metabolism through understanding phylogeny and evolution of this plant could be another breakthrough that can be achieved through the availability of the draft genome of *C. roseus*. In addition, genome-wide association studies (GWAS) could be utilized for pathway analysis and to find genetic variations associated with traits in *C. roseus*. GWAS is a new approach that uses statistical models to examine the link between different traits (SNP) and phenotypes. The data acquired from GWAS could be used in marker-assisted selection (MAS) for molecular breeding in *C. roseus*.

Although the *C. roseus* genome has provided a high degree of genetics-related research but to identify the good candidate for molecular breeding and detect genomic variations, a single reference genome sequence is not sufficient. Hence, a high-quality re-sequenced genome of *C. roseus* with automated phenotyping under controlled conditions is required. A re-sequenced reference genome would help to understand the conserved part of the related species. It would also be useful in improving our knowledge regarding the prevalence of variations (such as SNPs, InDels) within a species or a genus, phylogeny of a species or genus, and agronomic traits.

References

Aslam J, Khan SH, Siddiqui ZH, Fatima Z, Maqsood M, Bhat MA, Nasim SA, Ilah A, Ahmad IZ, Khan SA, Mujib A (2010) *Catharanthus roseus* (L.) G. Don. An important drug: it's applications and production. Pharmacie Globale (IJCP) 4(12):1–6

Ayora-Talavera T, Chappell J, Lozoya-Gloria E, Loyola-Vargas VM (2002) Overexpression in *Catharanthus roseus* hairy roots of a truncated hamster 3-hydroxy-3-methylglutaryl-CoA reductase gene. Appl Biochem Biotechnol 97(2):135–145. https://doi.org/10.1385/ABAB:97:2:135

Brun G, Dijoux MG, David B, Mariotte AM (1999) A new flavonol glycoside from *Catharanthus roseus*. Phytochemistry 50(1):167–169. https://doi.org/10.1016/S0031-9422(98)00501-9

Chung IM, Hong SB, Peebles CA, Kim JA, San KY (2007) Effect of the engineered indole pathway on accumulation of phenolic compounds in *Catharanthus roseus* hairy roots. Biotechnol Progr 23(2):327–332. https://doi.org/10.1021/bp060258e

Chung IM, Ahmad A, Ali M, Lee OK, Kim MY, Kim JH, Yoon DY, Peebles CAM, San KY (2009) Flavonoid glucosides from the hairy roots of *Catharanthus roseus*. J Nat Prod 72(4):613–620. https://doi.org/10.1021/np800378q

Champagne A, Rischer H, Oksman-Caldentey KM, Boutr M (2012) In-depth proteome mining of cultured

Catharanthus roseus cells identifies candidate proteins involved in the synthesis and transport of secondary metabolites. Proteomics 12:2536–2547. https://doi.org/10.1002/pmic.201200218

Collu G, Unver N, Peltenburg-Looman AM, van der Heijden R, Verpoorte R, Memelink J (2001) Geraniol 10-hydroxylase1, a cytochrome P450 enzyme involved in terpenoid indole alkaloid biosynthesis. FEBS Lett 508(2):215–220. https://doi.org/10.1016/S0014-5793(01)03045-9

Edwards D, Batley J (2010) Plant genome sequencing: applications for crop improvement. Plant Biotechnol J 8(1):2–9. https://doi.org/10.1111/j.1467-7652.2009.00459.x

Gantet P, Memelink J (2002) Transcription factors: tools to engineer the production of pharmacologically active plant metabolites. Trends Pharmacol Sci 23(12):563–569. https://doi.org/10.1016/s0165-6147(02)02098-9

Góngora-Castillo E, Childs KL, Fedewa G, Hamilton JP, Liscombe DK, Magallanes-Lundback M et al (2012) Development of transcriptomic resources for interrogating the biosynthesis of monoterpene indole alkaloids in medicinal plant species. PLoS ONE 7(12): e52506. https://doi.org/10.1371/journal.pone.0052506

Guimaraes G, Cardoso L, Oliveira H, Santos C, Duarte P, Sottomayor M (2012) Cytogenetic characterization and genome size of the medicinal plant *Catharanthus roseus* (L.) G. Don. AoB Plants. https://doi.org/10.1093/aobpla/pls002

Hao DC, Xiao PG (2015) Genomics and evolution in traditional medicinal plants: road to a healthier life. Evol Bioinform11:EBO-S31326. https://doi.org/10.4137/EBO.S31326

Hong SB, Peebles CA, Shanks JV, San KY, Gibson SI (2006) Expression of the Arabidopsis feedback-insensitive anthranilate synthase holoenzyme and tryptophan decarboxylase genes in *Catharanthus roseus* hairy roots. J Biotechnol 122(1):28–38. https://doi.org/10.1016/j.jbiotec.2005.08.008

Hu S, Li G, Yang J, Hou H (2017) Aquatic plant genomics: advances, applications, and prospects. Int J Genom 2017:6347874. https://doi.org/10.1155/2017/6347874

Hughes EH, Hong SB, Gibson SI, Shanks JV, San KY (2004a) Expression of a feedback-resistant anthranilate synthase in *Catharanthus roseus* hairy roots provides evidence for tight regulation of terpenoid indole alkaloid levels. Biotechnol Bioeng 86(6):718–727. https://doi.org/10.1002/bit.20081

Hughes EH, Hong SB, Gibson SI, Shanks JV, San KY (2004b) Metabolic engineering of the indole pathway in *Catharanthus roseus* hairy roots and increased accumulation of tryptamine and serpentine. Metab Eng 6(4):268-276. https://doi.org/10.1016/j.ymben.2004.03.002

Jaggi M, Kumar S, Sinha AK (2011) Overexpression of an apoplastic peroxidase gene *CrPrx* in transgenic hairy root lines of *Catharanthus roseus*. Appl Microb Biotechnol 90(3):1005–1016. https://doi.org/10.1007/s00253-011-3131-8

Jansen RK, Cai Z, Raubeso LA, Daniell H, Depamphilis CW, Leebens-Mack J, Müller KF, Guisinger-Bellian M, Haberle RC, Hansen AK, Chumley TW (2007) Analysis of 81 genes from 64 plastid genomes resolves relationships in angiosperms and identifies genome-scale evolutionary patterns. Proc Natl Acad Sci USA 104(49):19369–19374. https://doi.org/10.1073/pnas.0709121104

Kaul S, Koo HL, Jenkins J, Rizzo M, Rooney T, Tallon LJ, Feldblyum T, Nierman W, Benito MI, Lin X, Town CD (2000) Analysis of the genome sequence of the flowering plant *Arabidopsis thaliana*. Nature 408(6814):796–815. https://doi.org/10.1038/35048692

Kaushik SH, Tomar RS, Gupta MO, Mishra RK (2017) An overview of *Catharanthus roseus* and medicinal properties of their metabolites against important diseases. Eur Acad Res 5(2):1237–1247

Kellner F, Kim J, Clavijo BJ, Hamilton JP, Childs KL, Vaillancourt B, Cepela J, Habermann M, Steuernagel B, Clissold L, McLay K (2015) Genome-guided investigation of plant natural product biosynthesis. Plant J 82 (4):680–692. https://doi.org/10.1111/tpj.12827

Ku C, Chung WC, Chen LL, Kuo CH (2013) The complete plastid genome sequence of Madagascar periwinkle *Catharanthus roseus* (L.) G. Don: plastid genome evolution, molecular marker identification, and phylogenetic implications in asterids. PLoS One 8(6):68518. https://doi.org/10.1371/journal.pone.0068518

Kumar K, Kumar SR, Dwivedi V, Rai A, Shukla AK, Shanker K, Nagegowda DA (2015) Precursor feeding studies and molecular characterization of geraniol synthase establish the limiting role of geraniol in monoterpene indole alkaloid biosynthesis in *Catharanthus roseus* leaves. Plant Sci 239:56–66. https://doi.org/10.1016/j.plantsci.2015.07.007

Kumar SR, Rai A, Bomzan DP, Kumar K, Hemmerlin A, Dwivedi V, Godbole RC, Barvkar V, Shanker K, Shilpashree HB, Bhattacharya A (2020) A plastid-localized bona fide geranylgeranyl diphosphate synthase plays a necessary role in monoterpene indole alkaloid biosynthesis in *Catharanthus roseus*. Plant J 103(1):248–265. https://doi.org/10.1111/tpj.14725

Kumar SR, Shilpashree HB, Nagegowda DA (2018) Terpene moiety enhancement by overexpression of geranyl (geranyl) diphosphate synthase and geraniol synthase elevates monomeric and dimeric monoterpene indole alkaloids in transgenic *Catharanthus roseus*. Front Plant Sci 9:942. https://doi.org/10.3389/fpls.2018.00942

Kutchan TM (1995) Alkaloid biosynthesis—the basis for metabolic engineering of medicinal plants. Plant Cell 7:1059. https://doi.org/10.1105/tpc.7.7.1059

Li C, Zhang Y, Zheng L, Zheng Y, Huang P, Xiao X, Hao X, Chen Q, Hu Z, Li H, Wang X (2019) Assembly and annotation of a draft genome of the medicinal plant *Polygonum cuspidatum*. Front Plant Sci 10:1274. https://doi.org/10.3389/fpls.2019.01274

Liu DH, Ren WW, Cui LJ, Zhang LD, Sun XF, Tang KX (2011) Enhanced accumulation of catharanthine and vindoline in *Catharanthus roseus* hairy roots by overexpression of transcriptional factor *ORCA2*. Afr J Biotechnol 10(17):3260–3268. https://doi.org/10.5897/AJB10.1556

Liu Y, Patra B, Pattanaik S, Wang Y, Yuan L (2019) GATA and phytochrome interacting factor transcription factors regulate light-induced vindoline biosynthesis in *Catharanthus roseus*. Plant Physiol 180 (3):1336–1350. https://doi.org/10.1104/pp.19.00489

Magnotta M, Murata J, Chen J, De Luca V (2007) Expression of deacetylvindoline-4-O-acetyltransferase in *Catharanthus roseus* hairy roots. Phytochemistry 68(14):1922–1931. https://doi.org/10.1016/j.phytochem.2007.04.037

Menke FL, Champion A, Kijne JW, Memelink J (1999) A novel jasmonate-and elicitor-responsive element in the periwinkle secondary metabolite biosynthetic gene Str interacts with a jasmonate-and elicitor-inducible AP2-domain transcription factor, *ORCA2*. EMBO J 18 (16):4455–4463. https://doi.org/10.1093/emboj/18.16.4455

Moore MJ, Soltis PS, Bell CD, Burleigh JG, Soltis DE (2010) Phylogenetic analysis of 83 plastid genes further resolves the early diversification of eudicots. Proc Natl Acad Sci USA 107(10):4623–4628. https://doi.org/10.1073/pnas.0907801107

Moore BD, Andrew RL, Külheim C, Foley WJ (2014) Explaining intraspecific diversity in plant secondary metabolites in an ecological context. New Phytol 201 (3):733–750. https://doi.org/10.1111/nph.12526

Mustafa NR, Kim HK, Choi YH, Verpoorte R (2009) Metabolic changes of salicylic acid-elicited *Catharanthus roseus* cell suspension cultures monitored by NMR-based metabolomics. Biotechnol Lett 31 (12):1967–1974. https://doi.org/10.1007/s10529-009-0107-1

Murata J, Roepke J, Gordon H, De Luca V (2008) The leaf epidermome of *Catharanthus roseus* reveals its biochemical specialization. Plant Cell 20: 524–542. https://doi.org/10.1105/tpc.107.056630

O'Connor SE, Maresh JJ (2006) Chemistry and biology of monoterpene indole alkaloid biosynthesis. Nat Prod Rep 23(4):532–547. https://doi.org/10.1039/B512615K

Pan Q, Dai Y, Nuringtyas TR, Mustafa NR, Schulte AE, Verpoorte R, Choi YH (2014) Investigation of the chemomarkers correlated with flower colour in different organs of *Catharanthus roseus* using NMR-based metabolomics. Phytochem Anal 25(1):66–74. https://doi.org/10.1002/pca.2464

Pan Q, Mustafa NR, Tang K, Choi YH, Verpoorte R (2016) Monoterpenoid indole alkaloids biosynthesis and its regulation in *Catharanthus roseus*: a literature review from genes to metabolites. Phytochem Rev 5 (2):221–250. https://doi.org/10.1007/s11101-015-9406-4

Pan Q, Wang C, Xiong Z, Wang H, Fu X, Shen Q, Peng B, Ma Y, Sun X, Tang K (2019) *CrERF5*, an AP2/ERF transcription factor, positively regulates the biosynthesis of *bis*indole alkaloids and their precursors in *Catharanthus roseus*. Front Plant Sci 10:931. https://doi.org/10.3389/fpls.2019.00931

Pan Q, Wang Q, Yuan F, Xing S, Zhao J, Choi YH, Verpoorte R, Tian Y, Wang G, Tang K (2012) Overexpression of *ORCA3* and *G10H* in *Catharanthus roseus* plants regulated alkaloid biosynthesis and metabolism revealed by NMR-metabolomics. PLoS One 7(8):43038. https://doi.org/10.1371/journal.pone.0043038

Pathania S, Bagler G, Ahuja PS (2016) Differential network analysis reveals evolutionary complexity in secondary metabolism of *Rauvolfia serpentina* over *Catharanthus roseus*. Front Plant Sci 7:1229. https://doi.org/10.3389/fpls.2016.01229

Peebles CA, Hughes EH, Shanks JV, San KY (2009) Transcriptional response of the terpenoid indole alkaloid pathway to the overexpression of *ORCA3* along with jasmonic acid elicitation of *Catharanthus roseus* hairy roots over time. Metabol Eng 11(2):76–86. https://doi.org/10.1016/j.ymben.2008.09.002

Peebles CA, Sander GW, Hughes EH, Peacock R, Shanks JV, San KY (2011) The expression of 1-deoxy-D-xylulose synthase and geraniol-10-hydroxylase or anthranilate synthase increases terpenoid indole alkaloid accumulation in *Catharanthus roseus* hairy roots. Metab Eng 13(2):234–240. https://doi.org/10.1016/j.ymben.2010.11.005

Pomahačová B, Dušek J, Dušková J, Yazaki K, Roytrakul S, Verpoorte R (2009) Improved accumulation of ajmalicine and tetrahydroalstonine in Catharanthus cells expressing an ABC transporter. J Plant Physiol 166(13):1405–1412. https://doi.org/10.1016/j.jplph.2009.02.015

Rastogi S, Kalra A, Gupta V, Khan F, Lal RK, Tripathi AK, Parameswaran S, Gopalakrishnan C, Ramaswamy G, Shasany AK (2015) Unravelling the genome of Holy basil: an "incomparable" "elixir of life" of traditional Indian medicine. BMC Genom 16 (1):413. https://doi.org/10.1186/s12864-015-1640-z

Ravi V, Khurana JP, Tyagi AK, Khurana P (2008) An update on chloroplast genomes. Plant Syst Evol 271(1):101–122. https://doi.org/10.1007/s00606-007-0608-0

Samson N, Bausher MG, Lee SB, Jansen RK, Daniell H (2007) The complete nucleotide sequence of the coffee (*Coffea arabica* L.) chloroplast genome: organization and implications for biotechnology and phylogenetic relationships amongst angiosperms. Plant Biotechnol J 5(2):339–353. https://doi.org/10.1080/23802359.2019.1636729

Sen S, Chakraborty R, De B (2011) Challenges and opportunities in the advancement of herbal medicine: India's position and role in a global context. J Herb

Med 1(3–4):67–75. https://doi.org/10.1016/j.hermed.2011.11.001
Schweizer F, Colinas M, Pollier J, Van Moerkercke A, Bossche RV, De Clercq R, Goossens A (2018) An engineered combinatorial module of transcription factors boosts production of monoterpenoid indole alkaloids in *Catharanthus roseus*. Metabol Eng 48:150–162. https://doi.org/10.1016/j.ymben.2018.05.016
She J, Yan H, Yang J, Xu W, Su Z (2019) croFGD: *Catharanthus roseus* functional genomics database. Front Genet 10:238. https://doi.org/10.3389/fgene.2019.00238
Shen Q, Zhang L, Liao Z, Wang S, Yan T, Shi P, Liu M, Fu X, Pan Q, Wang Y, Lv Z (2018) The genome of *Artemisia annua* provides insight into the evolution of Asteraceae family and artemisinin biosynthesis. Mol Plant 11(6):776–788. https://doi.org/10.1016/j.molp.2018.03.015
Singh SK, Patra B, Paul P, Liu Y, Pattanaik S, Yuan L (2020). Revisiting the ORCA gene cluster that regulates terpenoid indole alkaloid biosynthesis in *Catharanthus roseus*. Plant Sci 293:110408. https://doi.org/10.1016/j.plantsci.2020.110408
Sonnante G (2019) The chloroplast genome. In: Portis E, Acquadro A, Lanteri S (eds) The globe artichoke genome. Compendium of plant genomes. Springer, Cham, Switzerland. https://doi.org/10.1007/978-3-030-20012-1_10
Stavrinides A, Tatsis EC, Foureau E, Caputi L, Kellner F, Courdavault V, O'Connor SE (2015) Unlocking the diversity of alkaloids in *Catharanthus roseus*: nuclear localization suggests metabolic channeling in secondary metabolism. Chem Biol 22(3):336–341. https://doi.org/10.1016/j.chembiol.2015.02.006
Sumner LW, Mendes P, Dixon RA (2003) Plant metabolomics: large-scale phytochemistry in the functional genomics era. Phytochemistry 62(6):817–836. https://doi.org/10.1016/s0031-9422(02)00708-2
Suttipanta N, Pattanaik S, Kulshrestha M, Patra B, Singh SK, Yuan L (2011) The transcription factor *CrWRKY1* positively regulates the terpenoid indole alkaloid biosynthesis in *Catharanthus roseus*. Plant Physiol 157(4):2081–2093. https://doi.org/10.1104/pp.111.181834
Toki K, Saito N, Irie Y, Tatsuzawa F, Shigihara A, Honda T (2008) 7-O-Methylated anthocyanidin glycosides from *Catharanthus roseus*. Phytochemistry 69 (5):1215–1219. https://doi.org/10.1016/j.phytochem.2007.11.005
Upadhyay AK, Chacko AR, Gandhimathi A, Ghosh P, Harini K, Joseph AP, Joshi AG, Karpe SD, Kaushik S, Kuravadi N, Lingu CS (2015) Genome sequencing of herb Tulsi (*Ocimum tenuiflorum*) unravels key genes behind its strong medicinal properties. BMC Plant Biol 15(1):212. https://doi.org/10.1186/s12870-015-0562-x
Van der Fits L, Zhang H, Menke FL, Deneka M, Memelink J (2000) A *Catharanthus roseus* BPF-1 homologue interacts with an elicitor-responsive region of the secondary metabolite biosynthetic gene Str and is induced by elicitor via a JA-independent signal transduction pathway. Plant Mol Biol 44(5):675–685. https://doi.org/10.1023/a:1026526522555
Van der Fits L, Memelink J (2000) ORCA3, a jasmonate-responsive transcriptional regulator of plant primary and secondary metabolism. Sci 289(5477):295–297. https://doi.org/10.1126/science.289.5477.295
Van Moerkercke A, Fabris M, Pollier J, Baart GJ, Rombauts S, Hasnain G, Rischer H, Memelink J, Oksman-Caldentey KM, Goossens A (2013) CathaCyc, a metabolic pathway database built from *Catharanthus roseus* RNA-Seq data. Plant Cell Physiol 54:673–685. https://doi.org/10.1093/pcp/pct039
Verma M, Ghangal R, Sharma R, Sinha AK, Jain M (2014) Transcriptome analysis of *Catharanthus roseus* for gene discovery and expression profiling. PLoS ONE 9(7):e103583. https://doi.org/10.1371/journal.pone.0103583
Wang CT, Liu H, Gao XS, Zhang HX (2010) Overexpression of G10H and ORCA3 in the hairy roots of *Catharanthus roseus* improves catharanthine production. Plant Cell Rep 29(8):887–894. https://doi.org/10.1007/s00299-010-0874-0
Whitmer S, van der Heijden R, Verpoorte R (2002a) Effect of precursor feeding on alkaloid accumulation by a tryptophan decarboxylase over-expressing transgenic cell line T22 of *Catharanthus roseus*. J Biotechnol 96(2):193–203. https://doi.org/10.1016/S0168-1656(02)00027-5
Whitmer S, Van der Heijden R, Verpoorte R (2002b) Effect of precursor feeding on alkaloid accumulation by a strictosidine synthase over-expressing transgenic cell line S1 of *Catharanthus roseus*. Plant Cell Tissue Org Cult 69(1):85–93. https://doi.org/10.1023/A:1015090224398
Xu H, Song J, Luo H, Zhang Y, Li Q, Zhu Y, Xu J, Li Y, Song C, Wang B, Sun W (2016) Analysis of the genome sequence of the medicinal plant *Salvia miltiorrhiza*. Mol Plant 9(6):949–952. https://doi.org/10.1016/j.molp.2016.03.010
You Q, Zhang L, Yi X, Zhang K, Yao D, Zhang X, Wang Q, Zhao X, Ling Y, Xu W, Li F (2016) Co-expression network analyses identify functional modules associated with development and stress response in *Gossypium arboreum*. Sci Rep 6(1):1–15. https://doi.org/10.1038/srep38436
Zhang H, Hedhili S, Montiel G, Zhang Y, Chatel G, Pré M, Gantet P, Memelink J (2011) The basic helix-loop-helix transcription factor *CrMYC2* controls the jasmonate-responsive expression of the *ORCA* genes that regulate alkaloid biosynthesis in *Catharanthus roseus*. Plant J 67(1):61–71. https://doi.org/10.1111/j.1365-313X.2011.04575.x
Zhou ML, Hou HL, Zhu XM, Shao JR, Wu YM, Tang YX (2011) Soybean transcription factor *Gm MYBZ2* represses catharanthine biosynthesis in hairy roots of *Catharanthus roseus*. Appl Microb Biotechnol 91(4):1095–1105. https://doi.org/10.1007/s00253-011-3288-1

Functional Genomics Approaches for Gene Discovery Related to Terpenoid Indole Alkaloid Biosynthetic Pathway in *Catharanthus roseus*

9

Sanjay Kumar Rai, Krishna Kumar Rai, Apoorva, Sushil Kumar, and Shashi Pandey Rai

Abstract

Catharanthus roseus is one of the most explored medicinal plants owing to its immense potential to synthesize various terpenoid indole alkaloids (TIAs) which are extensively used in the treatment of various chronic diseases. Extensive efforts have been devoted to decipher the metabolic pathways and the mechanisms behind the regulation of TIA accumulation, still less information is available regarding their structural and functional mode of transportation and corresponding modes of action. The elucidation of transportation of TIAs is extremely important as most of the TIAs are synthesized in different cellular compartments and exhibit function in another. In the recent decades, considerable progress has been achieved in unraveling transcriptional mediated regulation of TIA biosynthetic pathways along with the transcription factors, genes and proteins which regulate them. With the advent of omics-based tools along with the conjunction with next-generation sequencing technology, the metabolic engineering of efforts has been accelerated to a greater pace. Furthermore, the combined use of cell/tissue specific transcriptome with the metabolome together with deep bioinformatic tools can provide valuable insight into the TIA biosynthetic pathways in *C. roseus*. In addition, the functional genomics tools such as RNA interference and virus-induced gene silencing can help in identifying and characterizing novel genes of the metabolic pathways. So, the present chapter deals with the latest advancement made in the "omics" techniques that allow rapid and precise identification of biosynthetic genes in *C. roseus*. Further, how these genes/proteins can be systematically engineered to enhance the production of the important TIAs having pharmaceutical and therapeutic importance has also been discussed here.

S. K. Rai
Dr. Rajendra Prasad Central Agricultural University, PUSA, Samastipur, Bihar, India

K. K. Rai · Apoorva · S. P. Rai (✉)
Department of Botany, Center of Advance Study in Botany, Institute of Science, Banaras Hindu University (BHU), Varanasi 221005, India
e-mail: pshashi@bhu.ac.in

S. Kumar
SKA Institute for Research, Education and Development (SKAIRED), 4/11 Sarv Priya Vihar, New Delhi, India

9.1 Introduction

The perennial herb *Catharanthus roseus* L., commonly known as periwinkle, belongs to the family Apocynaceae. It is an important medicinal plant with immense pharmaceutical potential due

© The Author(s), under exclusive license to Springer Nature Switzerland AG 2022
C. Kole (ed.), *The Catharanthus Genome*, Compendium of Plant Genomes,
https://doi.org/10.1007/978-3-030-89269-2_9

to the presence terpenoid indole alkaloids (TIAs) (Pandey-Rai et al. 2006). The TIAs are the class of secondary metabolites which are generally synthesized in leaf and other cellular tissues of plants exposed to stressful conditions (Tolambiya and Mathur 2016). The synthesis of TIAs is regulated at different developmental stages and are also stimulated upon hormonal changes which are compartmentalized into different cellular organelles (Mangaiyarkarasi et al. 2014). Various researchers have reported that the site of synthesis of these secondary metabolites is entirely different from the site of their actual biological functions (Das et al. 2020). Furthermore, all TIAs are the precursors of strictosidine which is the synthesis product of tryptamine and secologanin moiety of shikimate and nonmevalonate pathway (Das et al. 2020). The secondary metabolite synthesis in *C. roseus* starts with reaction between tryptophan and monoterpenoid geraniol which involves at least 30 enzymes and reaction intermediates. Several lines of literature have well-documented idioblastic/laticifer cells as the major site of vindoline synthesis, whereas majority of biosynthesis of TIAs occurs in the epidermal layers of leaves and stems (Goswami 2013).

This plant produces over 50–100 different type of TIAs which play significant role in the treatment of various chronic diseases (Tolambiya and Mathur 2016). For instance, TIAs like vinblastine and vincristine are predominantly used in the treatment of cancer, whereas vindoline and catharanthine which are synthesized in trace amount have been reported to have antidiabetic, antidiuretic and antibacterial properties (Kotakadi et al. 2013). Various other TIAs identified and reported form *C. roseus* such as ajmalicine and serpentine are exceptionally involved in the treatment of hypertension and inflammatory disorders (Table 9.1) (Zhu et al. 2014). Numerous reports have revealed the genes/transcription factors (TFs) are coordinately involved in the biosynthesis of these TIAs in conjunction with the external and internal signals (Singh et al. 2020). TFs like Apetala2/Ethelene Responsive Factor (AP2/ERF), no apical meristem, Arabidopsis transcription activation factor and cup-shaped cotyledon (NAC) have been identified in *C. roseus* and overexpression of these TFs in the presence of methyl jasmonate have resulted in increased production of TIAs (Paul et al. 2017). Furthermore, other TFs have also been identified which stimulate the biosynthesis of TIAs after binding to promoter element in response to abiotic elicitors like jasmonic acid (Goldhaber-Pasillas et al. 2014).

Considerable progress has been made in understanding the metabolic pathway related to biosynthesis of TIAs, still these pharmacologically important TIAs are produced at extremely low amounts (Sharma et al. 2018). Due to their immense therapeutic potential, the biosynthesis of these TIAs needs systemic rewiring in order to meet the demand supply gap (Zhu et al. 2015). Nonetheless, the researchers are working extremely hard to improve the production of these valuable TIAs using large-scale culture systems, bioreactors and other metabolic tools and biotechnological interventions, but limited success has been achieved so far (Liu et al. 2017). Therefore, in order to delve further and deeper into the biochemical and molecular mechanisms related to biosynthesis of TIAs, it has become imperative to exploit state-of-the-art techniques to improve alkaloid synthesis in *C. roseus.* In this chapter, we will be describing about the latest advancement made in the "omics" techniques in the identification of genes related to biosynthesis of TIAs in *C. roseus.* Further, how these techniques could be exploited in *C. roseus* for its successful metabolic engineering are also deliberated.

9.2 Functional Biological Pathways Involved in the Biosynthesis of TIAs

The pathway of TIAs in *C. roseus* is complex biosynthetic pathway which involves coordinated action of more than 35 enzymes, regulators and intra- or intercellular transporters (Pan et al. 2015). In past recent years, extensive efforts have been made for deciphering this complex pathway by investigating biochemical and molecular

Table 9.1 List of some important TIAs synthesized by *C. roseus* and their potential functional properties

S. No	Alkaloids	Chemical structures	Pharmaceutical/therapeutic properties	Mode of action	Adverse effect	Mode of delivery	References
1.	Vindoline	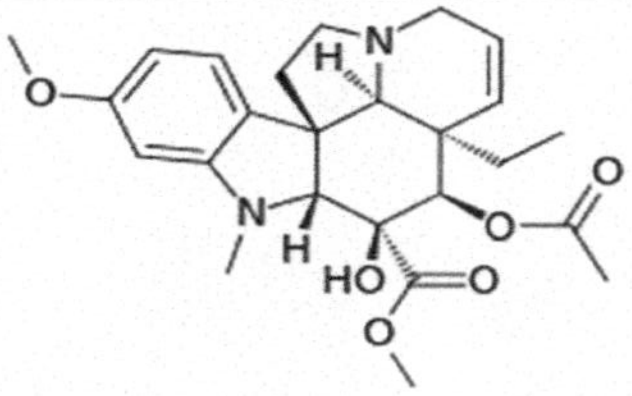	Brain cancer, melanoma	Blocking cell division	Headaches, constipation	Injection into a vein	Liu et al.
2	Vinblastine	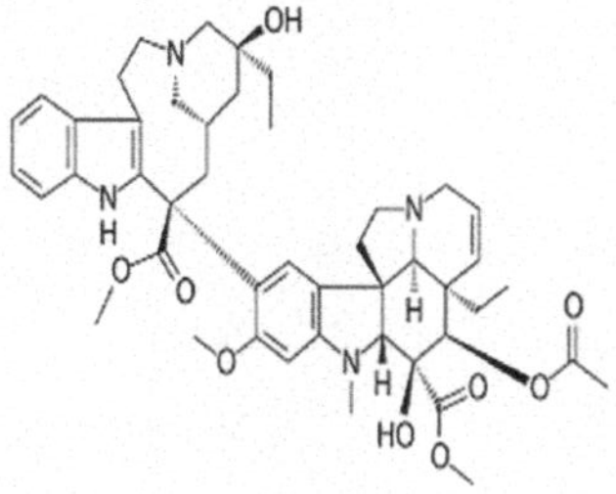	Lung cancer, bladder cancer, Hodgkin's lymphoma	Blocking cell division	Low blood cell counts, shortness of breath	Injection into a vein	Cooper and Deakin (2016)
3	Vincristine	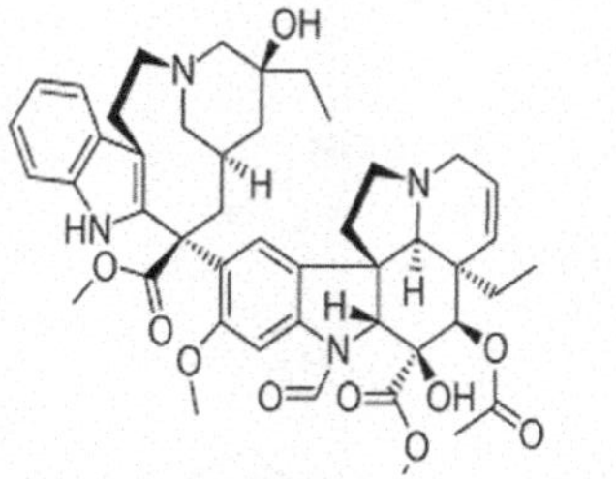	Leukaemia and other cancer diseases	Inhibition of microtubule formation during cell division	Jaw pain, difficulty in walking, drooping eyelids	Intravenously	WHO (2019)

(continued)

Table 9.1 (continued)

S. No	Alkaloids	Chemical structures	Pharmaceutical/therapeutic properties	Mode of action	Adverse effect	Mode of delivery	References
4	Horhammericine	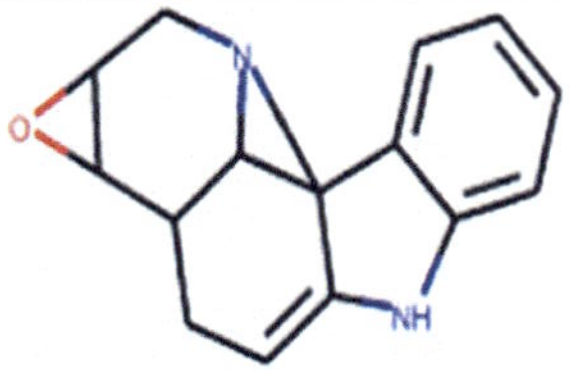	Anti-hypertension, Anti-tumour	Vasodilation and inhibiting cell division	Diarrhoea and dizziness	Orally and intravenously	Sharma et al. (2020)
5	Catharanthine	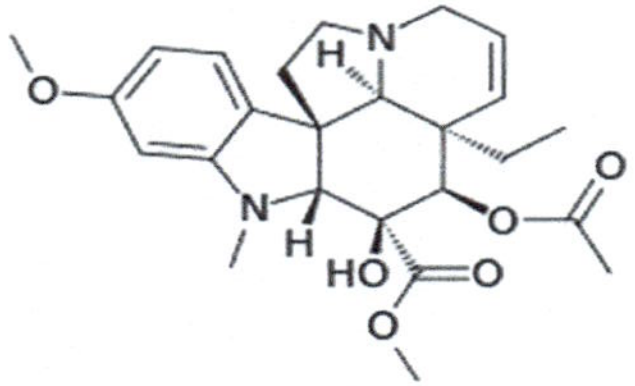	Anti-tumorigenic effect	Increases cAMP levels in neurons	Hyponatremia and hair loss	Injection into a vein	Verma et al. (2020)
6	Serpentine	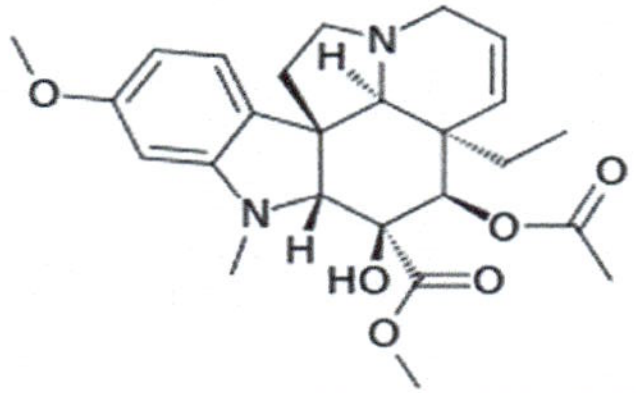	Treatment of circulatory disease,	Regulates the biosynthesis of catecholamines	Change in appetite and weight	Orally	Monforte-González et al. (1992)
7	Ajmalicine	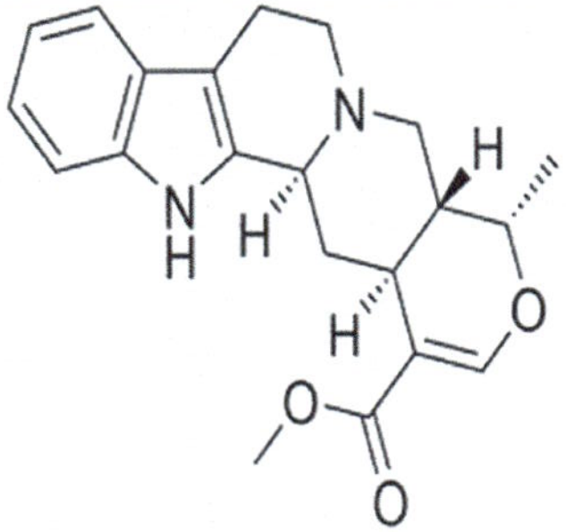	Maintenance of normal cerebral blood flow, Lowering of high blood pressure	Act as adrenergic receptor antagonist	Nightmares, drowsiness	Orally	Wink and Roberts (1998)

(continued)

Table 9.1 (continued)

S. No	Alkaloids	Chemical structures	Pharmaceutical/therapeutic properties	Mode of action	Adverse effect	Mode of delivery	References
8	Tryptamine	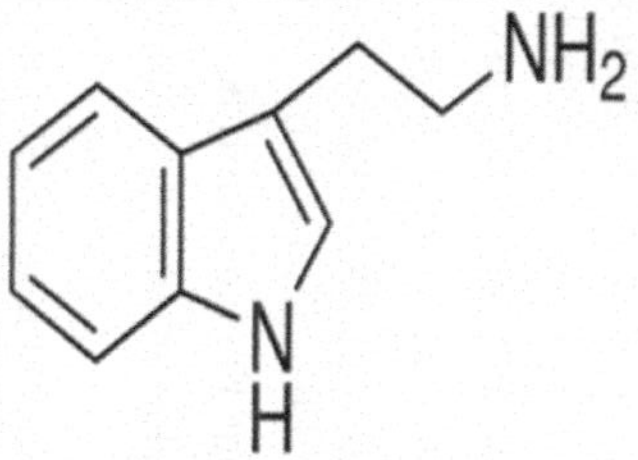	Used as psychotropic drug, Sleep, cognition and memory	Neurotransmitter, vasodilator	Anxiety nausea	Orally and intravenously	Khan and Waqas (2016)
9	Geraniol	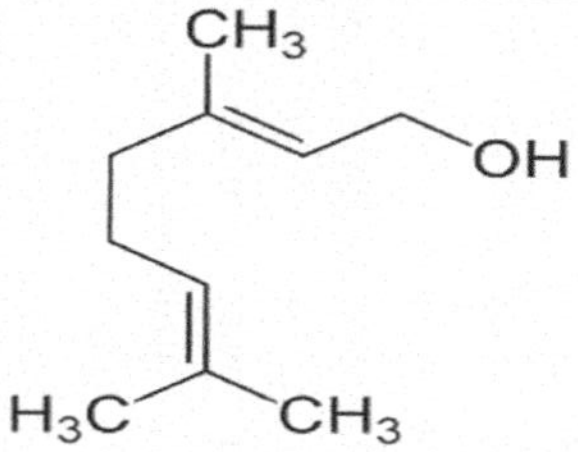	Treatment of various cancer, antibacterial, antiseptic and anti-inflammatory properties	Inhibition of cell division and growth	Depression, anxiety paranoia	Orally	Chen and Viljoen (2010)
10	Loganin	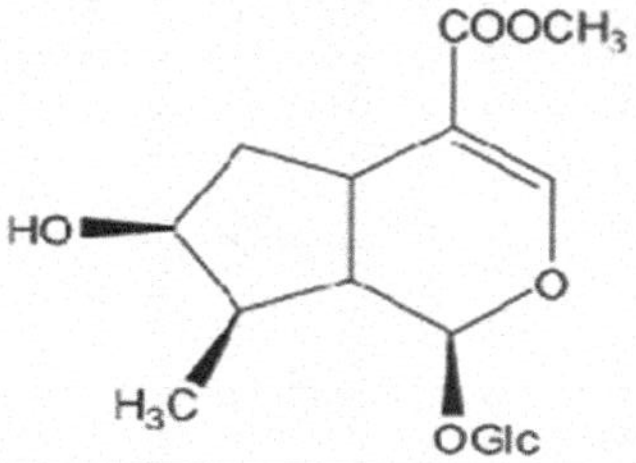	Treatment of insomnia and other neural disorders	Inhibit inflammation and kidney failure	Anxiety nausea	Orally	Kwon et al. (2009)
11	10-Hydroxygeraniol	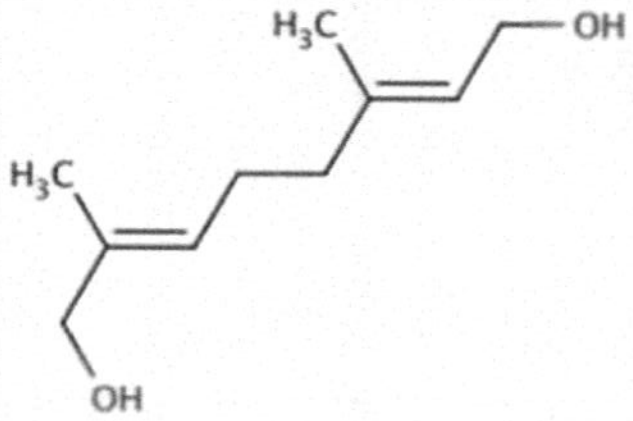	Treatment of various cancer, antibacterial, antiseptic and anti-inflammatory properties	Inhibition of cell division and growth	Depression, anxiety paranoia	Orally	Verma et al. (2020)

regulation of TIA biosynthesis and transport (Verma et al. 2012). TIAs in *C. roseus* are mainly synthesized from strictosidine derived by coupling tryptamine, indole which are generated via shikimate pathway and secologanin generated via nonmevalonic acid pathway (Saiman et al. 2014). Comprehensive research has indicated that secologanin is formed from isopentenyl diphosphate (IPP) through iridoid pathway or is regenerated itself through cytosolic mevalonic acid pathway or methyl-D-erythritol-4 phosphate (MEP) pathway in the plastid (Guirimand et al. 2020). Formerly, the IPP was thought to be originated from mevalonic acid pathway, but later the researchers confirm the involvement of MEP pathway in the biosynthesis of IPP to a greater extent (Zhang et al. 2018). Additionally, it has now been realized that mevalonic acid pathway is a minor pathway contributing to iridoid biosynthesis mainly through protein prenylation which regulates the expression of genes of MEP pathway (Commault et al. 2019).

The iridoid biosynthesis via MEP pathway (Fig. 9.1) is a multi-step process which involves seven enzymes to modulate the biosynthesis of intermediates within the pathway. Till date, the role of two enzymes, viz. 1-deoxy-D-xylulose-5-phosphate synthase (DXS) and 1-deoxy-D-xylulose-5-phosphate reductoisomerase (DXR), has been extensively studied and functionally characterized for their regulatory role in IPP biosynthesis (Fan et al. 2016). These two enzymes play significant role in the regulating the biosynthesis of TIAs in various plant species such as tomato, potato, Arabidopsis and also in *E. coli* (Henriquez et al. 2016). In *C. roseus,* enhanced activities/expression of DXS, DSR and 2-C-methyl-D-erythritol-2,4-cyclodiphosphate synthase (MECS) have been shown to stimulate TIA biosynthesis (Han et al. 2013). Nonetheless, enhanced expression of two enzymes, viz. anthranilate synthase (AS) and tryptophan decarboxylase (TDC), has been positively correlated with tryptamine biosynthesis; however, the mechanisms by which tryptamine is synthesized need further elucidation (Table 9.2).

In iridoid pathway, secologanin is produced from IPP in a multistep processes majority of which is unknown in *C. roseus*. In *C. roseus,* the enzyme geraniol-10-hydroxylase (G10H) which is a cytochrome P_{450} monooxygenase is involved in biosynthesis of TIAs (Sung et al. 2011; Pan et al. 2012). Cytochrome P_{450} reductase catalyzes the transfer of electron from NADPH which in turn hydroxylates geraniol at 10th position resulting in the formation of geranyl diphosphate (GPP) (Giddings et al. 2011). The GPP thus generated is further converted to geraniol-by-geraniol synthase (GES) through eight-step processes (Kumar et al. 2018). Scientists have also confirmed the regulatory role of G10H in seco-iridoid pathway in TIAs biosynthesis as they observed reversible inhibition of G10H by catharanthine. Furthermore, all enzymes involved in this pathway have been identified and characterized for their role in TIAs' biosynthesis. In another step, the central precursor strictosidine is hydrolyzed by strictosidine β-D-glucosidase (SGD) to produce strictosidine aglycoside which is then exploited by intermediates to synthesize various different types of TIAs.

In indole pathway, tryptophan derived from shikimate pathway initiates the biosynthesis of TIAs with two intermediates of carbohydrate metabolism pathway, i.e., phosphoenolpyruvate (PEP) and erythrose 4-phospahte (Idrees et al. 2013). The tryptophan derived is then converted by the combined action of tryptophan decarboxylase and anthranilate synthase enzymes to tryptamine (Peebles et al. 2011). This tryptamine then acts as the precursor for the synthesis of various TIAs in *C. roseus*. Overall, both the above pathways confirmed that the strictosidine and tryptamine are the central intermediates of biosynthesis pathway of TIAs in *C. roseus*. Furthermore, strictosidine β-D-glucosidase (SGD) also catalyze the formation of 4,2,1-dehydrogeissoschizine which in turn results in the formation of another class of TIAs like serpentine and tetrahydroalstonine while another branch leads to the synthesis of catharanthine and tabersonine (Carqueijeiro et al. 2020). Additionally, the tabersonine

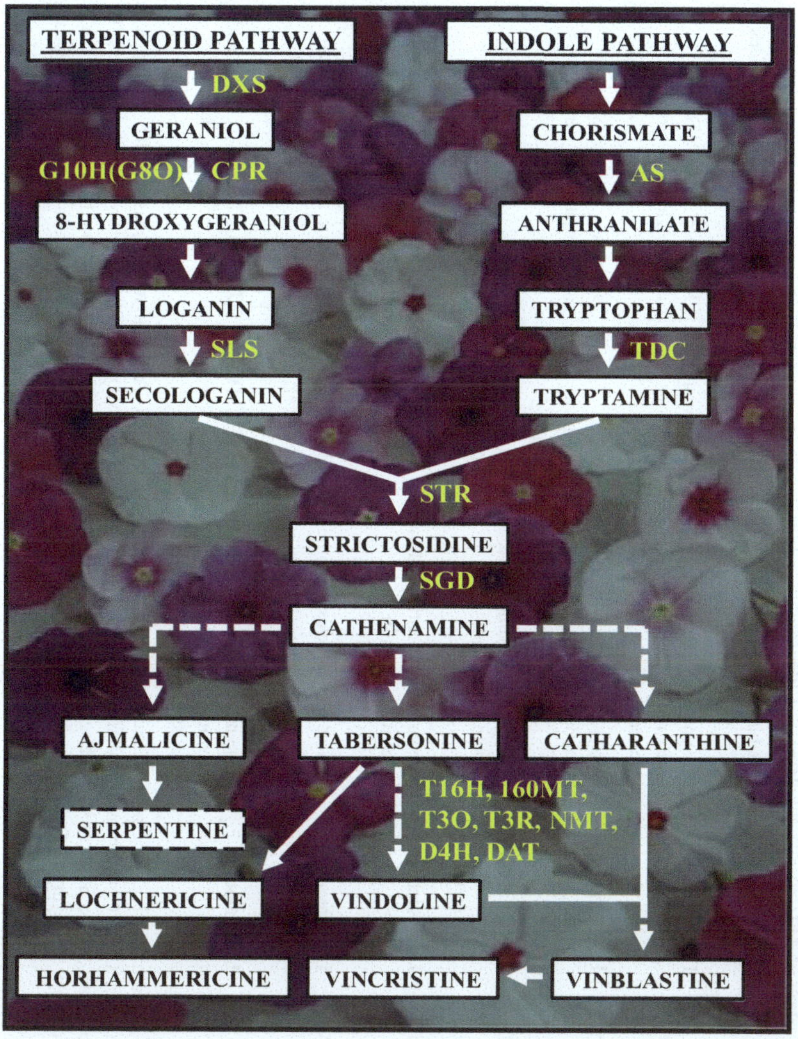

Fig. 9.1 Terpenoid and indole pathways involved in the biosynthesis of TIAs in *Catharanthus roseus.* AS, anthranilate synthase; TDC, tryptophan decarboxylase; DXS, 1-deoxy-D-xylulose 5-phosphate synthase; DXR, 1-deoxy-D-xylulose 5-phosphate reductoisomerase; MECS, 2C-methyl-D-erythritol-2,4,cyclodiphosphate synthase; G10H, geraniol 10-hydroxylase; CPR, cytochrome P450 reductase; LAMT, loganic acid methyltransferase; SCS, secologanin synthase; STR, strictosidine synthase; SGD, strictosidine β-D-glucosidase; T11H, tabersonine 11-hydroxylase; OMT, S-adenosyl-L-methionine:16 hydroxyl tabersonine O-methyl transferase; NMT, S-adenosyl-L-methionine: 2, 3-dihydro-3-hydroxytabersonine-N-methyltransferase; D17H, desacetoxy–vindoline-17-hydroxylase; DAT, acetyl CoA: 17-O-deacetylvindoline17-O-acetyltransferase; and MAT, acetyl CoA: minovincine-O-acetyltransferase

completes its biosynthetic pathway by generating vindoline, vincristine and vinblastine. Moreover, researchers have also corroborated above notion by analyzing expression of genes of above biosynthetic pathway where they reported the genes were downregulated by auxin and upregulated upon biotic, abiotic elicitors and phytohormone applications. List of potential TIAs with their functional properties has been mentioned in Table 9.1.

Table 9.2 List of transcription factors involved in the regulation of TIA biosynthetic genes in *Catharanthus roseus*

Transcription factors	Transcription factors family	Regulation of biosynthetic genes	Functional role	TIA biosynthesis	Pathway involved	Biotic/Abiotic elicitor	References
ORCA1, ORCA2 and ORCA3	AP2/ERF	*AS, TDC, DXS, CPR, G10H, SLS, STR, DGD and D4H*	Upregulation	Biosynthesis tabersonine hörhammericine, and lochnericine	Shikimate pathway	Jasmonic acid/fungal	Sharma et al. (2020)
CrBPF-1	P-binding factor	*STR, PRX1, T16H, DAT, MAT* and *NMT*	Upregulation	Biosynthesis tabersonine like compounds	MVA pathway	Jasmonic acid/fungal	Sharma et al. (2020)
CrMYC1 and CrMYC2	MYC/helix-loop-helix	*G10H, SLS, STR, DGD, GES*	Upregulation	Biosynthesis of ajmalicine, tryptamine, geraniol	Iridoid/MEP pathway	Jasmonic acid/fungal	Chen and Viljoen (2010)
ZCT1, ZCT2 and ZCT3	C_2H_2 zinc finger protein family	*STR, PRX1, T16H* and *SLS*	Upregulation	Biosynthesis of catharanthine, serpentine, loganin, 10-hydroxygeraniol	MVA pathway	Jasmonic acid/fungal	Verma et al. (2020)
GBF1 and GBF2	G box binding factors	*STR, PRX1, T16H, MVD*	Downregulation	Inhibit biosynthesis of some TIAs	Iridoid pathway	Jasmonic acid	Khan and Waqas (2016)
CrWRKY1 and CrWRKY2	WRKY	*AS, TDC, DXS, CPR, G10H, SGD, D4H*	Upregulation	Biosynthesis of catharanthine, serpentine, loganin, 10-hydroxygeraniol	MEP/MVA pathway	Jasmonic acid	Cooper and Deakin (2016)

ORCA, Octadecanoid-responsive Catharanthus AP2/ERF domain; ZCT, Zinc finger Catharanthus transcription factor; AS, Anthranilate synthase; TDC, Tryptophan decarboxylase; DXS, 1-deoxy-D-xylulose 5-phosphate synthase; CPR, Cytochrome P450 reductase; G10H, Geraniol-10-hydroxylase; SLS, Secologanin synthase; STR, Strictosidine synthase; D4H, Desacetoxyvindoline 4-hydroxylase; SGD, Strictosidine b-D-glucosidase; PRX1, Peroxidase 1; T16H, Tabersonine 16-hydroxylase; DAT, Deacetylvindoline 4-O-acetyltransferase; MAT, Minovincinine 19-hydroxy-O-acetyltransferase and NMT-N-methyltransferase; MVD, Mevalonate 5-diphosphate decarboxylase; GES, Geraniol-by-geraniol synthase

9.3 Regulatory Genes/Transcription Factors Involved in TIA Biosynthesis

A plethora of research has well documented about the involvement of transcription factor-mediated regulation of TIA biosynthetic genes in *C. roseus* (Gantet and Memelink 2002; Memelink and Gantet 2007). Researchers have also confirmed the putative role of various biotic (yeast, bacteria and fungi) and abiotic (phytohormone, stresses like drought, heat, etc.) elicitors in stimulating TIA biosynthesis in *C. roseus* (Sharma et al. 2019). The biotic and abiotic elicitor-mediated modulation of TIA biosynthesis is also under tight regulation of TFs/ genes such as TDC, G10H and STR (Fig. 9.2); however, the exact mechanism behind this regulation is complicated and poorly understood. Several lines of literature have confirmed that the TFs play significant role in stimulating production of secondary metabolites in plants. For example, researcher working on *Medicago truncatula* generated a transgenic line by overexpressing TSAR1/TSAR2 and basic helix-loop-helix genes noted significant upsurge in the triterpene biosynthetic genes and increased accumulation of triterpenes saponins. Furthermore, Apetala 2/Ethylene Responsive Factor (AP2/ERF) and

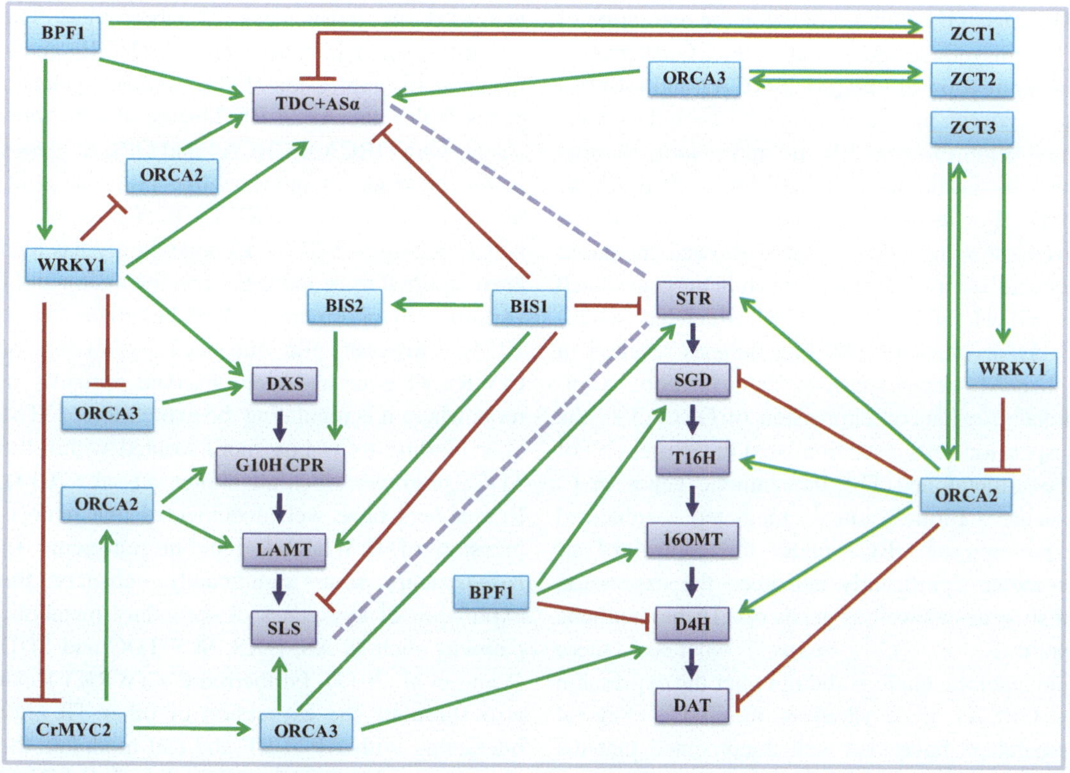

Fig. 9.2 Genes/transcription factors involved in the biosynthesis of TIAs in *Catharanthus roseus.* Transcription factor WRKYs control the expression of structural genes along with the expression of other TFs including MYC2, ORCAs, ZCTs, etc. TFs, Line marked in green represent positive regulation, whereas most TFs (marked in red) are negative regulator and line marked in purple represent TIA biosynthesis. Abbreviations: DXS, 1-deoxy-D-xylulose 5-phosphate synthase; G10H, geraniol 10-hydroxylase; CPR, cytochrome P450 reductase; LAMT, loganic acid methyltransferase; SLS, secologanin synthase; STR, strictosidine synthase; SGD, strictosidine β-D-glucosidase; CR, cathenamine reductases; T16H2, tabersonine 16-hydroxylase 2; 16OMT, 16-hydroxytabersonine-16-O-methyltransferase; NMT, N-methyltransferase; D4H, desacetoxyvindoline 4-hydroxylase; DAT, acetyl CoA: deacetylvindoline 4-O-acetyltransferase

MYB TFs have also been reported to stimulate synthesis of secondary metabolites in tobacco and *Papaver somniferum* in conjunction with biotic and abiotic elicitors (Table 9.2) (Srivastava and Kumar 2019). The MYB TFs have also been reported to induce the biosynthesis of both indolic and aliphatic glucosinolate (GSL) in cabbage (Chen et al. 2019). Transgenic *Artemisia annua* plants generated by overexpressing ERF1&2 TFs stimulated the accumulation of artemisinin and artemisinic acid up to twofold as compared to non-transformed plants (Shen et al. 2016).

In *C. roseus,* the TFs such as octadecanoid-responsive *Catharanthus* AP2/ERF (ORCAs) domain such as ORCA1, ORCA2 and ORCA3 are most recognized TFs which belong to AP2/ERF TF family involved in the regulation of TIA biosynthesis (Liu et al. 2011). Furthermore, the expression of ORCA1, ORCA2 and ORCA3 in *C. roseus* has been shown to be induced via application of several biotic and abiotic elicitors such as jasmonic acid and fungi (Paul et al. 2017). For instance, transgenic *C. roseus* plants overexpressing ORCA3 gene showed increased accumulation of distinct biosynthetic genes such as, G10H, SGD, DXS, TDC which are exclusively involved in TIA biosynthesis resulted in enhanced accumulation of TIAs in roots. Additionally, enhanced expression of ORCA3 in the suspension cells also stimulated the expression of above-mentioned TIA biosynthetic genes in *C. roseus.* Biotic/abiotic mediated enhanced expression of ORC2 under the control of str promoter significantly enhanced the expression of str genes as well as genes of TIA biosynthetic pathway in *C. roseus,* whereas these biotic/abiotic elicitors did not alter the expression of ORCA1 gene (Paul et al. 2017). Several researchers have also well documented that the overexpression of ORCA2 in the hairy roots of *C. roseus* induced compelling expression of TIA biosynthetic genes such as AS, G10H, SGD, DXS, TDC (Rizvi et al. 2016).

Another TFs which are extensively involved in the modulation of TIA biosynthesis are *C. roseus* G box binding factor (CrGBF1). The CrGBF1 TFs binds to str promoter but in reverse orientation as compared to ORCA genes (Rizvi et al. 2016) and regulate the expression of several TIA biosynthetic genes. Furthermore, MYC TFs such as MYC1 and MYC2 are the family of basic loop helix TFs which are also expressed under various biotic and abiotic stresses can stimulate the expression of str genes which ultimately activates the transcription of TIAs biosynthetic genes in *C. roseus.* Likewise, two AP2/ERF proteins were cloned, induced and were allowed to express in *C. roseus* hairy roots (Pan et al. 2019). The transformed hairy roots showed enhanced expression of TIA biosynthetic genes of both tryptophan and seco-iridoid pathways as well as increased the biosynthesis of tabersonine. The possible explanation by which AP2/ERF genes stimulate the expression of these TIA biosynthetic genes is that the AP2/ERF TFs are able to form cluster with ORCA3, ORCA4 and ORCA5 genes in the presence of inducers like jasmonic acid. Similarly, *C. roseus* WRKY1 (CrWRKY1) TF which belong to WRKY III superfamily has also been reported to be induced upon the treatment of various phytohormones (Schluttenhofer et al. 2014). However, the enhanced expression of CrWRKY1 gene has been observed primarily in roots where it is regulating the expression of TDC after binding to W-box motif located within the TDC promoter (Schluttenhofer et al. 2014). Researchers have well documented that overexpression of CrWRKY1 gene in transgenic *C. roseus* hairy roots significantly enhanced the expression of key genes of secondary metabolic pathway such as AS, DXS, SLS TDC and SGD (Yang et al. 2013). Furthermore, CrWRKY1 can also stimulate the expression of other TF after interacting with AP2/ERF and can modulate the expression of ORCA2, ORCA3 and CrMYC2 thereby increasing the accumulation of serpentine and catharanthine.

Another TF, viz. CrBPF1 which belongs to MYB TF superfamily, can also stimulate the expression of TIA biosynthetic genes after binding to BA motif of STR promoter (Colinas et al. 2020). Various studies have reported that overexpression of CrBPF1 in the hairy roots of *C. roseus* resulted in the up-accumulation of several other genes such as ORCA3, BIS1, CrMYC1 GBF2 and ZCTs and CrMYC2, whereas it downregulated the expressions of ORCA2 and CrWRKY1 (Sazegari et al. 2018). Furthermore, researchers have also used yeast two-hybrid system to decipher functional mechanisms behind the TFs binding activity of CrGBF1 gene by using G-box element as bait in the STR promoter of *C. roseus* (Sazegari et al. 2018). They observed that CrGBF1 acts as transcriptional repressor of NR element that specifically to STR promoter and activates the expression of TIA biosynthetic genes thereby suggesting that GBF genes are extensively involved in TIA accumulation in *C. roseus*.

Zinc finger transcription factor proteins, viz. Cys2/His2, have also been isolated and cloned in *C. roseus* reported to alter the expression of STR and TDC promoters as revealed by yeast one hybrid system using DB element as a bait (Pandey et al. 2016). Another group of researchers also confirm the repressible effect of ZCT1, ZCT2 and ZCT3 on STR and TDC promoter via trans-activation assays (Pauw et al. 2004). The possible reason behind this repressible effect of zinc finger transcription factor could be due LxLxL motif at C-terminal domain which may have inhibitory effect on such genes. Furthermore, studies have also well documented the inhibitory role of ZCT on ORCAs genes; however, the ZCT was able to upregulate the expression of ORCAs genes in the presence of methyl jasmonate (Patra et al. 2018). Recently, researchers have also created a knock-out of CrZCT genes with the notion that silencing of ZCT may enhance the expression of TIA biosynthetic genes. However, the researchers observed marginal increased in the TIAs accumulation in transgenic *C. roseus* thus indicating that ZCT genes play distinct role in modulating the expression of these metabolic pathway genes.

9.4 Functional Genes/TFs Involved in Transport of TIA in *C. roseus*

In *C. roseus*, the TIA transport mainly occurs through epidermal cells, laticifers, idioblastic cells and internal phloem-associated parenchymatic cells (IPAP, located on the periphery of the stem pith). In leaves, the IPAP is located in the chloroplast and near the plastid containing the enzymes of MEP pathway involved in strengthening the structural and functional properties of chloroplasts in green tissues (Fig. 9.3). Initially, G10H enzyme of seco-iridoid pathway catalyzes the conversion of geraniol into 10-oxogeranial in IPAP cell leaves and stem. Moreover, the enzymes such as 7-deoxyloganic acid hydroxylase, 7-deoxyganetic acid glucosyltransferase and iridoid oxidase have also been shown to exist in IPAP cells which produce loganic acid as an intermediate (Larsen et al. 2017). Further, the loganic acid is transported from IPAP cells to epidermis where it is converted into secologanin by loganic acid methyltransferase and secologanin synthase enzyme (Larsen et al. 2017). Meanwhile, enzymes STR and SGD catalyze the condensation of tryptamine and secologanin to yield strictosidine in epidermal cell of root, leaves and stems which act as precursor of some TIAs like ajmaline and catharanthine (Larsen et al. 2017). Several studies have well-documented roots as the primary site of ajmaline biosynthesis, whereas catharanthine is primarily synthesized on leaves in the presence of ATP-binding cassette (ABC) transporter which transports catharanthine from the epidermis to the surface of the leaves. Furthermore, desacetoxyvindoline, the precursor of vindoline, is also generated in epidermis, where tabersonine 16-hydroxylase 2, 16-hydroxytaberosine-16-O-methyltransferase and N-methyltransferase catalyze the biosynthesis of vindoline in epidermis (Murata and Luca 2005). Additionally, the enzymes, viz. deacetoxyvindoline-4-hydroxylase and acetyl-CoA-4-O-deacetylvindoline-4-O-acetyltransferase, are also involved in the biosynthesis of vindoline in a two-step process

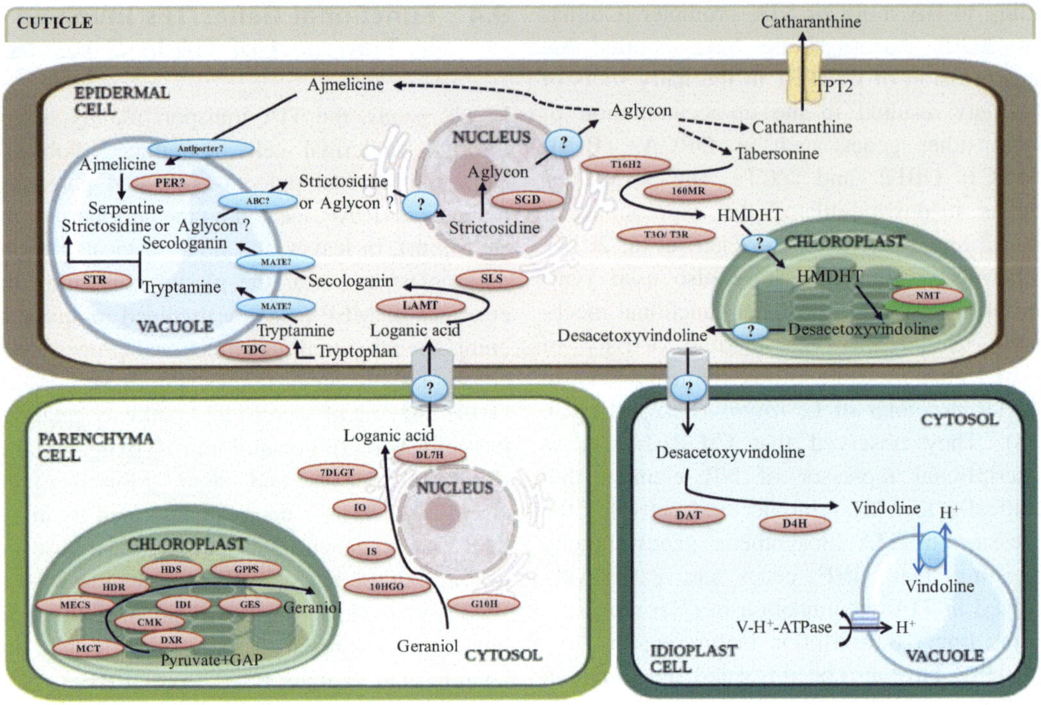

Fig. 9.3 Biosynthesis and transport of TIAs across different organelles in *Catharanthus roseus*. Abbreviations are: SGD, strictosidine β-D-glucosidase; STR, strictosidine synthase; TDC, tryptophan decarboxylase; DXS, 1-deoxy-D-xylulose 5-phosphate synthase; DXR, 1-deoxy-D-xylulose 5-phosphate reductoisomerase; MCT, MEP cytidylyltransferase; CMK, 4-(cytidine 5′-diphospho)-2-C-methyl-D-erythritol kinase; MECS, 2-C-methylerythritol 2,4-cyclodiphosphate synthase; HDS, hydroxymethylbutenyl 4-diphosphate synthase; HDR, hydroxymethylbutenyl 4-diphosphate reductase; IDI, IPP isomerase; GPPS, geranyl diphosphate synthase; GES, geraniol synthase; G10H, geraniol 10-hydroxylase; 10HGO, 10-hydroxygeraniol dehydrogenase; IDS, iridoid synthase; 7DLS, 7-deoxyloganetic acid synthase; DLGT, 7-deoxyloganetic acid glucosyltransferase; DL7H, 7-deoxyloganic acid 7-hydroxylase; LAMT, loganic acid methyltransferase; SLS, secologanin synthase; T16H2, tabersonine 16-hydroxylase 2; 16OMT, 16-hydroxytabersonine-16-O-methyltransferase; T3O, tabersonine 3-oxygenase; T3R, tabersonine 3-reductase; NMT, N-methyltransferase; D4H, desacetoxyvindoline 4-hydroxylase; DAT, acetyl CoA: deacetylvindoline 4-O-acetyltransferase

preferentially located in the idioblasts and laticiferic cells of leaves, stem and roots (Levac et al. 2008).

The MEP pathway also serves as one the potential route for the biosynthesis of various isoprenoids which occur in the stroma of plastids or IPAP cells. The enzyme isopentenyl diphosphate isomerase is involved in the synthesis of isopentenyl diphosphate and dimethyl diphosphate which act as the precursors for the biosynthesis of various TIAs which are then targeted to distinct organs such as chloroplast, mitochondria and peroxisomes for their biological functions (Guirimand et al. 2012). Geraniol which acts as the substrate for the synthesis 10-hydroxygeraniol is also synthesized in stroma of plastids which is then transported to cytosol of IPAP cells through some known metabolic flux (Simkin et al. 2013). The 10-hydroxygeraniol is then converted into loganic acid by 7-deoxyloganic acid hydroxylase enzyme within the cytosol of IPAP cells. Furthermore, other TIAs like tryptamine and secologanin are also transported from cytosol into the vacuole where they react with each other to yield strictosidine which is then transported form vacuole to cytosol for its actual biological function (Courdavault et al. 2014). Strictosidine-β-D-glucosidase forms a supramolecular complex in nucleus playing exceptional role in the biosynthesis of various other TIAs like catharanthine and tabersonine with in the cytosol. The TIAs like tabersonine and vindoline to other TIAs take place in various cellular compartments of leaf, stems and roots. The vindoline, catharanthine and other TIA biosynthetic enzymes are located within the cytosol and nucleus of laticifers and idioblasts cells which are then transported to vacuole by a proton-specific transporter in conjugation with V-H^+ ATPase (Mahroug et al. 2007). Interestingly, the catharanthine and vindoline are not synthesized in the same vacuole; however, they are transported and coupled by an external stimulus to form bis-indole alkaloids vinblastine and vincristine which are synthesized and accumulated with in the vacuoles for their characteristic functions.

9.5 Mining of Putative Genes/TFs Involved in TIA Biosynthesis

The alkaloid biosynthesis in *C. roseus* involves systemic production of various TIAs in different cellular compartments. For example, tabersonine is converted into horhammericine and lochnericine in root tissue, while in leaf and stem tissues, it is converted to vindoline (Mekky et al. 2018). The enzymes present in roots oxidize tabersonine into epoxide via a NADPH-dependent microsomal cytochrome P450 monooxygenase, whereas 19-O-acetyltransferase (MAT) gene located in the root tip catalyzed the formation of 6,7-dehyroecitovenine which acts as precursor for the biosynthesis of other TIA. Similarly, another root-specific TIA biosynthetic gene, viz. cytochrome P450 monooxygenase, has been cloned and characterized as CYP71BJ1 in *C. roseus* which causes hydroxylation of tabersonine and lochnericine. Likewise, the biosynthesis of vindoline is a five-step process which is catalyzed by tabersonine-16-hydroxyalse. Several researchers have reported that this gene is strongly induced by light causing hydroxylation of tabersonine (Besseau et al. 2013). Another gene, 16-hydroxytabersonine-16-O-methyltransferase, has been cloned and characterized which is involved in the synthesis of another TIA intermediate, viz. 16-methoxytabersonine. N-methyltransferase has also been cloned and functionally annotated to cause the conversion of 16-hydroxytabersonine to desacetoxyvindoline which is further hydroxylated by oxoglutarate-dependent dioxygenase gene to yield vindoline.

Researchers have constructed cDNA libraries using next-generation sequencing approach to get further insight in the TIA biosynthetic pathways and to identify putative genes of secondary metabolite pathway. For instance, Roepke et al. (2010) generated a loganic acid methyltransferase (LAMT) homology-based cloning that has been exploited using comparative bioinformatic studies to decipher transcriptome of *C. roseus* for unraveling novel genes/protein-involved TIA biosynthetic pathway. Additionally, expressed sequence tags (ESTs) library generated from leaf transcriptome of *C. roseus* resulted in the identification of novel putative gene that showed enhanced expression under methyl jasmonate treatment; however, its functional annotation still needs in-depth understanding before linking it with TIA biosynthetic pathway. Furthermore, this leaf transcriptome could also act as axillary resource for the identification of genes involved in catharanthine pathway which still remain to be elucidated. In the recent year, several researchers have generated large-scale transcriptomic data associated with TIA biosynthetic pathway and thus unraveling the function of uncharacterized genes (Shukla et al. 2006). Additionally, a researcher has also attempted to combine transcriptomic data with metabolomic data using cDNA/AFLP method to identify any links between biosynthetic gene and the product they produce as metabolites. Overall, the above studies successfully link various genes with their corresponding metabolites which are important intermediates of TIA biosynthetic pathway. Therefore, to delve further and deeper, several large-scale transcriptomes have been created and maintained by the Medicinal Plant Genomics Consortium which can further be exploited in conjunction with next-generation sequencing technology for identification of novel genes of TIA pathway in *C. roseus*.

Various researchers have also used RNA interference (RNAi) as a potential tool to induce or silence the expression of genes within the cell for their structural and functional characterization (Papon et al. 2004). For instance, Zárate and Verpoorte (2007) have used RNAi technology in hairy roots of *C. roseus* to alter the expression of target gene. The researchers observed that the suppression of target gene resulted in the enhance accumulation of TIA (Zhou et al. 2010). However, RNAi technique is very labor-intensive and time-consuming technique and required prior understanding of the target gene failure of which can result in the silencing of nearby genes as it has rate of off-target formation (Zhou et al. 2010). Another very effective tool which is being currently exploited by various researchers is virus-induced gene silencing (VIGS) for structural and functional annotation of uncharacterized genes. Several virus-mediated gene delivery systems have been developed and tested successfully in various plants (Van Moerkercke et al. 2015), for example, in *Nicotiana benthamiana* binary tobacco rattle virus (pTRV) to silence the gene of secondary metabolite pathway. VIGS has various advantages over RNAi as it is extremely versatile, easy and less time-consuming technique (Kumar et al. 2015). Additionally, VIGS does not require complete sequence of gene and could be achievable in those plant species in which transformation technique has not yet been well established (Liscombe and O'Connor 2011). In *Hyoscyamus niger*, VIGS was used to suppress the expression of CYP80F1 gene which is involved in the conversion of littorine to hyoscyamine. Additionally, VIGS has also been exploited in *C. roseus* to control the production of vindoline by silencing key gene involved in vindoline biosynthesis (Cacace et al. 2003). Similarly, N-methyltransferase gene has also been silenced in *C. roseus* using VIGS technique which resulted in enhanced accumulation of 16-methoxy-2,3-duhydro-3-hydroxytabersonine and decreased synthesis of vindoline (Liscombe and O'Connor 2011) using pTRV vector.

Various researchers are using VIGS technology for the deciphering structural and functional properties of genes and their corresponding proteins. In opium poppy, VIGS resulted in the suppression of gene responsible for the O-demethylation of thebaine to codeine and finally to morphine of benzylisoquinoline alkaloid pathway (Hagel and Facchini 2010). Further, the researchers also observed that silencing of demethylase resulted in the enhanced biosynthesis of benzylisoquinoline alkaloid pathway which was further confirmed by thin-layer chromatography technique. Overall, the VIGS is an effective and sophisticated reverse genetic tool for characterizing the function of uncharacterized genes of various metabolic pathways. However, their applicability is limited to certain plant species; still it can be used for the identification of candidate genes after combining/modifying VIGS technique using novel omics technologies.

9.6 Functional Genomic Tools for Engineering TIA Biosynthetic Pathway in *C. roseus*

Extensive research is being directed toward metabolic engineering of medicinally important plants, and in last one or two decades, considerable progress has been made to unravel genes/proteins which are responsible for the synthesis of important therapeutic metabolites (Zhao and Verpoorte 2007). With the advent of next-generation sequencing technologies, it has become easier to rewire the metabolic pathway in plants to increase the biosynthesis of valuable pharmaceutical metabolites (Zhou et al. 2009). Significant achievement has also been made toward the metabolic engineering of TIA metabolic pathway in *C. roseus* which has shed the light on the existence of several enzymes which are involved in the biosynthetic pathway (Zhao et al. 2013). Several studies have well documented that both tryptophan and tryptamine are the central regulatory step in the tryptophan biosynthetic pathway which are exclusively involved in the production of indole alkaloids (Zhao et al. 2013). In a study, researchers took *Arabidopsis* glucocorticoid-inducible promoter and overexpressed it in *C. roseus* hairy roots (Hong et al. 2006). They demonstrated that the α subunit of glucocorticoid-inducible promoter was compatible with the native β subunit which enhanced the production of both tryptophan and tryptamine up to a greater extent. They further recorded that while the level of tryptophan and tryptamine was increased, it did not significantly enhance the production of other TIAs except for lochnericine which showed twofold accumulation (Zhao et al. 2013).

Similarly, in another study, the researchers performed co-expression of TDC and Arabidopsis glucocorticoid-inducible promoter in *C. roseus* and observed compelling upsurge of TDC activity and serpentine accumulation with marginal or no increase in tryptamine levels (Hughes et al. 2004). Both the above findings clearly revealed that the levels of tryptamine can be modulated/regulated in *C. roseus* hairy roots using above-mentioned promoter; however, the complete engineering of the metabolic pathway requires extensive studies of other unknown genes (Suttipanta et al. 2011). Efforts were also diverted toward the engineering of TIA metabolic pathway using *Arabidopsis* As*β* subunit which was allowed to co-express along with the glucocorticoid-inducible promoter (Sun et al. 2016). The transgenic C. *roseus* plants showed enhanced expression of various TIA biosynthesis genes as well as greater accumulation of tryptophan and tryptamine. Furthermore, efforts have also been diverted toward engineering of iridoid pathway by overexpressing *Arabidopsis* DXS gene under the control of glucocorticoid-inducible promoter in *C. roseus* hairy roots (Peebles et al. 2011). The enhanced expression of DXS resulted in the increased accumulation of loganin, serpentine and ajmalicine levels while there was no change in the levels of tabersonine, lochnericine and horhammericine. Additionally, to increase the biosynthesis of TIAs, researchers also co-expressed G10H with Arabidopsis glucocorticoid-inducible promoter and recorded steep increase in ajmalicine, serpentine and lochnericine when G10H was expressed alone and also observed significant increase in tabersonine, ajmalicine upon co-expression (Han et al. 2013).

In *C. roseus*, the pathway for tabersonine and vindoline biosynthesis is less characterized. Nonetheless, various researchers have diverted their attention to device a method for engineering vindoline biosynthesis pathway (Sun et al. 2016). To achieve this goal, researchers tried to overexpress tabersonine 16-hydroxylase gene in the presence of cauliflower mosaic virus (CaMV35S) 35S promoter in *C. roseus* hairy roots and observed increase accumulation of lochnericine, albeit the overexpression decreases the accumulation of vindoline, strictosidine and catharanthine (Wang et al. 2010). Furthermore, the researchers have also exploited TFs such as ORCA2, ORCA3, CrMYC1 and CrMYC2for their overexpression analysis (Van Moerkercke et al. 2015). The transgenic *C. roseus* plants showed enhanced expression of TIA biosynthetic genes and accumulation of important metabolites of TIA metabolic pathway (Pan et al. 2012). Moreover, overexpression of CrWRKY genes along with Arabidopsis glucocorticoid-inducible promoter induced significant up-accumulation of serpentine, catharanthine and tabersonine (Schluttenhofer et al. 2014).

In the recent years, several researchers have exploited advanced tool and techniques to identify and quantify metabolites of secondary metabolite pathway in economically important medicinal plants. In this continuation, high-performance liquid chromatography (HPLC) linked with different detectors, for example, fluorescence detector, photo diode array and electron spray ionization, nuclear magnetic resonance and circular dichroism are being currently used to decipher secondary metabolite pathways and to provide in-depth understanding of mechanisms involved in the TIA biosynthesis

(Schluttenhofer et al. 2014). Owing to the presence of aromatic rings, the TIA metabolites and their corresponding intermediate can also be easily detected by UV absorbance. Further, the alkaloids are having pie electrons in their aromatic rings easily ionized and detected by electron spray ionization-mass spectrometry. The techniques like circular dichroism and nuclear magnetic resonance have become an indispensable tool for precise identification and quantification of TIA metabolites in *C. roseus*.

9.7 Conclusion and Future Outlook

In-depth understanding and elucidation of mechanisms behind the regulation of TIA biosynthesis pathway are of utmost importance in order to enhance the production of important metabolites in *C. roseus* as well as in other economically important medicinal plants. However, extensive research has been conducted for deciphering TIA metabolic pathways that have led to the identification of various genes, proteins and transcription factors which regulate the secondary metabolite production. In *C. roseus,* the genes involved in vindoline biosynthesis pathway have been well characterized in detail which have provided significant insight into its structural and functional role in stimulating the biosynthesis of other TIAs. However, till date very little has been known regarding various other genes which are involve in the biosynthetic process due to unavailability of complete genome sequence and presence of any gene library which make it quite difficult to decipher the metabolic pathway in *C. roseus.* The metabolic engineering has still not been exploited up to its full potential owing to the limitation that hampers its applicability to various systems which need a proper attention in the coming future. The major limitation behind the successful utilization of various techniques is the identification of suitable analyte due to lack of structural and biochemical information. Nonetheless, the techniques like NMR and HPLC and circular dichroism are being extensively used to identify and quantify important metabolites in various plants. In conclusion, mass spectrophotometry-based tools and techniques can be successfully utilized to develop in depth metabolic map which can provide comprehensive information for studying metabolic pathways in another complex system such as *C. roseus.*

References

Besseau S, Kellner F, Lanoue A, Thamm AM, Salim V, Schneider B, Geu-Flores F, Höfer R, Guirimand G, Guihur A, Oudin A (2013) A pair of tabersonine 16-hydroxylases initiates the synthesis of vindoline in an organ-dependent manner in *Catharanthus roseus*. Plant Physiol 163(4):1792–1803

Cacace S, Schröder G, Wehinger E, Strack D, Schmidt J, Schröder J (2003) A flavonol O-methyltransferase from *Catharanthus roseus* performing two sequential methylations. Phytochemistry 62(2):127–137

Carqueijeiro I, Koudounas K, de Bernonville TD, Sepúlveda LJ, Mosquera A, Bomzan DP, Oudin A, Lanoue A, Besseau S, Cruz PL, Kulagina N (2020) Alternative splicing creates a pseudo-strictosidine β-D-glucosidase modulating alkaloid synthesis in *Catharanthus roseus*. Plant Physiol 185(3):836–856

Chen C, Zhang K, Khurshid M, Li J, He M, Georgiev MI, Zhang X, Zhou M (2019) MYB transcription repressors regulate plant secondary metabolism. Crit Rev Plant Sci 38(3):159–170

Chen W, Viljoen AM (2010) Geraniol—a review of a commercially important fragrance material. S Afr J Bot 76(4):643–651

Colinas M, Pollier J, Vaneechoutte D, Malat DG, Schweizer F, De Clercq R, Guedes JG, Martinez-Cortes T, Hidalgo FJM, Sottomayor M, Vandepoele K (2020) A modular system regulates specialized metabolite pathway branch choice in the medicinal plant *Catharanthus roseus*. bioRxiv

Commault AS, Fabris M, Kuzhiumparambil U, Adriaans J, Pernice M, Ralph PJ (2019) Methyl jasmonate treatment affects the regulation of the 2-C-methyl-D-erythritol 4-phosphate pathway and early steps of the triterpenoid biosynthesis in *Chlamydomonas reinhardtii*. Algal Res 39:101462

Cooper R, Deakin JJ (2016) Africa's gift to the world. In: Cooper R, Deakin JF (eds) Botanical miracles: chemistry of plants that changed the world. CRC Press, Boca Raton, FL, pp 46–51

Courdavault V, Papon N, Clastre M, Giglioli-Guivarc'h N, St-Pierre B, Burlat V (2014) A look inside an alkaloid multisite plant: the *Catharanthus logistics*. Curr Opin Plant Biol 19:43–50

Das A, Sarkar S, Bhattacharyya S, Gantait S (2020) Biotechnological advancements in *Catharanthus roseus* (L.) G. Don. Appl Microbiol Biotechnol 104 (11):4811–4835

Fan H, Wu Q, Wang X, Wu L, Cai Y, Lin Y (2016) Molecular cloning and expression of 1-deoxy-d-xylulose-5-phosphate synthase and 1-deoxy-d-xylulose-5-phosphate reductoisomerase in *Dendrobium officinale*. Plant Cell Tissue Org Cult 125 (2):381–385

Gantet P, Memelink J (2002) Transcription factors: tools to engineer the production of pharmacologically active plant metabolites. Trends Pharmacol Sci 23(12):563–569

Giddings LA, Liscombe DK, Hamilton JP, Childs KL, DellaPenna D, Buell CR, O'Connor SE (2011) A stereoselective hydroxylation step of alkaloid biosynthesis by a unique cytochrome P450 in *Catharanthus roseus*. J Biol Chem 286(19):16751–16757

Goldhaber-Pasillas GD, Mustafa NR, Verpoorte R (2014) Jasmonic acid effect on the fatty acid and terpenoid indole alkaloid accumulation in cell suspension cultures of *Catharanthus roseus*. Molecules 19 (7):10242–10260

Goswami S (2013) Preliminary phytochemical screening and standardisation of leaves of *Catharanthus roseus* (L.) G. Don. Indian J Res Pharm Biotechnol 1(1):21

Guirimand G, Guihur A, Perello C, Phillips M, Mahroug S, Oudin A, Dugé de Bernonville T, Besseau S, Lanoue A, Giglioli-Guivarc'h N, Papon N (2020) Cellular and subcellular compartmentation of the 2C-methyl-D-erythritol 4-phosphate pathway in the Madagascar periwinkle. Plants 9 (4):462

Guirimand G, Guihur A, Phillips MA, Oudin A, Glévarec G, Melin C, Papon N, Clastre M, St-Pierre B, Rodríguez-Concepción M, Burlat V (2012) A single gene encodes isopentenyl diphosphate isomerase isoforms targeted to plastids, mitochondria and peroxisomes in *Catharanthus roseus*. Plant Mol Biol 79(4–5):443–459

Hagel J, Facchini P (2010) Biochemistry and occurrence of O-demethylation in plant metabolism. Front Physiol 1:14

Han M, Heppel SC, Su T, Bogs J, Zu Y, An Z, Rausch T (2013) Enzyme inhibitor studies reveal complex control of methyl-D-erythritol 4-phosphate (MEP) pathway enzyme expression in *Catharanthus roseus*. PLoS One 8(5):e62467

Henriquez MA, Soliman A, Li G, Hannoufa A, Ayele BT, Daayf F (2016) Molecular cloning, functional characterization and expression of potato (*Solanum tuberosum*) 1-deoxy-d-xylulose 5-phosphate synthase 1 (StDXS1) in response to *Phytophthora infestans*. Plant Sci 243:71–83

Hong SB, Peebles CA, Shanks JV, San KY, Gibson SI (2006) Expression of the Arabidopsis feedback-insensitive anthranilate synthase holoenzyme and tryptophan decarboxylase genes in *Catharanthus roseus* hairy roots. J Biotechnol 122(1):28–38

Hughes EH, Hong SB, Gibson SI, Shanks JV, San KY (2004) Metabolic engineering of the indole pathway in *Catharanthus roseus* hairy roots and increased accumulation of tryptamine and serpentine. Metabol Eng 6(4):268–276

Idrees M, Naeem M, Aftab T, Khan MMA (2013) Salicylic acid restrains nickel toxicity, improves antioxidant defence system and enhances the production of anticancer alkaloids in *Catharanthus roseus* (L.). J Hazard Mater 252:367–374

Khan MZ, Waqas N (2016) The emerging roles of human trace amines and human trace amine-associated receptors (hTAARs) in central nervous system. Biomed Pharmacother 83:439–449

Kotakadi VS, Rao YS, Gaddam SA, Prasad TNVKV, Reddy AV, Gopal DS (2013) Simple and rapid biosynthesis of stable silver nanoparticles using dried leaves of *Catharanthus roseus*. Linn. G. Donn and its anti-microbial activity. Colloids Surf B Biointerf 105:194–198

Kumar K, Kumar SR, Dwivedi V, Rai A, Shukla AK, Shanker K, Nagegowda DA (2015) Precursor feeding studies and molecular characterization of geraniol synthase establish the limiting role of geraniol in monoterpene indole alkaloid biosynthesis in *Catharanthus roseus leaves*. Plant Sci 239:56–66

Kumar SR, Shilpashree HB, Nagegowda DA (2018) Terpene moiety enhancement by overexpression of geranyl (geranyl) diphosphate synthase and geraniol synthase elevates monomeric and dimeric monoterpene indole alkaloids in transgenic *Catharanthus roseus*. Front Plant Sci 9:942

Kwon SH, Kim HC, Lee SY, Jang CG (2009) Loganin improves learning and memory impairments induced by scopolamine in mice. Eur J Pharmacol 619(1–3):44–49

Larsen B, Fuller VL, Pollier J, Van Moerkercke A, Schweizer F, Payne R, Colinas M, O'Connor SE, Goossens A, Halkier BA (2017) Identification of iridoid glucoside transporters in *Catharanthus roseus*. Plant Cell Physiol 58(9):1507–1518

Levac D, Murata J, Kim WS, De Luca V (2008) Application of carborundum abrasion for investigating the leaf epidermis: molecular cloning of *Catharanthus roseus* 16-hydroxytabersonine-16-O-methyltransferase. Plant J 53(2):225–236

Liscombe DK, O'Connor SE (2011) A virus-induced gene silencing approach to understanding alkaloid metabolism in *Catharanthus roseus*. Phytochemistry 72(16):1969–1977

Liu DH, Ren WW, Cui LJ, Zhang LD, Sun XF, Tang KX (2011) Enhanced accumulation of catharanthine and vindoline in *Catharanthus roseus* hairy roots by overexpression of transcriptional factor ORCA2. Afr J Biotechnol 10(17):3260–3268

Liu J, Cai J, Wang R, Yang S (2017) Transcriptional regulation and transport of terpenoid indole alkaloid in *Catharanthus roseus*: exploration of new research directions. Intl J Mol Sci 18(1):53

Mahroug S, Burlat V, St-Pierre B (2007) Cellular and sub-cellular organisation of the monoterpenoid indole alkaloid pathway in *Catharanthus roseus*. Phytochem Rev 6(2–3):363–381

Mangaiyarkarasi R, Girija M, Gnanamurthy S, Don CLG (2014) Mutagenic effectiveness and efficiency of gamma rays and ethyl methane sulphonate in *Catharanthus roseus*. Int J Curr Microbiol Appl Sci 3:881–889

Mekky H, Al-Sabahi J, Abdel-Kreem MFM (2018) Potentiating biosynthesis of the anticancer alkaloids vincristine and vinblastine in callus cultures of *Catharanthus roseus*. S Afr J Bot 114:29–31

Memelink J, Gantet P (2007) Transcription factors involved in terpenoid indole alkaloid biosynthesis in *Catharanthus roseus*. Phytochem Rev 6(2):353–362

Monforte-González M, Ayora-Talavera T, Maldonado-Mendoza IE, Loyola-Vargas VM (1992) Quantitative analysis of serpentine and ajmalicine in plant tissues of *Catharanthus roseus* and hyoscyamine and scopolamine in root tissues of *Datura stramonium* by thin layer chromatography-densitometry. Phytochem Anal 3(3):117

Murata J, Luca VD (2005) Localization of tabersonine 16-hydroxylase and 16-OH tabersonine-16-O-methyltransferase to leaf epidermal cells defines them as a major site of precursor biosynthesis in the vindoline pathway in *Catharanthus roseus*. Plant J 44 (4):581–594

Pan Q, Wang C, Xiong Z, Wang H, Fu X, Shen Q, Peng B, Ma Y, Sun X, Tang K (2019) CrERF5, an AP2/ERF transcription factor, positively regulates the biosynthesis of bisindole alkaloids and their precursors in *Catharanthus roseus*. Front Plant Sci 10:931

Pan Q, Wang Q, Yuan F, Xing S, Zhao J, Choi YH, Verpoorte R, Tian Y, Wang G, Tang K (2012) Overexpression of ORCA3 and G10H in *Catharanthus roseus* plants regulated alkaloid biosynthesis and metabolism revealed by NMR-metabolomics. PLoS One 7(8):e43038

Pan YJ, Liu J, Guo XR, Zu YG, Tang ZH (2015) Gene transcript profiles of the TIA biosynthetic pathway in response to ethylene and copper reveal their interactive role in modulating TIA biosynthesis in *Catharanthus roseus*. Protoplasma 252(3):813–824

Pandey SS, Singh S, Babu CV, Shanker K, Srivastava NK, Shukla AK, Kalra A (2016) Fungal endophytes of *Catharanthus roseus* enhance vindoline content by modulating structural and regulatory genes related to terpenoid indole alkaloid biosynthesis. Sci Rep 6 (1):1–14

Pandey-Rai S, Mallavarapu GR, Naqvi AA, Yadav A, Rai SK, Srivastava S, Singh D, Mishra R, Kumar S (2006) Volatile components of leaves and flowers of periwinkle *Catharanthus roseus* (L.) G. Don from New Delhi. Flavour Fragrance J 21(3):427–430

Papon N, Vansiri A, Gantet P, Chénieux JC, Rideau M, Crèche J (2004) Histidine-containing phosphotransfer domain extinction by RNA interference turns off a cytokinin signalling circuitry in *Catharanthus roseus* suspension cells. FEBS Lett 558(1–3):85–88

Patra B, Pattanaik S, Schluttenhofer C, Yuan L (2018) A network of jasmonate-responsive bHLH factors modulate monoterpenoid indole alkaloid biosynthesis in *Catharanthus roseus*. New Phytol 217(4):1566–1581

Paul P, Singh SK, Patra B, Sui X, Pattanaik S, Yuan L (2017) A differentially regulated AP 2/ERF transcription factor gene cluster acts downstream of a MAP kinase cascade to modulate terpenoid indole alkaloid biosynthesis in *Catharanthus roseus*. New Phytol 213 (3):1107–1123

Pauw B, Hilliou FA, Martin VS, Chatel G, de Wolf CJ, Champion A, Pré M, van Duijn B, Kijne JW, van der Fits L, Memelink J (2004) Zinc finger proteins act as transcriptional repressors of alkaloid biosynthesis genes in *Catharanthus roseus*. J Biol Chem 279 (51):52940–52948

Peebles CA, Sander GW, Hughes EH, Peacock R, Shanks JV, San KY (2011) The expression of 1-deoxy-D-xylulose synthase and geraniol-10-hydroxylase or anthranilate synthase increases terpenoid indole alkaloid accumulation in *Catharanthus roseus* hairy roots. Metabol Eng 13(2):234–240

Rizvi NF, Weaver JD, Cram EJ, Lee-Parsons CW (2016) Silencing the transcriptional repressor, ZCT1, illustrates the tight regulation of terpenoid indole alkaloid biosynthesis in *Catharanthus* roseus hairy roots. PLoS One 11(7):e0159712

Roepke J, Salim V, Wu M, Thamm AM, Murata J, Ploss K, Boland W, De Luca V (2010) Vinca drug components accumulate exclusively in leaf exudates of Madagascar periwinkle. Proc Natl Acad Sci USA 107(34):15287–15292

Saiman MZ, Mustafa NR, Pomahočová B, Verberne M, Verpoorte R, Choi YH, Schulte AE (2014) Analysis of metabolites in the terpenoid pathway of *Catharanthus roseus* cell suspensions. Plant Cell Tissue Org Cult 117(2):225–239

Sazegari S, Niazi A, Shahriari-Ahmadi F, Moshtaghi N, Ghasemi Y (2018) CrMYC1'transcription factor overexpression promotes the production of low abundance terpenoid indole alkaloids in'*Catharanthus roseus*. Plant Omics 11(1):30–36

Schluttenhofer C, Pattanaik S, Patra B, Yuan L (2014) Analyses of *Catharanthus roseus* and Arabidopsis thaliana WRKY transcription factors reveal involvement in jasmonate signaling. BMC Genomics 15 (1):1–20

Sharma A, Amin D, Sankaranarayanan A, Arora R, Mathur AK (2020) Present status of *Catharanthus roseus* monoterpenoid indole alkaloids engineering in homo-and hetero-logous systems. Biotechnol Lett 42 (1):11–23

Sharma A, Mathur AK, Ganpathy J, Joshi B, Patel P (2019) Effect of abiotic elicitation and pathway precursors feeding over terpenoid indole alkaloids production in multiple shoot and callus cultures of *Catharanthus roseus*. Biologia 74(5):543–553

Sharma A, Verma P, Mathur A, Mathur AK (2018) Genetic engineering approach using early Vinca alkaloid biosynthesis genes led to increased tryptamine and terpenoid indole alkaloids biosynthesis in differentiating cultures of *Catharanthus roseus*. Protoplasma 255(1):425–435

Shen Q, Yan T, Fu X, Tang K (2016) Transcriptional regulation of artemisinin biosynthesis in *Artemisia annua* L. Sci Bull 61(1):18–25

Shukla AK, Shasany AK, Gupta MM, Khanuja SP (2006) Transcriptome analysis in *Catharanthus roseus* leaves and roots for comparative terpenoid indole alkaloid profiles. J Exp Bot 57(14):3921–3932

Simkin AJ, Miettinen K, Claudel P, Burlat V, Guirimand G, Courdavault V, Papon N, Meyer S, Godet S, St-Pierre B, Giglioli-Guivarc'h N (2013) Characterization of the plastidial geraniol synthase from Madagascar periwinkle which initiates the monoterpenoid branch of the alkaloid pathway in internal phloem associated parenchyma. Phytochemistry 85:36–43

Singh SK, Patra B, Paul P, Liu Y, Pattanaik S, Yuan L (2020) Revisiting the ORCA gene cluster that regulates terpenoid indole alkaloid biosynthesis in *Catharanthus roseus*. Plant Sci 293:110408

Srivastava R, Kumar R (2019) The expanding roles of APETALA2/Ethylene responsive factors and their potential applications in crop improvement. Brief Funct Genom 18(4):240–254

Sun J, Manmathan H, Sun C, Peebles CA (2016) Examining the transcriptional response of overexpressing anthranilate synthase in the hairy roots of an important medicinal plant *Catharanthus roseus* by RNA-seq. BMC Plant Biol 16(1):1–14

Sung PH, Huang FC, Do YY, Huang PL (2011) Functional expression of geraniol 10-hydroxylase reveals its dual function in the biosynthesis of terpenoid and phenylpropanoid. J Agric Food Chem 59(9):4637–4643

Suttipanta N, Pattanaik S, Kulshrestha M, Patra B, Singh SK, Yuan L (2011) The transcription factor CrWRKY1 positively regulates the terpenoid indole alkaloid biosynthesis in *Catharanthus roseus*. Plant Physiol 157(4):2081–2093

Tolambiya P, Mathur S (2016) A study on potential phytopharmaceuticals assets in *Catharanthus roseus* L. (Alba). Int J Life Sci Biotechnol Pharm Res 5(1):1–6

Van Moerkercke A, Steensma P, Schweizer F, Pollier J, Gariboldi I, Payne R, Bossche RV, Miettinen K, Espoz J, Purnama PC, Kellner F (2015) The bHLH transcription factor BIS1 controls the iridoid branch of the monoterpenoid indole alkaloid pathway in *Catharanthus roseus*. Proc Natl Acad Sci USA 112 (26):8130–8135

Verma P, Mathur AK, Srivastava A, Mathur A (2012) Emerging trends in research on spatial and temporal organization of terpenoid indole alkaloid pathway in *Catharanthus roseus*: a literature update. Protoplasma 249(2):255–268

Verma P, Singh N, Khan SA, Mathur AK, Sharma A, Jamal F (2020) TIAs pathway genes and associated miRNA identification in Vinca minor: supporting aspidosperma and eburnamine alkaloids linkage via transcriptomic analysis. Physiol Mol Biol Plants 26 (8):1695–1711

Wang CT, Liu H, Gao XS, Zhang HX (2010) Overexpression of G10H and ORCA3 in the hairy roots of *Catharanthus roseus* improves catharanthine production. Plant Cell Rep 29(8):887–894

Wink M, Roberts MW (1998) Alkaloids: biochemistry, ecology, and medicinal applications. Plenum Press, New York

World Health Organization (2019) World Health Organization model list of essential medicines: 21st list 2019. Geneva: World Health Organization. hdl:10665/325771. WHO/MVP/EMP/IAU/2019.06. License: CC BY-NC-SA 3.0 IGO

Yang Z, Patra B, Li R, Pattanaik S, Yuan L (2013) Promoter analysis reveals cis-regulatory motifs associated with the expression of the WRKY transcription factor CrWRKY1 in *Catharanthus roseus*. Planta 238 (6):1039–1049

Zárate R, Verpoorte R (2007) Strategies for the genetic modification of the medicinal plant *Catharanthus roseus* (L.) G. Don. Phytochem Rev 6(2–3):475–491

Zhang XN, Liu J, Liu Y, Wang Y, Abozeid A, Yu ZG, Tang ZH (2018) Metabolomics analysis reveals that ethylene and methyl jasmonate regulate different branch pathways to promote the accumulation of terpenoid indole alkaloids in *Catharanthus roseus*. J Nat Prod 81(2):335–342

Zhao J, Verpoorte R (2007) Manipulating indole alkaloid production by *Catharanthus roseus* cell cultures in bioreactors: from biochemical processing to metabolic engineering. Phytochem Rev 6(2):435–457

Zhao L, Sander GW, Shanks JV (2013) Perspectives of the Metabolic engineering of terpenoid indole alkaloids in *Catharanthus roseus* hairy roots. In: Doran P (ed) Biotechnology of hairy root systems. Advances in biochemical engineering/biotechnology. Springer, Berlin, Heidelberg, pp 23–54

Zhou ML, Shao JR, Tang YX (2009) Production and metabolic engineering of terpenoid indole alkaloids in cell cultures of the medicinal plant *Catharanthus roseus* (L.) G. Don (Madagascar periwinkle). Biotechnol Appl Biochem 52(4):313–323

Zhou ML, Zhu XM, Shao JR, Wu YM, Tang YX (2010) Transcriptional response of the catharanthine biosynthesis pathway to methyl jasmonate/nitric oxide elicitation in *Catharanthus roseus* hairy root culture. Appl Microbiol Biotechnol 88(3):737–750

Zhu J, Wang M, Wen W, Yu R (2015) Biosynthesis and regulation of terpenoid indole alkaloids in *Catharanthus roseus*. Pharmacog Rev 9(17):24

Zhu X, Zeng X, Sun C, Chen S (2014) Biosynthetic pathway of terpenoid indole alkaloids in *Catharanthus roseus*. Front Med 8(3):285–293

Future Prospects for Research on *Catharanthus roseus* 10

Maneesha Mall, Pooja Singh, Seema Yadav, Velusamy Sundaresan, Daya N. Mani, Ajit K. Shasany, and Ashutosh K. Shukla

Abstract

Catharanthus roseus represents a reservoir of over 130 medicinally important alkaloids, many of which are currently in therapeutic use. It is one of the most extensively investigated medicinal plants and has acquired the status of a model non-model plant due to many decades of active research conducted upon it by various groups across the globe. Nevertheless, the plant still remains an enigma due to gaps in the understanding of the biosynthesis of pharmacologically important terpenoid indole alkaloids (TIAs) coupled with extremely low content of the potent anticancer bisindole alkaloids, vincristine and vinblastine. Genomic and transcriptomic resources have been generated for the plant, but their full exploitation is yet to be achieved. This chapter will provide an overview of the research potential with respect to *C. roseus* and also elaborate upon the future prospects for research on the plant in terms of the gaps to be filled.

10.1 Introduction

Catharanthus roseus (L.) G. Don (family Apocynaceae; common Indian name 'Sadabahar') produces over 130 terpenoid indole alkaloids (TIAs) and is a natural reservoir of medicinally important phytomolecules. The plant has acquired an exalted status among medicinal plants due to long-term consolidated research efforts carried out on it globally by various groups (Shukla and Khanuja 2013). The plant derives fame from the costly and scarce shoot-specific bisindole alkaloids, vinblastine and vincristine that are highly potent anticancer agents and are derived by the condensation of monomeric TIAs, vindoline and catharanthine. Vindoline, one of the monomeric precursors of the bisindoles, is produced only in the aerial green tissues of the plant and represents the major bottleneck in the semi-synthetic commercial production of the bisindole alkaloids.

Maneesha Mall, Pooja Singh, and Seema Yadav have made equal contribution as first author

M. Mall · P. Singh · S. Yadav · D. N. Mani · A. K. Shasany · A. K. Shukla (✉)
CSIR-Central Institute of Medicinal and Aromatic Plants, P.O. CIMAP, Lucknow 226015, India
e-mail: ashutoshshukla@cimap.res.in

V. Sundaresan
CSIR-Central Institute of Medicinal and Aromatic Plants, Research Centre, Allalasandra, GKVK Post, Bengaluru 560065, India

© The Author(s), under exclusive license to Springer Nature Switzerland AG 2022
C. Kole (ed.), *The Catharanthus Genome*, Compendium of Plant Genomes,
https://doi.org/10.1007/978-3-030-89269-2_10

C. roseus has been described in Ayurveda for treating *Kapha-Vata doshas* and *Rakta-Arbud* (Leukemia) (Sharma 2005). The plant is also mentioned for treatment of diabetes, but unlike its anticancer activity, the antidiabetic property is yet to be thoroughly investigated. The antidiabetic and antioxidant properties of TIAs from *C. roseus* were studied by Tiong et al. (2013), and vindogentianine (a vindoline-gentianine dimer), a new alkaloid from the plant leaves, was found to show hypoglycemic property in *in vitro* studies through enhancement of glucose uptake and protein tyrosine phosphatase-1B (PTP-1B) inhibition, indicating its therapeutic potential against type 2 diabetes (Tiong et al. 2015).

Although other species like *C. trychophyllus*, *C. ovalis*, *C. longifolius*, *C. coriaceus*, *C. lanceus*, *C. scitulus* and *C. pusillus* are also present in the genus *Catharanthus*, *C. roseus* is unique due to the presence of TIAs like vindoline, vincristine and vinblastine.

10.2 Availability of Various Omics Resources for *C. roseus*

With the arrival of the omics era, many omics-based studies were initiated on *C. roseus*. Initial transcriptomic studies employed ESTs (Murata et al. 2006; Shukla et al. 2006; Murata et al. 2008) and cDNA-AFLP (Rischer et al. 2006; Shukla et al. 2012) tools, but with the advent of NGS, many transcriptomes of the plant were generated. Presently, many omics resources dedicated to *C. roseus* have become available (Table 10.1). The genome sequence of the plant is now available, and the genome size is 522.654 Mb. A comprehensive genome assembly for *C. roseus* was provided by Kellner et al. (2015) after Ku et al. (2013) assembled the complete plastome (plastid genome sequence) of the plant.

10.3 Future Prospects for Research

The future direction to be adopted for research on *C. roseus* was delineated earlier (Shukla and Khanuja 2013). Since then, many of the aspects were worked out by various groups, but many of them still remain to be addressed. It is now evident that the plant cannot produce large amounts of bisindole alkaloids due to problems of cytotoxicity that hinder their sequestration and storage. For commercial production of the bisindoles, the best option is to procure the monomeric precursors, vindoline and catharanthine, from the plant and couple them semi-synthetically. Towards this end, the major obstacle is sourcing vindoline, which is derived only from the green aerial tissues of the plant unlike catharanthine, which is produced throughout the plant as well as in cell suspension cultures. This makes vindoline a key metabolite of *C. roseus* (Mall et al. 2019a).

10.3.1 Sequestration and Transportation of TIA Biosynthetic Intermediates

TIA biosynthesis is spread over various cell types and organelles (Murata et al. 2008). A good number of TIA biosynthetic pathway steps have been characterized till date, but the information about the transporters and sequestration of the intermediate metabolites is mostly unavailable. Also, there are many gaps in the understanding of the biosynthesis of yet unexplored TIAs. Interestingly, the TIA biosynthetic genes are not clustered in the plant genome (Caputi et al. 2018). The post-strictosidine modifications are still an enigma as this phytochemical chameleon molecule has the potential to give rise to several TIA structures in *C. roseus* as well as related species. Besides, many of the regulators of TIA biosynthesis may still be unknown. The major cell types involved in TIA biosynthesis are the internal phloem-associated parenchyma (IPAP), epidermis and laticifers/idioblasts. The TIA biosynthetic intermediates need to be translocated in both ways, intracellular and intercellular (Fig. 10.1). Champagne et al. (2012) adopted a proteomics-based approach and suggested that the transporters possibly involved in the transport of TIA intermediates could be belonging to the

Table 10.1 *C. roseus* omics resources available in the public domain

S. No.	Omics resource	Salient feature	Reference
1	croFGD database (http://bioinformatics.cau.edu.cn/croFGD/)	Co-expression network exploring multi-dimensional gene expression	She et al. (2019)
2	Medicinal Plant Genomics Resource (http://www.medicinalplantgenomics.msu.edu/) (https://datadryad.org/stash/dataset/doi:10.5061/dryad.hs593)	Genome sequence of variety Sunstorm Apricot using whole-genome shotgun sequencing; 33,829 genes annotated	Kellner et al. (2015)
3	CathaCyc (http://www.cathacyc.org)	Metabolic pathway database built from RNA-seq data	Van Moerkercke et al. (2013)
4	MAPslnc database (https://lncrnapipe.cimap.res.in)	Repository of long non-coding RNAs (lncRNAs)	Shukla et al. (2021)
5	Genetic linkage map	Integrated genetic linkage map based on different types of molecular and morphological markers; total map length 1131.9 cM	Gupta et al. (2007)
6	Plastome (plastid genome)	The complete plastid genome sequence is 154,950 bp long	Ku et al. (2013)
7	Proteome	Comparative analysis of two distinct cell lines using two-dimensional difference gel electrophoresis (2D-DIGE) revealed differentially expressed proteins that could yield novel insights	Champagne et al. (2012)
8	miRNAome	Comprehensive account of *C. roseus* miRNAs that were mapped to the reference genome of the plant	Shen et al. (2017)
9	Developmental methylome	Whole-genome bisulfite sequencing (WGBS) along with RNA-seq was employed for identifying differentially methylated and expressed genes. Indicated an epigenetic regulation of TIA metabolism	Dugé de Bernonville et al. (2020)

ATP-binding cassette (ABC), multidrug and toxic compound extrusion (MATE) and the nitrate/peptide families of transporters. Since then, a few transporters have already been characterized, like the cytosol-to-cuticle catharanthine transporter TPT2, belonging to the pleiotropic drug resistance (PDR) family of ABC transporters (Yu and De Luca 2013) and the vacuole-to-cytosol strictosidine transporter NPF2.9, belonging to the nitrate/peptide family of transporters (Payne et al. 2017). However, most of the transporters are yet to be characterized. This could be facilitated by the availability of huge amount of transcriptomic data on the plant. The problem is compounded by the fact that there is a lack of consensus sequences enabling the prediction of substrates for transporters available in the transcriptomics/proteomics resources. For elucidating transporter–substrate combinations, physical cloning of the genes followed by functional transport assays is required to be carried out (Champagne et al. 2012). An elegant study by Carqueijeiro et al. (2013) suggested that TIAs like vindoline, catharanthine and α-3′,4′- anhydrovinblastine (AVLB) that accumulate in the plant leaves are

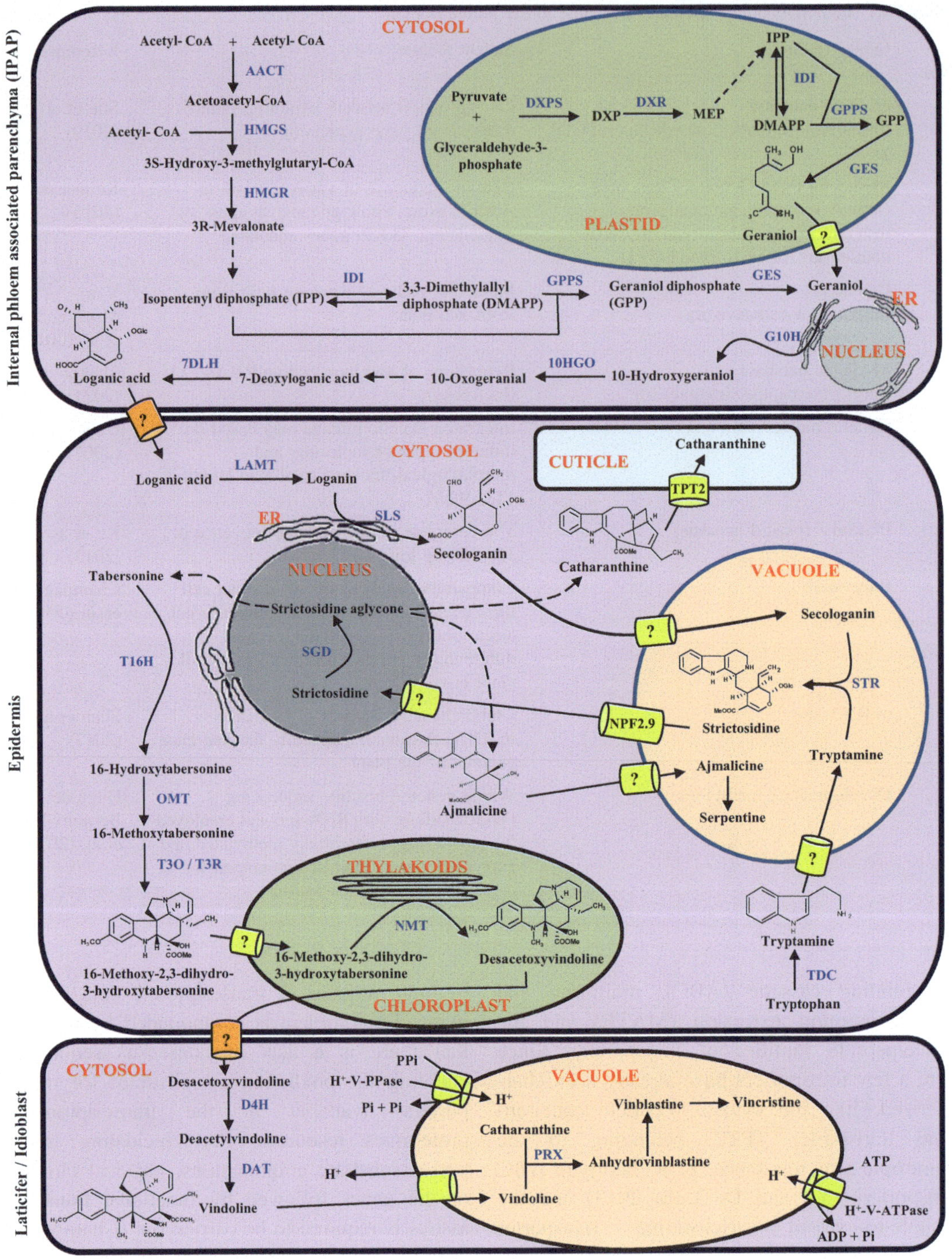
Internal phloem associated parenchyma (IPAP)
CYTOSOL
Acetyl- CoA + Acetyl- CoA
AACT
Acetoacetyl-CoA
Acetyl- CoA
HMGS
3S-Hydroxy-3-methylglutaryl-CoA
HMGR
3R-Mevalonate
Isopentenyl diphosphate (IPP)
IDI
3,3-Dimethylallyl diphosphate (DMAPP)
GPPS
Geraniol diphosphate (GPP)
GES
Geraniol
ER
G10H
NUCLEUS
Pyruvate
+
Glyceraldehyde-3-phosphate
DXPS
DXP
DXR
MEP
IPP
IDI
DMAPP
GPPS
GPP
GES
Geraniol
PLASTID
?
10-Hydroxygeraniol
10HGO
10-Oxogeranial
7-Deoxyloganic acid
7DLH
Loganic acid
Epidermis
CYTOSOL
Loganic acid
LAMT
Loganin
SLS
ER
Secologanin
CUTICLE
Catharanthine
TPT2
Catharanthine
NUCLEUS
Strictosidine aglycone
SGD
Strictosidine
Tabersonine
T16H
VACUOLE
Secologanin
STR
Strictosidine
NPF2.9
Tryptamine
Ajmalicine
Serpentine
Ajmalicine
16-Hydroxytabersonine
OMT
16-Methoxytabersonine
T3O / T3R
16-Methoxy-2,3-dihydro-3-hydroxytabersonine
THYLAKOIDS
NMT
16-Methoxy-2,3-dihydro-3-hydroxytabersonine
Desacetoxyvindoline
CHLOROPLAST
Tryptamine
TDC
Tryptophan
Laticifer / Idioblast
CYTOSOL
Desacetoxyvindoline
D4H
Deacetylvindoline
DAT
Vindoline
PPi
H⁺-V-PPase
Pi + Pi
H⁺
VACUOLE
Vinblastine
Vincristine
Catharanthine
PRX
Anhydrovinblastine
Vindoline
H⁺
H⁺
ATP
H⁺-V-ATPase
ADP + Pi

◀ **Fig. 10.1** A comprehensive model of TIA biosynthesis in *Catharanthus roseus* depicting the inter-cellular and intra-cellular transport of metabolites as best understood through currently available literature. Various types of cells like, internal phloem associated parenchyma (IPAP), epidermis, laticifers and idioblasts are involved in the process. Solid black arrows indicate a single step, whereas dashed black arrows indicate multiple steps; uncharacterized transportation steps are marked with '?'; yellow transporters indicate intra-cellular transport, whereas orange transporters indicate inter-cellular transport. AACT, acetoacetyl-CoA thiolase; DAT, deacetylvindoline 4-*O*-acetyltransferase; D4H, desacetoxyvindoline 4-hydroxylase; DXPS, 1-deoxy-D-xylulose-5-phosphate synthase; DXR, 1-deoxy-D-xylulose-5-phosphate reductoisomerase; GES, geraniol synthase; G10H, geraniol 10-hydroxylase; GPPS, geranyl pyrophosphate synthase; 10HGO, 10-hydroxygeraniol oxidoreductase; HMGR, hydroxymethylglutaryl-CoA reductase; HMGS, hydroxymethylglutaryl-CoA synthase; IDI, isopentenyl diphosphate isomerase; LAMT, loganic acid *O*-methyltransferase; NMT, 16-methoxy-2,3-dihydro-3-hydroxy-tabersonine-*N*-methyltransferase; NPF2.9, nitrate/peptide family strictosidine transporter 2.9; OMT, 16-hydroxytabersonine *O*-methyltransferase; PRX, peroxidase; SGD, strictosidine β-D-glucosidase; SLS, secologanin synthase; STR, strictosidine synthase; TDC, tryptophan decarboxylase; T16H, tabersonine 16-hydroxylase; T3O, tabersonine 3-oxygenase; TPT2, catharanthine transporter (pleiotropic drug resistance family of ABC transporters); T3R, tabersonine 3-reductase; V, vacuolar pump.

actively transported into vacuoles through specific proton-driven antiport system and not through the action of ABC transporters or ion-trap mechanism. The transtonoplast pH gradient is presumably generated by a yet unidentified vacuolar pump (vacuolar H^+-ATPase or vacuolar H^+-pyrophosphatase), and it remains to be characterized whether there is one vacuolar transporter showing varying degree of affinities toward various TIAs or there are dedicated vacuolar transporters for each TIA (Carqueijeiro et al. 2013). The other putative transportation steps where future studies must be focused are the transport of geraniol from the plastid to the cytosol, loganic acid from cytosol of IPAP cells to that of the epidermis, tryptamine and secologanin from cytosol into the vacuole, strictosidine from cytosol to the nucleus, 16-methoxy-2,3-dihydro-3-hydroxytabersonine from cytosol to chloroplast, desacetoxyvindoline from chloroplast of epidermal cells to cytosol of laticifer/idioblast cells, ajmalicine from cytosol to vacuole and the proton-driven antiporters for the active transport of specific TIAs like vindoline from the cytosol to the vacuoles (Fig. 10.1).

Apart from the transporters, many structural and regulatory genes involved in TIA metabolism still remain to be elucidated. Despite the identification and characterization of many transcription factors regulating TIA metabolism, there may still be some unexplored ones. Although the structural genes involved in vindoline biosynthesis have been mostly worked out, there is a lack of information on the dedicated steps involved in the biosynthesis of other TIAs.

10.3.2 Future Scope for Biotechnological Improvements

Synthetic biology (Qu et al. 2015; Kulagina et al. 2021; Lemos Cruz et al. 2021) as well as transgenic (Kumar et al. 2018) strategies have been attempted for enhanced production of *C. roseus* TIAs like vindoline. However, due to large number of complex steps, metabolite trafficking and compartmentation, carefully designed breeding still seems to be highly promising (Mall et al. 2021a). Since the genome of the plant is now available (Kellner et al. 2015), genome-editing technologies can be attempted in the plant. However, it is well known that *C. roseus* is recalcitrant to established transformation and gene expression methods. In future, *de novo* induction of gene-edited meristems in the plant (Maher et al. 2020) or utilizing plant viruses to obtain gene-edited plants (Ellison et al. 2020; Zhu et al. 2020) could be employed for circumventing tissue culture, which is inefficient, especially in case of *C. roseus*. For favorably modulating TIA metabolic pathways, the down-regulation of specific genes could be a desirable approach to circumvent the biosynthesis of competitive intermediate metabolites or to completely shut the metabolic pathway branches leading toward the undesired metabolites. One such example is provided by the root-specific minovincinine-19-*O*-acetyltransferase (MAT) which is somewhat similar to the shoot-specific deacetylvindoline-4-*O*-acetyltransferase (DAT) but has different substrate specificity (Laflamme et al. 2001), whereby the shoot-specific pathway channelizing the metabolic flux toward vindoline is the preferred one (Shukla et al. 2012). However, any sort of biotechnological intervention for improving content of TIAs like vindoline must be tried in elite genotypes like CIM-Sushil (Gupta et al. 2020) to get a booster effect.

10.3.3 Potential Role of Micronutrients in Modulating TIA Content in the Plant

Micronutrients like boron (B), chlorine (Cl), copper (Cu), iron (Fe), manganese (Mn), molybdenum (Mo) and zinc (Zn) are essential elements required by plants in small quantities (DuPont, Pioneer, website). Most of them are enzyme cofactors that are easily available through the soil. Apart from influencing plant growth, development and primary metabolism, the micronutrients are also known to modulate plant secondary metabolism (Poschenrieder et al. 2008). However, this aspect has not been explored in depth in relation to TIA accumulation in *C. roseus*. In a few cases, micronutrients have been implicated in biosynthesis/accumulation of specific secondary metabolites. For example, Zn has been shown to stimulate tryptophan biosynthesis, which is a precursor for TIAs (Lovkova et al. 2005). Recently, a genotype of *C. roseus* having higher-than-usual level of vindoline was found to be rich in Zn, Fe and Mn but deficient in B as compared to the usual genotypes (Mall et al. 2021a). Earlier, Zn finger proteins have been shown to act as transcriptional repressors of some TIA biosynthesis-related genes in *C. roseus* (Pauw et al. 2004). Although many genes have been characterized in *C. roseus*, till date there has been no concerted effort to study the genes involved in micronutrient-based modulation of the biology of the plant, including its secondary metabolism. Since there are preliminary indications that certain micronutrients may be affecting the *in planta* TIA accumulation, the exact cause–effect relationship is required to be established so that it could be beneficially exploited for TIA sourcing. Thus, future studies should be aimed at advancing the current understanding of the role of specific micronutrients in TIA metabolism.

10.3.4 Key Areas Requiring Focus in Future

- Creating genotypic diversity in *C. roseus* germplasm for enhanced content of specific TIAs must be an important target. These efforts could be either through conventional plant breeding or through biotechnological approaches including genome editing. A combination of both the approaches is also possible. However, keeping in view the recalcitrance of the plant toward genetic transformation, conventional plant breeding approaches seem to be more attractive in case of *C. roseus*. An effort for achieving enhanced vindoline yield in the plant has recently fructified in the form of the mutation breeding-derived high vindoline-yielding variety CIM-Sushil (Gupta et al. 2020; Mall et al. 2021a). Such high-yielding varieties should be the starting material for any biotechnological intervention to achieve a multiplier effect for specific TIA content.
- Elucidation of pathway genes for various TIAs, other than vindoline and bisindole alkaloids, is yet to be completed. Unless all the structural and regulatory components of TIA biosynthesis are completely known, successful metabolic engineering will be a distant dream.
- Analysis of metabolic flux channeling into various branches originating from strictosidine (the central intermediate) is crucial for optimization of production of key TIAs.

- The complete TIA arsenal of *C. roseus* is yet to be completely elucidated. Novel alkaloids with unique therapeutic properties present in the plant may still be waiting to be isolated and characterized. As an example, a unique bisindole alkaloid named cathachunine was isolated from the plant and found to be showing potent antitumor effect toward human leukemia cells (Wang et al. 2016). Additionally, novel structures formed from known alkaloids may be investigated for enhanced therapeutic capability as shown by Zhang et al. (2019) through the design of a dimer molecule from double vindoline-bonded (DVB) that was suggested to have stronger anticancer cytotoxicity than vinblastine.
- Since microRNAs (miRNAs) are crucial regulators of gene expression, further investigations are needed to elucidate the regulatory roles of *C. roseus* miRNAs in TIA biosynthesis in the plant. Similarly, the long non-coding RNAs (lncRNAs) operating in *C. roseus* need to be characterized.
- To circumvent the recalcitrance of *C. roseus* to established transformation/gene expression methods, recently, Mortimer et al. (2020) have exploited the replication system of a *C. roseus*-infecting geminivirus toward development of an autonomously replicating viral expression system tailored for *C. roseus*, which consistently yields high gene expression in the plant leaves. This and other similar platforms could be used for expressing candidate genes that need to be functionally characterized, like, for example, transporters.
- The influence of seasonal and ontogenic parameters on the pattern of key TIA accumulation in *C. roseus* has been well established now (Mall et al. 2019b). An agro-advisory could be suggested to the commercial growers of the crop for appropriate harvesting (time and mode) for specific TIAs.
- Although cellular differentiation, elicitation (Shukla et al. 2010) and biotic/abiotic stress factors (Mall 2017; Mall et al. 2021b) are known to influence TIA biosynthetic steps, a holistic picture still eludes us. The exact mode of action of each type of stress factor needs to be worked out for better understanding of TIA metabolism.
- Comparative analysis of *C. roseus* must be carried out with other species in the genus *Catharanthus* so as to identify the similarities and differences. Attempts may be made to explore whether any of the other species is more amenable to genetic transformation and could be used to improve specific TIA production, which may be a limiting factor in *C. roseus*.
- *C. roseus* has high potential for phytoremediation of toxic metals in sewage sludge (Barbosa et al. 2020). This dimension of the plant needs to be explored in depth, and the metal accumulation/transportation machinery needs to be elucidated.
- *C. roseus* acts as a good host for phytoplasma infections. It could thus serve the purpose for making a repository of various phytoplasma candidates for studying their properties.
- *C. roseus* leaf extract offers a good choice for synthesizing zinc oxide (ZnO) nanoparticles, which could be used in miscellaneous applications such as antimicrobial agents (Gupta et al. 2018).

Acknowledgements The authors are grateful to Director, CSIR-CIMAP, Lucknow, for encouragement and research facilities. The authors' research is supported through a grant (CRG/2020/002970) from the Science and Engineering Research Board (SERB), India. Maneesha Mall and Pooja Singh thank UGC, India for their fellowships, and Seema Yadav thanks CSIR, India for her fellowship.

References

Barbosa ÉS, Cacique AP, de Pinho GP, Silvério FO (2020) *Catharanthus roseus* potential for phytostabilizing metals in sewage sludge. J Environ Sci Health Part A Toxic/Hazard Subst Environ Eng 55:209–215

Caputi L, Franke J, Farrow SC, Chung K, Payne RME, Nguyen TD, Dang TTT, Carqueijeiro IST, Koudounas K, de Bernonville TD, Ameyaw B, Jones DM, Vieira IJC, Courdavault V, O'Connor SE (2018) Missing enzymes in the biosynthesis of the anticancer drug vinblastine in Madagascar periwinkle. Science 360:1235–1239

Carqueijeiro I, Noronha H, Duarte P, Gerós H, Sottomayor M (2013) Vacuolar transport of the medicinal alkaloids from *Catharanthus roseus* is mediated by a proton-driven antiport. Plant Physiol 162:1486–1496

Champagne A, Rischer H, Oksman-Caldentey KM, Boutry M (2012) In-depth proteome mining of cultured *Catharanthus roseus* cells identifies candidate proteins involved in the synthesis and transport of secondary metabolites. Proteomics 12:3536–3547

Dugé de Bernonville T, Maury S, Delaunay A, Daviaud C, Chaparro C, Tost J, O'Connor SE, Courdavault V (2020) Developmental methylome of the medicinal plant *Catharanthus roseus* unravels the tissue-specific control of the monoterpene indole alkaloid pathway by DNA methylation. Int J Mol Sci 21:6028

DuPont, Pioneer. https://www.pioneer.com/us/agronomy/micronutrients_crop_production.html

Ellison EE, Nagalakshmi U, Gamo ME, Huang P-J, Dinesh-Kumar S, Voytas DF (2020) Multiplexed heritable gene editing using RNA viruses and mobile single guide RNAs. Nat Plants 6:620–624

Gupta AK, Shukla AK, Singh P, Mall M, Yadav S, Shasany AK, Shanker K, Baskaran K, Sundaresan V, Talha M, Srivastava A, Gupta S, Khare P, Mani DN, Samad A (2020) CIM-Sushil: a high vindoline yielding variety of *Catharanthus roseus* (L.) G. Don. J Med Aromat Plant Sci 42:51–63

Gupta M, Tomar RS, Kaushik S, Mishra RK, Sharma D (2018) Effective antimicrobial activity of green ZnO nano particles of *Catharanthus roseus*. Front Microbiol 9:2030

Gupta S, Pandey-Rai S, Srivastava S, Naithani SC, Prasad M, Kumar S (2007) Construction of genetic linkage map of the medicinal and ornamental plant *Catharanthus roseus*. J Genet 86:259–268

Kellner F, Kim J, Clavijo BJ, Hamilton JP, Childs KL, Vaillancourt B, Cepela J, Habermann M, Steuernagel B, Clissold L, McLay K, Buell CR, O'Connor SE (2015) Genome-guided investigation of plant natural product biosynthesis. Plant J 82:680–692

Ku C, Chung WC, Chen LL, Kuo CH (2013) The complete plastid genome sequence of Madagascar Periwinkle *Catharanthus roseus* (L.) G. Don: plastid genome evolution, molecular marker identification, and phylogenetic implications in asterids. PLoS One 8:e68518

Kulagina N, Guirimand G, Melin C, Lemos-Cruz P, Carqueijeiro I, De Craene J-O, Oudin A, Heredia V, Koudounas K, Unlubayir M, Lanoue A, Imbault N, St-Pierre B, Papon N, Clastre M, Giglioli-Guivarc'h N, Marc J, Besseau S, Courdavault V (2021) Enhanced bioproduction of anticancer precursor vindoline by yeast cell factories. Microbial Biotechnology. https://doi.org/10.1111/1751-7915.13898

Kumar SR, Shilpashree HB, Nagegowda DA (2018) Terpene moiety enhancement by overexpression of geranyl(geranyl) diphosphate synthase and geraniol synthase elevates monomeric and dimeric monoterpene indole alkaloids in transgenic *Catharanthus roseus*. Front Plant Sci 9:942

Laflamme P, St-Pierre B, De Luca V (2001) Molecular and biochemical analysis of a Madagascar periwinkle root-specific minovincinine-19-hydroxy-*O*-acetyltransferase. Plant Physiol 125:189–198

Lemos Cruz P, Kulagina N, Guirimand G, De Craene J-O, Besseau S, Lanoue A, Oudin A, Giglioli-Guivarc'h N, Papon N, Clastre M, Courdavault V (2021) Optimization of tabersonine methoxylation to increase vindoline precursor synthesis in yeast cell factories. Molecules 26:3596

Lovkova MY, Buzuk GN, Sokolova SM, Buzuk LN (2005) Role of elements and physiologically active compounds in the regulation of synthesis and accumulation of indole alkaloids in *Catharanthus roseus* L. Appl Biochem Microbiol 41:299–305

Maher MF, Nasti RA, Vollbrecht M, Starker CG, Clark MD, Voytas DF (2020) Plant gene editing through *de novo* induction of meristems. Nat Biotechnol 38:84–89

Mall M (2017) Molecular studies related to key terpenoid indole alkaloid biosynthesis in *Catharanthus roseus*. PhD thesis, Jawaharlal Nehru University, New Delhi, India

Mall M, Singh P, Kumar R, Shanker K, Gupta AK, Khare P, Shasany AK, Khatoon S, Sundaresan V, Baskaran K, Yadav S, Shukla AK (2021a) Phenotypic, genetic and expression profiling of a vindoline-rich genotype of *Catharanthus roseus*. S Afr J Bot 139:50–57

Mall M, Shanker K, Samad A, Kalra A, Sundaresan V, Shukla AK (2021b) Stress responsiveness of vindoline accumulation in *Catharanthus roseus* leaves is mediated through co-expression of allene oxide cyclase with pathway genes. Protoplasma. https://doi.org/10.1007/s00709-021-01701-6

Mall M, Singh P, Yadav S, Shukla AK (2019a) Why is vindoline such an important metabolite? Presented at the UGC-SAP (DRS-II), DST-SERB & CSIR sponsored National Seminar on Biotechnology Research in India: Current Status and Future Prospects, organized by the Department of Biotechnology, School of Chemical and Life Sciences, Jamia Hamdard, New Delhi, during 26–27 March, 2019, p 111

Mall M, Verma RK, Gupta MM, Shasany AK, Khanuja SPS, Shukla AK (2019b) Influence of seasonal and ontogenic parameters on the pattern of key terpenoid indole alkaloids biosynthesized in the leaves of *Catharanthus roseus*. S Afr J Bot 123:98–104

Mortimer C, Dugdale B, Waterhouse P (2020) Development of an autonomously replicating viral expression system tailored for *Catharanthus roseus*. Plant Biotechnol J 18:1115–1117

Murata J, Bienzle D, Brandle JE, Sensen CW, De Luca V (2006) Expressed sequence tags from Madagascar periwinkle (*Catharanthus roseus*). FEBS Lett 580:4501–4507

Murata J, Roepke J, Gordon H, De Luca V (2008) The leaf epidermome of *Catharanthus roseus* reveals its biochemical specialization. Plant Cell 20:524–542

Pauw B, Hilliou FAO, Martin VS, Chatel G, de Wolf CJF, Champion A, Pre M, van Duijn B, Kijne JW, van der Fits L, Memelink J (2004) Zinc finger proteins act as transcriptional repressors of alkaloid biosynthesis genes in *Catharanthus roseus*. J Biol Chem 279:52940–52948

Payne RME, Xu D, Foureau E, Carqueijeiro MIST, Oudin A, de Bernonville TD, Novak V, Burow M, Olsen CE, Jones DM, Tatsis EC, Pendle A, Halkier BA, Geu-Flores F, Courdavault V, Nour-Eldin HH, O'Connor SE (2017) An NPF transporter exports a central monoterpene indole alkaloid intermediate from the vacuole. Nat Plants 3:16208

Poschenrieder C, Allue′ J, Tolra` R, Llugany M, Barcelo` J (2008) Trace elements and plant secondary metabolism: quality and efficacy of herbal products. In: Prasad MNV (ed) Trace elements: nutritional benefits, environmental contamination, and health implications. John Wiley & Sons, Inc

Qu Y, Easson MLAE, Froese J, Simionescu R, Hudlicky T, De Luca V (2015) Production of anticancer drugs in yeast. Proc Natl Acad Sci USA 112:6224–6229

Rischer H, Oresic M, Seppänen-Laakso T, Katajamaa M, Lammertyn F, Ardiles-Diaz W, Van Montagu MCE, Inzé D, Oksman-Caldentey KM, Goossens A (2006) Gene-to-metabolite networks for terpenoid indole alkaloid biosynthesis in *Catharanthus roseus* cells. Proc Natl Acad Sci USA 103:5614–5619

Sharma PV (2005) Dravyaguna-Vijnana Volume II [Audbhid (Plant origin)]. Chaukhambha Bharti Academy, Varanasi, India, pp 831–832

She J, Yan H, Yang J, Xu W, Su Z (2019) croFGD: *Catharanthus roseus* functional genomics database. Front Genet 10:238

Shen EM, Singh SK, Ghosh JS, Patra B, Paul P, Yuan L, Pattanaik S (2017) The miRNAome of *Catharanthus roseus*: identification, expression analysis, and potential roles of microRNAs in regulation of terpenoid indole alkaloid biosynthesis. Sci Rep 7:43027

Shukla AK, Khanuja SPS (2013) *Catharanthus roseus*: the metabolome that represents a unique reservoir of medicinally important alkaloids under precise genomic regulation. In: Barh D (ed) OMICS applications in crop science. CRC Press (Taylor & Francis Group), Boca Raton, pp 325–384

Shukla AK, Shasany AK, Gupta MM, Khanuja SPS (2006) Transcriptome analysis in *Catharanthus roseus* leaves and roots for comparative terpenoid indole alkaloid profiles. J Exp Bot 57:3921–3932

Shukla AK, Shasany AK, Khanuja SPS (2012) cDNA-AFLP-based numerical comparison of leaf and root organ cDNAs in *Catharanthus roseus*. OMICS: J Integr Biol 16:397–401

Shukla AK, Shasany AK, Verma RK, Gupta MM, Mathur AK, Khanuja SPS (2010) Influence of cellular differentiation and elicitation on intermediate and late steps of terpenoid indole alkaloid biosynthesis in *Catharanthus roseus*. Protoplasma 242:35–47

Shukla B, Gupta S, Srivastava G, Sharma A, Shukla AK, Shasany AK (2021) lncRNADetector: a bioinformatics pipeline for long non-coding RNA identification and MAPslnc: a repository of medicinal and aromatic plant lncRNAs. RNA Biol. https://doi.org/10.1080/15476286.2021.1899673

Tiong SH, Looi CY, Arya A, Wong WF, Hazni H, Mustafa MR, Awang K (2015) Vindogentianine, a hypoglycemic alkaloid from *Catharanthus roseus* (L.) G. Don (Apocynaceae). Fitoterapia 102:182–188

Tiong SH, Looi CY, Hazni H, Arya A, Paydar M, Wong WF, Cheah SC, Mustafa MR, Awang K (2013) Antidiabetic and antioxidant properties of alkaloids from *Catharanthus roseus* (L.) G. Don. Molecules 18:9770–9784

Van Moerkercke A, Fabris M, Pollier J, Baart GJ, Rombauts S, Hasnain G, Rischer H, Memelink J, Oksman-Caldentey KM, Goossens A (2013) CathaCyc, a metabolic pathway database built from *Catharanthus roseus* RNA-Seq data. Plant Cell Physiol 54:673–685

Wang XD, Li CY, Jiang MM, Li D, Wen P, Song X, Chen JD, Guo LX, Hu XP, Li GQ, Zhang J, Wang CH, He ZD (2016) Induction of apoptosis in human leukemia cells through an intrinsic pathway by cathachunine, a unique alkaloid isolated from *Catharanthus roseus*. Phytomedicine 23:641–653

Yu F, De Luca V (2013) ATP-binding cassette transporter controls leaf surface secretion of anticancer drug components in *Catharanthus roseus*. Proc Natl Acad Sci USA 110:15830–15835

Zhang Z, Lu C, Wang P, Li A, Zhang H, Xu S (2019) Structural basis and mechanism for vindoline dimers interacting with α, β-tubulin. ACS Omega 4:11938–11948

Zhu H, Li C, Gao C (2020) Applications of CRISPR–Cas in agriculture and plant biotechnology. Nat Rev Mol Cell Biol 21:661–677

The manufacturer's authorised representative in the EU is Springer Nature Customer Service Centre GmbH, Europaplatz 3, 69115 Heidelberg, Germany. If you have any concerns regarding our products, please contact ProductSafety@springernature.com

Printed and bound by CPI Group (UK) Ltd, Croydon, CR0 4YY
30/09/2026
02225699-0018